Lecture Notes in Computer Science 4996

Commenced Publication in 1973
Founding and Former Series Editors:
Gerhard Goos, Juris Hartmanis, and Jan van Leeuwen

Hans Kleine Büning   Xishun Zhao (Eds.)

Theory and Applications of Satisfiability Testing – SAT 2008

11th International Conference, SAT 2008
Guangzhou, China, May 12-15, 2008
Proceedings

 Springer

Volume Editors

Hans Kleine Büning
University of Paderborn
Department of Computer Science
33095 Paderborn, Germany
E-mail: kbcsl@upb.de

Xishun Zhao
Sun Yat-sen University
Institute of Logic and Cognition
510275 Guangzhou, P.R. China
E-mail: hsszxs@mail.sysu.edu.cn

Library of Congress Control Number: 2008925418

CR Subject Classification (1998): F.4.1, I.2.3, I.2.8, I.2, F.2.2, G.1.6

LNCS Sublibrary: SL 1 – Theoretical Computer Science and General Issues

ISSN   0302-9743
ISBN-10   3-540-79718-1 Springer Berlin Heidelberg New York
ISBN-13   978-3-540-79718-0 Springer Berlin Heidelberg New York

Springer is a part of Springer Science+Business Media

springer.com

© Springer-Verlag Berlin Heidelberg 2008
Printed in Germany

Typesetting: Camera-ready by author, data conversion by Scientific Publishing Services, Chennai, India
Printed on acid-free paper   SPIN: 12265787   06/3180   5 4 3 2 1 0

Preface

This volume contains the papers presented at the 11th International Conference on Theory and Applications of Satisfiability Testing (SAT 2008).

The series of International Conferences on Theory and Applications of Satisfiability Testing (SAT) has evolved from a first workshop on SAT in 1996 to an annual international conference which is a platform for researchers studying various aspects of the propositional satisfiability problem and its applications. In the past, the SAT conference venue alternated between Europe and North America. For the first time, the conference venue was in Asia, more precisely at the Zhudao Guest House, near Sun Yat-Sen University in Guangzhou, P.R. China.

Many hard combinatorial problems can be encoded into SAT. Therefore improvements on heuristics on the practical side, as well as theoretical insights into SAT apply to a large range of real-world problems. More specifically, many important practical verification problems can be rephrased as SAT problems. This applies to verification problems in hardware and software. Thus SAT is becoming one of the most important core technologies to verify secure and dependable systems. The topics of the conference span practical and theoretical research on SAT and its applications and include but are not limited to proof systems, proof complexity, search algorithms, heuristics, analysis of algorithms, hard instances, randomized formulae, problem encodings, industrial applications, solvers, simplifiers, tools, case studies, and empirical results. SAT is interpreted in a rather broad sense: besides propositional satisfiability, it includes, for example, the domain of quantified Boolean formulae (QBF) and satisfiability modulo theories (SMT).

The Program Committee selected 25 papers out of 70 submissions. Submissions were rigorously reviewed by three Program Committee members. The committee decided to accept 17 regular papers with a page limit of 14 pages and 8 short papers with a page limit of 6 pages.

The conference program included two invited presentations. The first, by Alasdair Urquhart, addressed the exponential separation between regular and unrestricted resolution. The second, by Kazuo Iwama, dealt with recent developments on the CNF satisfiability problem: deterministic algorithms for k-SAT, inapproximability of MAX-3SAT and related problems, and proof complexity of unsatisfiable formulas.

An integral part of the SAT conferences are the competitions and evaluations. SAT 2008 featured a SAT Race in the spirit of the SAT Competitions, a competitive QBF Evaluation and a Max-SAT Evaluation. The SAT Race was organized by Carsten Sinz (Chair), Nina Amla, Toni Jussila, Daniel Le Berre, Panagiotis Manolios, and Lintao Zhang. For the first time there were special tracks for parallel SAT solvers and structural SAT solvers. The QBF Evaluation was organized by Massimo Narizzano, Luca Pulina, and Armando Tacchella. Participants of this evaluation could contribute by submitting implementations of QBF solvers as well as submitting hard instances of QBF formulas. The Max-SAT Evaluation was organized by Josep Argelich, Chu Min Li, Felip Manyà, and Jordi Planes. The evaluation was divided in four categories: Max-SAT, Weigthed Max-SAT, Partial Max-SAT, and Weighted Partial Max-SAT.

The organizers of this year's conference are deeply indebted to the large number of people who contributed to its preparation: the local organizers Shier Ju and Minghui Xiong in Guangzhou; Uwe Bubeck and Theo Lettmann in Paderborn; the organizers of the affiliated events as mentionend above.

We thank the authors for their contributions and we thank the Program Committee and the additional reviewers for their careful and thorough work. Without their efforts, it would not have been possible for us to put together such an excellent conference program. In particular, we are grateful to Andrei Voronkov for his EasyChair system. EasyChair is a very helpful tool for the organization of paper submissions, the reviewing process, Program Committee discussions, and assembly of the proceedings.

Financial and organizational support was generously provided by the National Natural Science Foundation of China, the Sun Yat-Sen University (Guangzhou, P.R. China), especially the Institute of Logic and Cognition, and the University of Paderborn (Paderborn, Germany), especially the Department of Computer Science.

Finally, we would like to thank the sponsors for their generous support of the SAT 2008 conference: Hexin Technology (Guangzhou), Intel Design Technologies and Solutions (Haifa), K.C. Wong Education Foundation (Hong Kong), Light Engineering (Guangzhou), Microsoft Research (Mountain View), NEC Labs America (Princeton), and Potevio (Beijing).

May 2008   Hans Kleine Büning
                                                                 Xishun Zhao

Organization

Conference Chairs

Hans Kleine Büning
Xishun Zhao

Local Chairs

Shier Ju
Minghui Xiong
Lin Xu
Uwe Bubeck
Theo Lettmann

Technical Program Committee

Fahiem Bacchus
Armin Biere
Nadia Creignou
Adnan Darwiche
Leonardo de Moura
Decheng Ding
John Franco
Ian Gent
Enrico Giunchiglia
Aarti Gupta
Ziyad Hanna
Holger Hoos

Henry Kautz
Oliver Kullmann
Daniel Le Berre
Chu-Min Li
Ines Lynce
Panagiotis Manolios
Joao Marques-Silva
David Mitchell
Stefan Porschen
Steve Prestwich
Karem Sakallah
Uwe Schöning

Roberto Sebastiani
Bart Selman
Laurent Simon
Ewald Speckenmeyer
Ofer Strichman
Stefan Szeider
Allen Van Gelder
Hans van Maaren
Toby Walsh
Jian Zhang
Lintao Zhang

External Reviewers

A. Anbulagan
Ralph Becket
Roderick Bloem
Sebastian Brand
Roberto Bruttomesso
Uwe Bubeck
Benjamin Chambers
Baiqiang Chen
Sylvain Darras
Hervé Daudé

Jed Davis
Gilles Dequen
Laure Devendeville
Jianfeng Du
Anders Franzén
Oded Fuhrmann
Alberto Griggio
Christine Hang
Warwick Harvey
Miki Hermann

Frank Hutter
Xiangxue Jia
George Katsirelos
Mark Liffiton
Sheng Liu
Feifei Ma
Michael Maher
Vasco Manquinho
Paolo Marin
Hannes Moser

Moritz Müller   Vadim Ryvchin   Stefano Tonetta
Massimo Narizzano   Marko Samer   Aaron Turon
Peter Nightingale   Tatjana Schmidt   Michele Vescovi
Luca Pulina   Ilka Schnoor   Wanxia Wei
Olivier Roussel   Martina Seidl   Lin Xu
Bert Randerath   Ohad Shacham   Christian Bessière
Igor Razgon   Yuping Shen
Emanuele Di Rosa   Dave Tompkins

Sponsoring Institutions

National Natural Science Foundation of China
Sun Yat-Sen University of Guangzhou
Institute of Logic and Cognition, Sun Yat-Sen University
University of Paderborn
Department of Computer Science, University of Paderborn

Hexin Technology (Guangzhou)
Intel Design Technologies and Solutions (Haifa)
K. C. Wong Education Foundation (Hongkong)
Light Engineering (Guangzhou)
Microsoft Research (Mountain View)
NEC Labs America (Princeton)
Potevio (Beijing)

Table of Contents

Modelling Max-CSP as Partial Max-SAT[*]

Josep Argelich[1], Alba Cabiscol[1], Inês Lynce[2], and Felip Manyà[1]

[1] Computer Science Department
Universitat de Lleida
Jaume II, 69, E-25001 Lleida, Spain
[2] IST/INESC-ID
Technical University of Lisbon
Rua Alves Redol 9, 1000-029 Lisboa, Portugal

Abstract. We define a number of original encodings that map Max-CSP instances into partial Max-SAT instances. Our encodings rely on the well-known direct and support encodings from CSP into SAT. Then, we report on an experimental investigation that was conducted to compare the performance profile of our encodings on random binary Max-CSP instances. Moreover, we define a new variant of the support encoding from CSP into SAT which produces fewer clauses than the standard support encoding.

1 Introduction

In the last years, there has been an increasing interest in the Boolean Max-SAT problem. Taking into account the success of SAT on solving NP-complete decision problems, the SAT community investigates how to transfer the technology created for SAT to Max-SAT with the aim of developing fast Max-SAT solvers, which can be used to solve NP-hard optimization problems via their reduction to Max-SAT.

The most recent and relevant results for Max-SAT can be summarized as follows: (i) there exist solvers like Clone [21], Lazy [1], MaxSatz [18], MiniMaxSat [12], ms4 [19], SR(w) [22] and Max-DPLL [15] which solve many instances that are beyond the reach of the solvers existing just five years ago; (ii) resolution refinements, which preserve the number of unsatisfied clauses, have been incorporated into Max-SAT solvers [14,15,18], as well as good quality under-estimations of the lower bound [16,17,21,22], (iii) a resolution-style calculus for Max-SAT has been proven to be complete [5,6], (iv) formalisms like Partial Max-SAT have been investigated for solving problems with soft constraints [2,8,12,3], and (v) two evaluations of Max-SAT solvers have been performed for the first time as a co-located event of the International Conference on Theory and Applications of Satisfiability Testing (SAT-2006 and SAT-2007).

[*] This research was funded by MEC research projects TIN2006-15662-C02-02, TIN2007-68005-C04-04 and Acción Integrada HP2005-0147, and FCT research projects SATPot (POSC/EIA/61852/2004) and SHIPs (PTDC/EIA/64164/2006).

H. Kleine Büning and X. Zhao (Eds.): SAT 2008, LNCS 4996, pp. 1–14, 2008.

In this paper, we define a number of original encodings that map Max-CSP instances into partial Max-SAT instances. Our encodings rely on the well-known direct and support encodings from CSP into SAT. Then, we report on an experimental investigation that was conducted to compare the performance profile of our encodings on random binary Max-CSP instances. Interestingly, we also define a new variant of the support encoding from CSP into SAT which produces fewer clauses than the standard support encoding. Our new encoding, called *minimal support encoding*, eliminates redundant clauses.

The objective of our research is to show that different Max-SAT encodings for a same optimization problem may produce substantial differences on performance, as well as to identify features of the encodings that lead to better performance profiles. To the best of our knowledge, this is the first paper that addresses the question of how to encode Max-CSP into Max-SAT, and analyzes the impact of modelling on the performance of Max-SAT solvers.

The structure of the paper is as follows. Section 2 contains preliminary definitions about Max-SAT and Max-CSP. Section 3 surveys the support and direct encodings from CSP into SAT, and defines a new encoding that we call minimal support encoding. Section 4 defines a number of original encodings from Max-CSP into partial Max-SAT. Section 5 reports and analyses the experimental investigation. Section 6 presents the conclusions and future research directions.

2   Preliminaries

2.1   Max-SAT Definitions

In propositional logic a variable x_i may take values 0 (for false) or 1 (for true). A literal l_i is a variable x_i or its negation $\bar{x}_i$. A clause is a disjunction of literals, and a CNF formula is a multiset of clauses.

An assignment of truth values to the propositional variables satisfies a literal x_i if x_i takes the value 1 and satisfies a literal $\bar{x}_i$ if x_i takes the value 0, satisfies a clause if it satisfies at least one literal of the clause, and satisfies a CNF formula if it satisfies all the clauses of the formula. An empty clause, denoted by $\square$, contains no literals and cannot be satisfied.

The Max-SAT problem for a CNF formula ϕ is the problem of finding an assignment of values to propositional variables that maximizes the number of satisfied clauses. In this sequel we often use the term Max-SAT meaning Min-UNSAT. This is because, with respect to exact computations, finding an assignment that minimizes the number of unsatisfied clauses is equivalent to finding an assignment that maximizes the number of satisfied clauses.

We also consider the extension of Max-SAT known as Partial Max-SAT because it is more well-suited for representing and solving NP-hard problems. A Partial Max-SAT instance is a CNF formula in which some clauses are *relaxable* or *soft* and the rest are *non-relaxable* or *hard*. Solving a Partial Max-SAT instance amounts to finding an assignment that satisfies all the hard clauses and the maximum number of soft clauses. Hard clauses are represented between square brackets, and soft clauses are represented between round brackets.

2.2   Max-CSP Definitions

Definition 1. *A* Constraint Satisfaction Problem (CSP) *instance is defined as a triple* $\langle \mathcal{X}, \mathcal{D}, \mathcal{C} \rangle$, *where* $\mathcal{X} = \{X_1, \ldots, X_n\}$ *is a set of variables,* $\mathcal{D} = \{d(X_1), \ldots, d(X_n)\}$ *is a set of finite domains containing the values the variables may take, and* $\mathcal{C} = \{C_1, \ldots, C_m\}$ *is a set of constraints. Each constraint* $C_i = \langle S_i, R_i \rangle$ *is defined as a relation* R_i *over a subset of variables* $S_i = \{X_{i_1}, \ldots, X_{i_k}\}$, *called the* constraint scope. *The relation* R_i *may be represented extensionally as a subset of the Cartesian product* $d(X_{i_1}) \times \cdots \times d(X_{i_k})$.

Definition 2. *An* assignment v *for a CSP instance* $\langle \mathcal{X}, \mathcal{D}, \mathcal{C} \rangle$ *is a mapping that assigns to every variable* $X_i \in \mathcal{X}$ *an element* $v(X_i) \in d(X_i)$. *An assignment* v *satisfies a constraint* $\langle \{X_{i_1}, \ldots, X_{i_k}\}, R_i \rangle \in C$ *iff* $\langle v(X_{i_1}), \ldots, v(X_{i_k}) \rangle \in R_i$.

Definition 3. *The Constraint Satisfaction Problem (CSP) for a CSP instance* P *consists in deciding whether there exists an assignment that satisfies* P.

In the sequel we assume that all CSPs are unary and binary; i.e., the scope of all the constraints has cardinality at most two.

Definition 4. *A CSP is* node consistent, *if for every variable* X_i, *every value of the domain of* X_i *is allowed for the unary constraints on* X_i. *A CSP is* arc consistent, *if for every constraint on two variables* X_i *and* Y_j, *for all* $a \in d(X_i)$, *there exists* $b \in d(Y_j)$, *such that* (a, b) *is in the constraint.*

Definition 5. *The Max-CSP problem for a CSP instance* $\langle \mathcal{X}, \mathcal{D}, \mathcal{C} \rangle$ *is the problem of finding an assignment that minimizes (maximizes) the number of violated (satisfied) constraints.*

3   Encoding CSP into SAT

Mappings of binary CSPs into SAT is an area of research that has been investigated by several authors [4,9,10,11,13,23]. They have proposed a number of encodings having different performance profiles and achieving different degrees of local propagation on SAT solvers. Among them, the most well-known are the direct encoding and the support encoding. In the rest of this section, we first define the direct encoding and the support encoding, and then define a new encoding from CSP into SAT called *minimal support encoding*.

3.1   Direct Encoding and Support Encoding

In the *direct encoding*, we associate a Boolean variable x_{ij} with each value j that can be assigned to the CSP variable X_i. Assuming that X_i has a domain of size m, the direct encoding contains clauses that ensure that each CSP variable X_i is given a value: for each i, $x_{i1} \vee \cdots \vee x_{im}$ (called *at-least-one* clauses), and contains clauses that rule out any binary nogoods. For example, if $X_1 = 2$ and $X_3 = 1$ is not allowed, then the clause $\overline{x}_{12} \vee \overline{x}_{31}$ (called *conflict* clause) is added.

We consider the version of the direct encoding that adds clauses that ensure that each CSP variable X_i takes no more than one value: for each i, j, k with $j < k$, $\overline{x}_{ij} \vee \overline{x}_{ik}$ (called *at-most-one* clauses). These clauses are redundant, but are considered in the literature in order to maintain a one-to-one mapping between CSP models and SAT models.

In the *support encoding*, the idea is to encode into clauses the *support* for a value instead of encoding conflicts. The support for a value j of a CSP variable X_i across a constraint is the set of values of the other variable in the constraint which allow $X_i = j$. If $v_1, v_2, \ldots, v_k$ are the supporting values of variable X_l for $X_i = j$, we add the clause $\overline{x}_{ij} \vee x_{lv_1} \vee x_{lv_2} \vee \cdots \vee x_{lv_k}$ (called *support* clause). There is one support clause for each pair of variables X_i, X_l involved in a constraint, and for each value in the domain of X_i. Unlike conflict clauses, a clause in each direction is used in the literature, one for the pair X_i, X_l and one for X_l, X_i. The support clauses on their own do not provide a correct encoding of CSPs into SAT. To complete an encoding using support clauses we need to add the at-least-one and at-most-one clauses for each CSP variable to ensure that each CSP variable takes exactly one value of its domain.

Example 1. The direct encoding of the CSP $\langle \mathcal{X}, \mathcal{D}, \mathcal{C} \rangle = \langle \{X, Y\}, \{d(X) = \{1, 2, 3\}, d(Y) = \{1, 2, 3\}\}, \{X \leq Y\} \rangle$ contains the following clauses:

at-least-one	$x_1 \vee x_2 \vee x_3$	$y_1 \vee y_2 \vee y_3$	
at-most-one	$\overline{x}_1 \vee \overline{x}_2$	$\overline{x}_1 \vee \overline{x}_3$	$\overline{x}_2 \vee \overline{x}_3$
	$\overline{y}_1 \vee \overline{y}_2$	$\overline{y}_1 \vee \overline{y}_3$	$\overline{y}_2 \vee \overline{y}_3$
conflict	$\overline{x}_2 \vee \overline{y}_1$	$\overline{x}_3 \vee \overline{y}_1$	$\overline{x}_3 \vee \overline{y}_2$

and the support encoding for that CSP contains the at-least-one clauses, the at-most-one clauses, and the following support clauses:

support	$\overline{x}_2 \vee y_2 \vee y_3$	$\overline{y}_1 \vee x_1$
	$\overline{x}_3 \vee y_3$	$\overline{y}_2 \vee x_1 \vee x_2$

The support clause for x_1 is missing because it is subsumed by $y_1 \vee y_2 \vee y_3$, and the support clause for y_3 is missing because it is subsumed by $x_1 \vee x_2 \vee x_3$.

3.2   Minimal Support Encoding

Our first contribution in this paper is to give a new version of the support encoding, which we call *minimal support encoding*. Our definition follows from the observation that the support encoding contains redundant clauses. More precisely, given a binary constraint C_k with scope $\{X, Y\}$, it is enough to add the support clauses either for the values of X or for the values of Y; it is not necessary to add a clause in each direction. Despite of the number of papers dealing with the support encodings, this fact has gone unnoticed so far.

Definition 6. *The* minimal support encoding *is like the support encoding except for the fact that, for every constraint C_k with scope $\{X, Y\}$, we only add either the support clauses for all the domain values of the CSP variable X or the support clauses for all the domain values of the CSP variable Y.*

Example 2. A minimal support encoding for the CSP instance from Example 1 contains the following clauses:

at-least-one	$x_1 \vee x_2 \vee x_3$	$y_1 \vee y_2 \vee y_3$	
at-most-one	$\overline{x}_1 \vee \overline{x}_2$	$\overline{x}_1 \vee \overline{x}_3$	$\overline{x}_2 \vee \overline{x}_3$
	$\overline{y}_1 \vee \overline{y}_2$	$\overline{y}_1 \vee \overline{y}_3$	$\overline{y}_2 \vee \overline{y}_3$
support	$\overline{x}_2 \vee y_2 \vee y_3$	$\overline{x}_3 \vee y_3$	

Proposition 1. *The* minimal support encoding *is correct.*

PROOF: We assume, without loss of generality, that we add the support clauses for all the domain values of the CSP variable X for every constraint C_k with scope $\{X, Y\}$. Given a CSP assignment, we construct its corresponding Boolean assignment by setting the variable x_i to true if the CSP assignment assigns the value i to X; otherwise, we set the variable x_i to false. Given a Boolean assignment that satisfies the minimal support encoding of a CSP, we construct its corresponding CSP assignment by assigning to the CSP variable X the value i if x_i is true. Note that there is exactly one x_i for each CSP variable X which is true because the minimal support encoding contains the at-least-one and at-most-one clauses. So, it is a valid CSP assignment.

We prove first that if a CSP assignment satisfies all the constraints of a CSP instance, then its corresponding Boolean assignment satisfies its minimal encoding. Since a CSP assignment assigns exactly one value to each CSP variable, the Boolean assignment satisfies the at-least-one and at-most-one clauses. For every constraint C_k with scope $\{X, Y\}$, the CSP assignment assigns a value i to X and a value j to Y. Since $(X = i, Y = j)$ is an allowed combination, among the clauses encoding that constraint, there is a clause of the form $\overline{x}_i \vee y_j \vee \cdots$ which is satisfied by the Boolean encoding because y_j is true. The remaining clauses are also satisfied by the Boolean assignment because they are of the form $\overline{x}_l \vee \cdots$, where $l \neq i$, and the Boolean assignment assigns the value false to all variables x_l with $l \neq i$.

We prove now that if a Boolean assignment satisfies the minimal support encoding of a CSP instance P, then its corresponding CSP assignment satisfies P. Assume that the CSP assignment does not satisfy P. Therefore, there exists a constraint C_k of P with scope $\{X, Y\}$ which is violated because the CSP assignment assigns a value i to X and a value j to Y which corresponds to a forbidden combination. In this case, there is exactly one support clause of the form $\overline{x}_i \vee y_{j_1} \vee \cdots \vee y_{j_k}$ among the support clauses encoding C_k which is not satisfied by the Boolean assignment because x_i is true and $y_{j_1} \neq y_j, \ldots, y_{j_k} \neq y_j$. The rest of support clauses encoding C_k are satisfied by the Boolean assignment because it assigns the value false to all variables x_l with $l \neq i$. ∎

Unlike the support encoding [11,13], the minimal support encoding does not maintain arc consistency through unit propagation. Recall that the direct encoding does not maintain arc consistency too.

Proposition 2. *The minimal support encoding does not maintain arc consistency through unit propagation.*

PROOF: We give a counterexample to prove the proposition. Given the CSP instance $\langle \mathcal{X}, \mathcal{D}, \mathcal{C} \rangle$, where $\mathcal{X} = \{X, Y\}, d(X) = d(Y) = \{1, 2, 3\}, \mathcal{C} = \{C_{XY}\} = \{\{(1,1),(2,2),(3,3)\}\}$ with the following minimal support encoding:

at-least-one	$x_1 \lor x_2 \lor x_3$	$y_1 \lor y_2 \lor y_3$	
at-most-one	$\overline{x}_1 \lor \overline{x}_2$	$\overline{x}_1 \lor \overline{x}_3$	$\overline{x}_2 \lor \overline{x}_3$
	$\overline{y}_1 \lor \overline{y}_2$	$\overline{y}_1 \lor \overline{y}_3$	$\overline{y}_2 \lor \overline{y}_3$
support	$\overline{x}_1 \lor y_1$	$\overline{x}_2 \lor y_2$	$\overline{x}_3 \lor y_3,$

if x_1 is set to false, then $\overline{y}_1$ is not derived by unit propagation, and the domain of Y is not arc consistent. Observe that if the support clauses are $\overline{y}_1 \lor x_1, \overline{y}_2 \lor x_2, \overline{y}_3 \lor x_3$, then $\overline{y}_1$ is derived by unit propagation, and the domain of Y becomes arc consistent. However, if y_1 is set to false, then arc consistency is not maintained in the last case. ■

4   Encoding Max-CSP into Partial Max-SAT

4.1   Direct Encoding for Partial Max-SAT

Given a CSP instance P, our goal is to define a version of the direct encoding that produces a partial Max-SAT instance ϕ such that the minimum number of constraints of P that are violated by a CSP assignment is exactly the same as the minimum number of clauses of ϕ that are falsified by a Boolean assignment.

Definition 7. *The* direct encoding *of a Max-CSP instance $\langle \mathcal{X}, \mathcal{D}, \mathcal{C} \rangle$ is the Partial Max-SAT instance that contains as hard clauses the corresponding at-least-one and at-most-one clauses for every CSP variable in $\mathcal{X}$, and contains a soft clause $\overline{x}_i \lor \overline{y}_j$ for every nogood $(X = i, Y = j)$ of every constraint of $\mathcal{C}$ with scope $\{X, Y\}$.*

Example 3. The Partial Max-SAT direct encoding for the Max-CSP problem of the CSP instance from Example 1 is as follows:

at-least-one	$[x_1 \lor x_2 \lor x_3]$	$[y_1 \lor y_2 \lor y_3]$	
at-most-one	$[\overline{x}_1 \lor \overline{x}_2]$	$[\overline{x}_1 \lor \overline{x}_3]$	$[\overline{x}_2 \lor \overline{x}_3]$
	$[\overline{y}_1 \lor \overline{y}_2]$	$[\overline{y}_1 \lor \overline{y}_3]$	$[\overline{y}_2 \lor \overline{y}_3]$
conflict	$(\overline{x}_2 \lor \overline{y}_1)$	$(\overline{x}_3 \lor \overline{y}_1)$	$(\overline{x}_3 \lor \overline{y}_2)$

Proposition 3. *Solving a Max-CSP instance is equivalent to solving the Partial Max-SAT problem of its direct encoding.*

PROOF: The hard clauses ensure that exactly one of the Boolean variables that encode a CSP variable is true and the rest are false in a feasible assignment. Therefore, we have that there is a one-to-one mapping between the set of CSP assignments and the set of feasible assignments of the Partial Max-SAT instance and, moreover, at most one of the conflict clauses that encode a certain constraint can be falsified by a feasible assignment. If the CSP assignment satisfies a constraint, then the corresponding Boolean assignment also satisfies the conflict clauses that encode that constraint because there is no clause forbidding allowed values. If the CSP assignment violates a constraint, then the corresponding Boolean assignment does not satisfy the conflict clause that encodes the forbidden values of the two variables involved in the constraint, and satisfies the remaining clauses.   ∎

There are other options for defining the direct encoding which amount to introducing auxiliary variables. For example, you can add all the clauses representing nogoods as hard clauses by adding an auxiliary literal c_i to every clause encoding a nogood of every constraint $C_i \in \mathcal{C}$, and adding the unit clause $\bar{c}_i$ as a soft clause. Nevertheless, we do not consider this encoding because we realized that its performance profile is worse than the performance profile of the direct encoding (at least for the benchmarks considered in our empirical evaluation).

4.2   Support Encoding for Partial Max-SAT

The support encoding for Partial Max-SAT may be defined by adapting the minimal support encoding from CSP into SAT:

Definition 8. *The* minimal support encoding *of a Max-CSP instance* $\langle \mathcal{X}, \mathcal{D}, \mathcal{C} \rangle$ *is the Partial Max-SAT instance that contains as hard clauses the corresponding at-least-one and at-most-one clauses for every CSP variable in* $\mathcal{X}$*, and contains as soft clauses the support clauses of the minimal support encoding from CSP into SAT.*

Example 4. A minimal Partial Max-SAT support encoding for the Max-CSP problem of the CSP instance from Example 1 contains the following clauses:

at-least-one	$[x_1 \vee x_2 \vee x_3]$	$[y_1 \vee y_2 \vee y_3]$	
at-most-one	$[\overline{x}_1 \vee \overline{x}_2]$	$[\overline{x}_1 \vee \overline{x}_3]$	$[\overline{x}_2 \vee \overline{x}_3]$
	$[\overline{y}_1 \vee \overline{y}_2]$	$[\overline{y}_1 \vee \overline{y}_3]$	$[\overline{y}_2 \vee \overline{y}_3]$
support	$(\overline{x}_2 \vee y_2 \vee y_3)$	$(\overline{x}_3 \vee y_3)$	

Proposition 4. *Solving a Max-CSP instance is equivalent to solving the Partial Max-SAT problem of its minimal support encoding.*

PROOF: Proposition 1 proves that there is one unsatisfied clause for every violated constraint. Since the minimal support encoding is correct, and the hard clauses ensure a one-to-one mapping between Max-CSP and feasible Partial Max-SAT assignments, the optimal solutions of Max-CSP are exactly the same as the optimal solutions of Partial Max-SAT.   ∎

We now define how to adapt to Partial Max-SAT the support encoding from CSP into SAT.

Definition 9. *The* support encoding *of a Max-CSP instance* $\langle \mathcal{X}, \mathcal{D}, \mathcal{C} \rangle$ *is the Partial Max-SAT instance that contains as hard clauses the corresponding* *at-least-one and at-most-one clauses for every CSP variable in $\mathcal{X}$, and contains, for every constraint $C_k \in \mathcal{C}$ with scope $\{X,Y\}$, a soft clause of the form $S_{X=j} \vee c_k$ for every support clause $S_{X=j}$ encoding the support for the value j of the CSP variable X, where c_k is an auxiliary variable , and contains a soft clause of the form $S_{Y=m} \vee \overline{c}_k$ for every support clause $S_{Y=m}$ encoding the support for the value m of the CSP variable Y.*

Observe that we introduce an auxiliary variable for every constraint. This is due to the fact that there are two unsatisfied soft clauses for every violated constraint of the Max-CSP instance if we do not introduce auxiliary variables. It is particularly important to have one unsatisfied clause for every violated constraints when mapping weighted Max-CSP instances into weighted Max-SAT instances.[1] In this case, all the clauses encoding a certain constraint have as weight the weight associated to that constraint. When a constraint is violated with weight w, this guarantees that there is exactly one unsatisfied clause with weight w.

Example 5. The Partial Max-SAT support encoding for the Max-CSP problem of the CSP instance from Example 1 is as follows:

at-least-one $[x_1 \vee x_2 \vee x_3]$ $[y_1 \vee y_2 \vee y_3]$
at-most-one $[\overline{x}_1 \vee \overline{x}_2]$ $[\overline{x}_1 \vee \overline{x}_3]$ $[\overline{x}_2 \vee \overline{x}_3]$
$[\overline{y}_1 \vee \overline{y}_2]$ $[\overline{y}_1 \vee \overline{y}_3]$ $[\overline{y}_2 \vee \overline{y}_3]$
support $(\overline{x}_2 \vee y_2 \vee y_3 \vee c_1)$ $(\overline{y}_1 \vee x_1 \vee \overline{c}_1)$
$(\overline{x}_3 \vee y_3 \vee c_1)$ $(\overline{y}_2 \vee x_1 \vee x_2 \vee \overline{c}_1)$

Proposition 5. *Solving a Max-CSP instance is equivalent to solving the Partial Max-SAT problem of its support encoding.*

PROOF: By introducing auxiliary variables we ensure that the optimal solutions of Max-CSP are exactly the same as the optimal solutions of Partial Max-SAT. The auxiliary variables allow to violate exactly one clause for every violated constraint. ∎

In the following proposition we assume that Partial Max-SAT solvers incorporate the rule that replaces any two complementary unit clauses with an empty clause. Actually, most of the solvers we know implement such a rule.

Proposition 6. *When solving a Max-CSP instance with the support encoding on a Partial Max-SAT solver, it is not necessary to branch on the auxiliary variables.*

PROOF: For every violated constraint C_k with scope $\{X,Y\}$, there is exactly one unsatisfied support clause of the form $\overline{x}_i \vee y_{j_1} \vee \cdots \vee y_{j_k}$ and one unsatisfied support clause of the form $\overline{y}_l \vee x_{m_1} \vee \cdots \vee x_{m_s}$ in the support encoding from

---

[1] In weighted Max-CSP (Max-SAT), each constraint (clause) has a weight and the goal is to minimize the sum of the weights of the violated constraints (falsified clauses).

CSP into SAT. Therefore, these clauses will produce the derivation of the two complementary unit clauses in the support encoding from Max-CSP into Partial Max-SAT: c_k (from $\overline{x}_i \vee y_{j_1} \vee \cdots \vee y_{j_k} \vee c_k$) and $\overline{c}_k$ (from $\overline{y}_l \vee x_{m_1} \vee \cdots \vee x_{m_s} \vee \overline{c}_k$). The solver will then derive a contradiction from these two clauses. If C_k is satisfied, both support clauses are satisfied and the fact of adding an extra literal does not affect their satisfaction.   ∎

On the solved benchmarks we did not see significant differences between branching including auxiliary variables and branching without including them. So, we only report results for branching including auxiliary variables. However, there may exist differences on other types of instances and solvers.

5   Experimental Results

We conducted an empirical evaluation to assess the impact of the defined encodings on the performance of two of the best performing Partial Max-SAT solvers: MiniMaxSat [12] and PMS [3]. Moreover, we compared the support encoding and the minimal support encoding when solving SAT-encoded CSP instances with MiniSat [7] and zChaff [20]. The evaluation was conducted on a cluster with 160 2 GHz AMD Opteron 248 Processors with 1 GB of memory.

As benchmarks we considered binary CSPs, which were obtained with a generator of uniform random binary CSPs[2] —designed and implemented by Frost, Bessière, Dechter and Regin— that implements the so-called model B: in the class $\langle n, d, p_1, p_2 \rangle$ with n variables of domain size d, we choose a random subset of exactly $p_1 n(n-1)/2$ constraints (rounded to the nearest integer), each with exactly $p_2 d^2$ conflicts (rounded to the nearest integer); p_1 may be thought of as the *density* of the problem and p_2 as the *tightness* of constraints. The difficulty of the instances depends on the selected values for n, d, p_1 and p_2. We selected values that allowed to solve the instances in a reasonable amount of time in each solver.

We used the following encodings: the direct encoding (`dir`), the support encoding (`supxy`), and three variants of the minimal support encoding (`supx`, `supl`, `supc`). The encoding `supx` refers to the minimal support encoding of a binary CSP containing only the support clauses for the CSP variable X and not for the variable Y for every constraint with scope $\{X, Y\}$; we do not show results for the encoding containing only support clauses for the CSP variable Y because its behaviour is very close to `supx` for the solved random instances. The encoding `supl` refers to the minimal support encoding containing, for each constraint, the support clauses for the variable that produces a smaller total number of literals. The encoding `supc` refers to the minimal support encoding containing, for each constraint, the support clauses for the variable that produces smaller size clauses; we give a score of 16 to unit clauses, a score of 4 to binary clauses and a score of 1 to ternary clauses, and choose the variable with higher sum of scores. For instance, given the CSP instance $\langle X, D, C \rangle$, where $X = \{X, Y\}, d(X) = d(Y) = \{1, 2, 3, 4\}, C = \{C_{XY}\} = \{\{(1, 2), (1, 3), (1, 4)\}\}$,

---

[2] http://www.lirmm.fr/~bessiere/generator.html

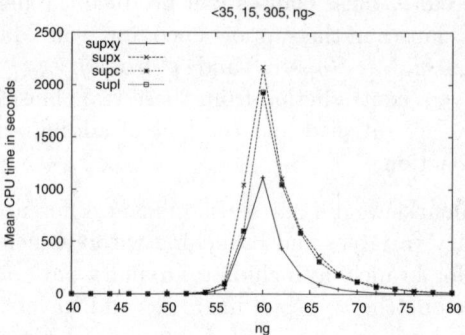

Fig. 1. Experimental results for MiniSat

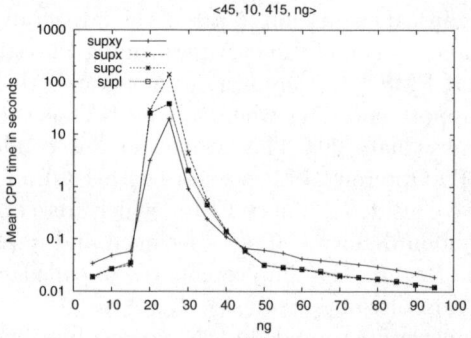

Fig. 2. Experimental results for zChaff (log scale)

supc prefers three binary support clauses $x_1 \vee \overline{y}_2, x_1 \vee \overline{y}_3, x_1 \vee \overline{y}_4$ rather than the quaternary support clause $\overline{x}_1 \vee y_2 \vee y_3 \vee y_4$, while supl prefers $\overline{x}_1 \vee y_2 \vee y_3 \vee y_4$.

In the first experiment we solved 100 CSP instances with MiniSat and zChaff for each data point. We compared all the support encodings from CSP into SAT. With MiniSat, we solved CSP instances with 35 variables, domains of 15 elements, 305 constraints and variable tightness (we vary the number of nogoods (ng)). The obtained results are shown in Figure 1. With zChaff, we solved CSP instances with 45 variables, domains of 10 elements, 415 constraints and variable tightness. The obtained results are shown in Figure 2. We observe that the support encoding outperforms the three variants of the minimal support encoding. We believe that this is due to the fact that the support encoding, unlike the minimal support encoding, maintains arc consistency through unit propagation.

In the second experiment we solved 100 Max-CSP instances with MiniMaxSat for each data point; the instances had 22 variables, domains of 4 elements, 231 constraints and variable tightness. We compared all the defined encodings of Max-CSP into Partial Max-SAT. The obtained results are shown in Figure 3.

In the third experiment we solved 100 Max-CSP instances with MiniMaxSat for each data point; the instances had 25 variables, domains of 5 elements, 150

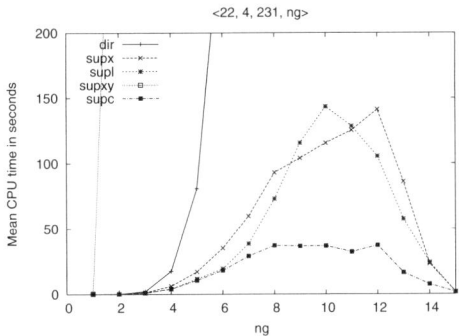

Fig. 3. Experimental results for MiniMaxSat

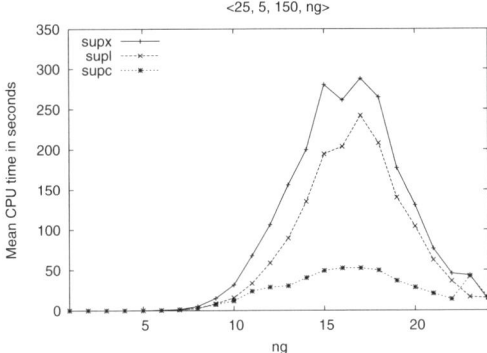

Fig. 4. Experimental results for MiniMaxSat

constraints and variable tightness. We compared all the defined encodings of Max-CSP into Partial Max-SAT. The obtained results are shown in Figure 4. We omit the results for the encodings dir and supxy because they are not competitive.

In the fourth experiment we solved 100 Max-CSP instances with PMS for each data point; the instances had 15 variables, domains of 4 elements, 120 constraints and variable tightness. We compared all the defined encodings of Max-CSP into Partial Max-SAT. The obtained results are shown in Figure 5.

In the fifth experiment we solved 100 Max-CSP instances with PMS for each data point; the instances had 14 variables, domains of 5 elements, 91 constraints and variable tightness. We compared all the defined encodings of Max-CSP into Partial Max-SAT. The obtained results are shown in Figure 6.

We observe that support encodings from Max-CSP into Partial Max-SAT, which have been introduced for the first time in this paper, outperform the direct encoding for both solvers. In MiniMaxSat, the best performing encoding is the minimal support encoding. Among the different versions of the minimal support encoding, we observe that supc is up to 6 times faster than the other

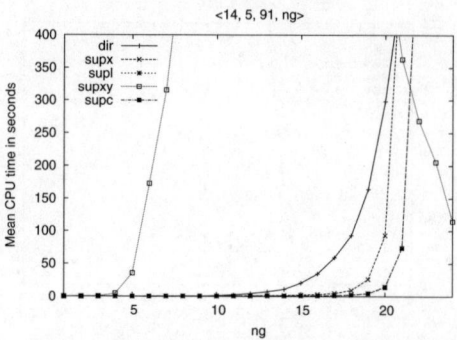

Fig. 5. Experimental results for PMS

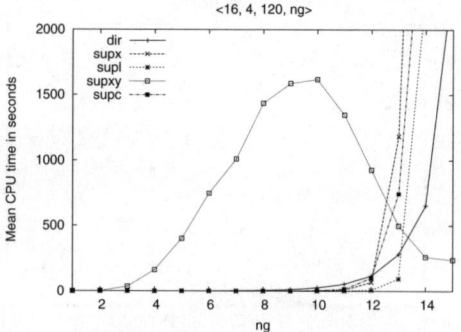

Fig. 6. Experimental results for PMS

Table 1. Number of clauses for different encodings

Parameters	dir	supxy	supx	supc	supl
$\langle 25, 5, 150, 2 \rangle$	575	824	551	549	524
$\langle 25, 5, 150, 4 \rangle$	875	1201	738	720	685
$\langle 25, 5, 150, 6 \rangle$	1175	1445	861	826	804
$\langle 25, 5, 150, 8 \rangle$	1475	1602	939	905	890
$\langle 25, 5, 150, 10 \rangle$	1775	1690	983	959	950
$\langle 25, 5, 150, 12 \rangle$	2075	1739	1007	993	991
$\langle 25, 5, 150, 14 \rangle$	2375	1762	1019	1012	1012
$\langle 25, 5, 150, 16 \rangle$	2675	1771	1023	1021	1021
$\langle 25, 5, 150, 18 \rangle$	2975	1774	1025	1024	1024
$\langle 25, 5, 150, 20 \rangle$	3275	1775	1025	1025	1025

two encodings (supx and supl). In PMS, the best encoding for high values of tightness is the support encoding while the best encodings for lower values are supc and supl in Figure 5, and supl in Figure 6.

Finally, in Table 1 we show, for each different encoding, the average number of clauses of some sets of Max-CSP instances solved in the third experiment (Figure 4) in order to illustrate the differences on the number of clauses among all the defined encodings from Max-CSP into Partial Max-SAT. Observe that encodings supx, supl and supc produce instances with a similar number of clauses.

6   Conclusions

We have defined the minimal support encoding, which is a new encoding from CSP into SAT, and a number of original encodings (dir, supxy, supx, supl, supc) that map Max-CSP instances into partial Max-SAT instances, and have provided experimental evidence that different Max-SAT encodings for a given optimization problem may produce substantial differences on the performance of a solver. Since our mappings produce one unsatisfied clause for every violated constraints, they can be easily extended to mappings from weighted Max-CSP instances into weighted Max-SAT instances; all the clauses encoding a certain constraint should have as weight the weight associated to that constraint.

To the best of our knowledge, this is the first paper that addresses the question of how to encode Max-CSP into Max-SAT, and analyzes the impact of modelling on the performance of Max-SAT solvers. Future research directions include analyzing the degree of soft local consistency achieved by each encoding, conducting an experimental investigation with benchmarks other than random binary Max-CSP instances, and generalizing our results to n-ary constraints.

References

1. Alsinet, T., Manyà, F., Planes, J.: Improved exact solver for weighted Max-SAT. In: Bacchus, F., Walsh, T. (eds.) SAT 2005. LNCS, vol. 3569, pp. 371–377. Springer, Heidelberg (2005)
2. Argelich, J., Manyà, F.: Exact Max-SAT solvers for over-constrained problems. Journal of Heuristics 12(4–5), 375–392 (2006)
3. Argelich, J., Manyà, F.: Partial Max-SAT solvers with clause learning. In: Marques-Silva, J., Sakallah, K.A. (eds.) SAT 2007. LNCS, vol. 4501, pp. 28–40. Springer, Heidelberg (2007)
4. Bessière, C., Hebrard, E., Walsh, T.: Local consistencies in SAT. In: Giunchiglia, E., Tacchella, A. (eds.) SAT 2003. LNCS, vol. 2919, pp. 299–314. Springer, Heidelberg (2004)
5. Bonet, M.L., Levy, J., Manyà, F.: A complete calculus for Max-SAT. In: Biere, A., Gomes, C.P. (eds.) SAT 2006. LNCS, vol. 4121, pp. 240–251. Springer, Heidelberg (2006)
6. Bonet, M.L., Levy, J., Manyà, F.: Resolution for Max-SAT. Artificial Intelligence 171(8–9), 240–251 (2007)
7. Een, N., Sorensson, N.: An extensible SAT-solver. In: Giunchiglia, E., Tacchella, A. (eds.) SAT 2003. LNCS, vol. 2919, pp. 502–518. Springer, Heidelberg (2004)
8. Fu, Z., Malik, S.: On solving the partial MAX-SAT problem. In: Biere, A., Gomes, C.P. (eds.) SAT 2006. LNCS, vol. 4121, pp. 252–265. Springer, Heidelberg (2006)

9. Gavanelli, M.: The log-support encoding of CSP into SAT. In: Bessière, C. (ed.) CP 2007. LNCS, vol. 4741, pp. 815–822. Springer, Heidelberg (2007)
10. Génisson, R., Jégou, P.: Davis and Putnam were already checking forward. In: Proceedings of the ECAI-1996, pp. 180–184 (1996)
11. Gent, I.P.: Arc consistency in SAT. In: Proceedings of ECAI-2002, pp. 121–125 (2002)
12. Heras, F., Larrosa, J., Oliveras, A.: MiniMaxSat: A new weighted Max-SAT solver. In: Proceedings of SAT-2007 (2007)
13. Kasif, S.: On the parallel complexity of discrete relaxation in constraint satisfaction networks. Artificial Intelligence 45, 275–286 (1990)
14. Larrosa, J., Heras, F.: Resolution in Max-SAT and its relation to local consistency in weighted CSPs. In: Proceedings of IJCAI-2005, Edinburgh, Scotland, pp. 193–198. Morgan Kaufmann, San Francisco (2005)
15. Larrosa, J., Heras, F., de Givry, S.: A logical approach to efficient max-sat solving. Artificial Intelligence 172(2–3), 204–233 (2008)
16. Li, C.M., Manyà, F., Planes, J.: Exploiting unit propagation to compute lower bounds in branch and bound Max-SAT solvers. In: van Beek, P. (ed.) CP 2005. LNCS, vol. 3709, pp. 403–414. Springer, Heidelberg (2005)
17. Li, C.M., Manyà, F., Planes, J.: Detecting disjoint inconsistent subformulas for computing lower bounds for Max-SAT. In: Proceedings of AAAI-2006, pp. 86–91 (2006)
18. Li, C.M., Manyà, F., Planes, J.: New inference rules for Max-SAT. Journal of Artificial Intelligence Research 30, 321–359 (2007)
19. Marques-Silva, J., Planes, J.: Algorithms for Maximum Satisfiability using Unsatisfiable Cores. In: Proceedings of DATE-2008 (2008)
20. Moskewicz, M., Madigan, C., Zhao, Y., Zhang, L., Malik, S.: Chaff: Engineering an efficient SAT solver. In: 39th Design Automation Conference (2001)
21. Pipatsrisawat, K., Darwiche, A.: Clone: Solving weighted max-sat in a reduced search space. In: 20th Australian Joint Conference on Artificial Intelligence, AI-2007, pp. 223–233 (2007)
22. Ramírez, M., Geffner, H.: Structural relaxations by variable renaming and their compilation for solving MinCostSAT. In: Bessière, C. (ed.) CP 2007. LNCS, vol. 4741, pp. 605–619. Springer, Heidelberg (2007)
23. Walsh, T.: SAT v CSP. In: Dechter, R. (ed.) CP 2000. LNCS, vol. 1894, pp. 441–456. Springer, Heidelberg (2000)

A Preprocessor for Max-SAT Solvers[*]

Josep Argelich[1], Chu Min Li[2], and Felip Manyà[1]

[1] Computer Science Department
Universitat de Lleida
Jaume II, 69, E-25001 Lleida, Spain
[2] LaRIA, Université de Picardie Jules Verne
33 Rue Saint Leu, 80039 Amiens Cedex 01, France

Abstract. We describe a preprocessor that incorporates a variable saturation procedure for Max-SAT, and provide empirical evidence that it improves the performance of some of the most successful state-of-the-art solvers on several partial (weighted) Max-SAT instances of the 2007 Max-SAT Evaluation.

1 Introduction

In the last years, there has been an increasing interest in Max-SAT formalisms such as (weighted) Max-SAT and partial (weighted) Max-SAT. Among the most relevant results we highlight the following ones: (i) there exist solvers like ChaffBS [6], Clone [14], Lazy [1], MaxSatz [12], MiniMaxSat [8], ms4 [13], PMS [3], Sat4Jmaxsat, SR(w) [15] and Toolbar [7] which solve many instances that are beyond the reach of the solvers existing just five years ago; (ii) resolution refinements, which preserve the number of unsatisfied clauses, have been incorporated into Max-SAT solvers [9,7,12], as well as good quality underestimations of the lower bound [10,11,14,15], (iii) a resolution-style calculus for Max-SAT has been proven to be complete [4,5], (iv) formalisms like Partial Max-SAT have been investigated for solving problems with soft constraints [2,6,8,3], and (v) two evaluations of Max-SAT solvers have been performed for the first time.

In this paper we present a preprocessor that can be applied to solvers for Max-SAT formalisms, including Max-SAT, weighted Max-SAT, partial Max-SAT and partial weighted Max-SAT solvers. Our preprocessor implements a variable saturation procedure defined in [4,5]. Moreover, we provide empirical evidence that it improves the performance of MiniMaxSat, SR(w) and PMS on several partial (weighted) Max-SAT instances of the 2007 Max-SAT Evaluation. The preprocessor applies the variable saturation procedure defined in [4,5] to a limited number of variables which are selected heuristically, and transforms the input instance into an equivalent instance which does not contain the saturated variables. To the best of our knowledge, this is the first paper that investigates the practical usefulness of the notion of variable saturation in the Max-SAT context.

---

[*] This research was funded by the MEC research projects TIN2006-15662-C02-02 and TIN2007-68005-C04-04, and Acción Integrada HP2005-0147.

H. Kleine Büning and X. Zhao (Eds.): SAT 2008, LNCS 4996, pp. 15–20, 2008.

The structure of the paper is as follows. Section 2 contains preliminary definitions about Max-SAT. Section 3 introduces the Max-SAT resolution rule and the notion of variable saturation. Section 4 reports and analyses the experiments.

2   Preliminaries

In propositional logic a variable x_i may take values 0 (false) or 1 (true). A literal l_i is a variable x_i or its negation $\bar{x}_i$. A clause is a disjunction of literals, and a CNF formula is a multiset of clauses. A weighted clause is a pair (C_i, w_i), where C_i is a disjunction of literals and w_i, its weight, is a positive number, and a weighted CNF formula is a multiset of weighted clauses. An assignment of truth values to the propositional variables satisfies a literal x_i ($\bar{x}_i$) if it takes the value 1 (0), satisfies a clause if it satisfies at least one literal of the clause, and satisfies a CNF formula if it satisfies all the clauses of the formula.

The Max-SAT problem for a CNF formula ϕ is the problem of finding an assignment that maximizes (minimizes) the number of satisfied (unsatisfied) clauses. The weighted Max-SAT problem for a weighted CNF formula ϕ is the problem of finding an assignment that minimizes the sum of weights of unsatisfied clauses. A Partial Max-SAT instance is a CNF formula in which some clauses are *relaxable* or *soft* and the rest are *non-relaxable* or *hard*. Solving a Partial Max-SAT instance amounts to find an assignment that satisfies all the hard clauses and the maximum number of soft clauses. The *weighted Partial Max-SAT* problem is the combination of weighted Max-SAT and Partial Max-SAT.

3   Resolution in Max-SAT

The *Max-SAT resolution* rule is defined as follows:

$$
\begin{array}{l}
x \vee a_1 \vee \cdots \vee a_s \\
\bar{x} \vee b_1 \vee \cdots \vee b_t \\
\hline
a_1 \vee \cdots \vee a_s \vee b_1 \vee \cdots \vee b_t \\
x \vee a_1 \vee \cdots \vee a_s \vee \overline{b_1} \\
x \vee a_1 \vee \cdots \vee a_s \vee b_1 \vee \overline{b_2} \\
\cdots \\
x \vee a_1 \vee \cdots \vee a_s \vee b_1 \vee \cdots \vee b_{t-1} \vee \overline{b_t} \\
\bar{x} \vee b_1 \vee \cdots \vee b_t \vee \overline{a_1} \\
\bar{x} \vee b_1 \vee \cdots \vee b_t \vee a_1 \vee \overline{a_2} \\
\cdots \\
\bar{x} \vee b_1 \vee \cdots \vee b_t \vee a_1 \vee \cdots \vee a_{s-1} \vee \overline{a_s}
\end{array}
$$

This inference rule is applied to multisets of clauses, and replaces the premises of the rule by its conclusions. We say that the rule *cuts* the variable x, and the tautologies concluded by the rule are removed from the resulting multiset. In partial Max-SAT, the hard clauses remain and the clauses subsumed by the hard clause are removed (see [3] for details). For the sake of clarity, we did not define the weighted version of the rule (see [5] for details).

Definition 1. *A multiset of clauses C is said to be* saturated *w.r.t. x if for every pair of clauses $C_1 = x \vee A$ and $C_2 = \overline{x} \vee B$ of C, there is a literal l such that l is in A and $\overline{l}$ is in B. A multiset of clauses C' is a* saturation *of C w.r.t. x if C' is saturated w.r.t. x and $C \vdash_x C'$; i.e., C' can be obtained from C applying Max-SAT resolution cutting x finitely many times.*

Lemma 1. *[5] For every multiset of clauses C and variable x, there exists a multiset C' such that C' is a saturation of C w.r.t. x. Moreover, this multiset C' can be computed by applying Max-SAT resolution to any pair of clauses $x \vee A$ and $\overline{x} \vee B$ with the restriction that $A \vee B$ is not a tautology, using any ordering of the literals, until we can not apply Max-SAT resolution any longer.*

The completeness proof of Max-SAT resolution [4,5] states that we can get a complete algorithm by successively saturating w.r.t. all the variables as follows: we saturate w.r.t x_1 and then remove all the clauses containing x_1, saturate w.r.t x_2 and then remove all the clauses containing x_2, etc. After saturating this way w.r.t. all the variables we get as many empty clauses as the minimum number of unsatisfied clauses in the original formula.

Solving a Max-SAT instance by successively saturating w.r.t. all the variables is clearly not competitive with solving it with a modern branch and bound solver. Nevertheless, we thought that it would make sense to saturate w.r.t. a limited number of variables as a preprocessing in order to simplify the formula. We select the variables to be saturated, depending on a parameter k, iteratively as follows: We build a graph whose nodes are the Boolean variables occurring in the instance, and add an edge between two vertices if the variables of the vertices occur in the same clause. We select a variable whose vertex has minimal degree if its degree is smaller than k. This process is repeated until no more variables can be selected. The idea is to saturate variables in which the application of variable saturation is not very costly in terms of time and space. We also tried to saturate the variables with a low number of occurrences, but the results were not so good.

4   Experimental Results

To assess the impact of the preprocessor on the performance of branch and bound Max-SAT solvers, we solved partial Max-SAT instances[1] of the 2007 Max-SAT Evaluation (with a timeout of 30 minutes as in the evaluation) on three of the most successful and representative state-of-the-art solvers: MiniMaxSat [8], PMS [3], and SR(w) [15]. For MiniMaxSat and SR(w) we used the same versions as in the evaluation. For PMS we used an improved version (PMS v1.3). We executed the preprocessor with $k = 6, 10, 14$ (remind that k is the parameter for selecting variables). All the experiments were performed on a Linux Cluster where the nodes have a 2GHz AMD Opteron processor with 1Gb of RAM.

---

[1] We solved only the instances in which the preprocessor detected variables that could be saturated. We also solved (weighted) Max-SAT instances, but the speed-ups were not so good as in (weighted) Partial Max-SAT.

Tables 1 and 2 show the experimental results for PMS. The instances are divided into sets. The first column is the name of the set, the second column shows the number of instances in each set, the third column shows the results for the solver without preprocessing, and the rest of columns show the results with preprocessing for $k = 6, 10, 14$. We display the mean time (in seconds) of the solved instances, as well as the number of solved instances (in brackets). We observe that PMS with preprocessing solves more instances in 5 sets, and reduces considerably the CPU time in most of the other sets. The best improvements are achieved for MaxClique (random), where the preprocessing allows to solve 8 additional instances, and for Auctions (paths), where the preprocessing allows to solve 9 additional instances.

Tables 3 and 4 show the results for MiniMaxSat. In this case, the gains are not so significant as for PMS, although the preprocessing allows to solve 1 additional instance for MaxClique (structured) and for WCSP (spot5 dir).

Tables 5 and 6 show the experimental results for SR(w). In this case, we solve an additional instance for 3 sets (MaxCSP (dense loose), MaxCSP (w-queens) and Auctions (scheduling)), and 185 additional instances for Pseudo (factor). The latter is the best improvement achieved with our preprocessor.

Table 1. Partial Max-SAT benchmarks with PMS

Instance set	#	PMS	PMS(6)	PMS(10)	PMS(14)
MaxClique (random)	96	43.69(80)	69.30(83)	61.04(85)	**53.85(88)**
MaxClique (structured)	62	175.27(23)	183.30(24)	178.03(24)	**171.13(25)**
MaxOne (3-SAT)	50	261.95(50)	122.08(50)	**62.06(50)**	328.07(48)
MaxOne (structured)	60	**177.84(58)**	234.56(56)	223.76(42)	6.57(1)
MaxCSP (dense loose)	20	5.50(20)	5.26(20)	**3.39(20)**	8.31(20)
MaxCSP (dense tight)	20	9.76(20)	9.76(20)	**7.83(20)**	12.95(20)
MaxCSP (sparse loose)	20	16.39(20)	9.18(20)	**4.77(20)**	36.51(19)
MaxCSP (sparse tight)	20	24.02(20)	21.70(20)	**18.07(20)**	84.81(20)
WCSP (w-queens)	7	72.22(6)	72.19(6)	**72.17(6)**	72.18(6)

Table 2. Weighted Partial Max-SAT benchmarks with PMS

Instance set	#	PMS	PMS(6)	PMS(10)	PMS(14)
Auctions (paths)	88	233.56(71)	**178.50(80)**	127.72(77)	266.47(63)
Auctions (regions)	84	**5.24(84)**	5.30(84)	5.52(84)	5.62(84)
Auctions (scheduling)	84	89.70(84)	89.62(84)	89.66(84)	**89.61(84)**
Pseudo (factor)	186	**11.00(186)**	11.64(186)	226.88(186)	924.37(2)
Pseudo (miplib)	16	1.94(4)	**0.96(4)**	190.93(4)	2.34(1)
QCP	25	**199.31(15)**	199.36(15)	199.46(15)	199.52(15)
WCSP (planning)	71	**13.96(71)**	21.97(71)	63.65(70)	233.29(42)
WCSP (spot5 dir)	21	14.86(2)	6.59(5)	**57.96(6)**	13.27(5)
WCSP (spot5 log)	21	18.95(2)	91.03(3)	2.55(4)	**1.46(4)**

Table 3. Partial Max-SAT benchmarks with MiniMaxSat

Instance set	#	MiniMS	MiniMS(6)	MiniMS(10)	MiniMS(14)
MaxClique (random)	96	**2.41(96)**	2.44(96)	2.67(96)	4.38(96)
MaxClique (structured)	62	85.22(36)	82.15(37)	**67.94(37)**	66.43(36)
MaxOne (3-SAT)	50	**0.37(50)**	0.40(50)	0.43(50)	8.87(50)
MaxOne (structured)	60	**31.35(60)**	20.57(54)	65.88(42)	0.78(1)
MaxCSP (dense loose)	20	**0.65(20)**	0.71(20)	0.87(20)	5.11(20)
MaxCSP (dense tight)	20	**0.69(20)**	0.70(20)	0.70(20)	2.87(20)
MaxCSP (sparse loose)	20	**0.35(20)**	0.36(20)	0.57(20)	21.20(20)
MaxCSP (sparse tight)	20	**0.85(20)**	0.87(20)	0.94(20)	27.05(20)
WCSP (w-queens)	7	55.47(7)	55.28(7)	**54.56(7)**	179.13(7)

Table 4. Weighted Partial Max-SAT benchmarks with MiniMaxSat

Instance set	#	MiniMS	MiniMS(6)	MiniMS(10)	MiniMS(14)
Auctions (paths)	88	29.82(88)	**19.44(88)**	13.52(84)	78.21(75)
Auctions (regions)	84	1.63(84)	**1.55(84)**	**1.55(84)**	1.56(84)
Auctions (scheduling)	84	**46.14(84)**	46.24(84)	46.28(84)	46.16(84)
Pseudo (factor)	186	**1.16(186)**	1.79(186)	5.53(186)	905.51(183)
Pseudo (miplib)	16	**41.35(5)**	84.90(5)	398.55(5)	1.43(1)
QCP	25	25.00(20)	26.71(20)	25.28(20)	**24.65(20)**
WCSP (planning)	71	**9.97(71)**	10.11(71)	22.12(71)	235.45(47)
WCSP (spot5 dir)	21	2.63(3)	11.82(3)	8.18(4)	**6.99(4)**
WCSP (spot5 log)	21	**9.07(4)**	5.69(2)	152.16(3)	323.82(4)

Table 5. Partial Max-SAT benchmarks with SR(w)

Instance set	#	SR-W	SR-W(6)	SR-W(10)	SR-W(14)
MaxClique (random)	96	244.85(55)	219.40(55)	224.65(55)	**218.38(55)**
MaxClique (structured)	62	21.18(9)	**17.56(9)**	22.67(8)	20.17(8)
MaxOne (3-SAT)	50	386.23(41)	**338.69(41)**	718.76(22)	758.61(1)
MaxOne (structured)	60	**471.72(22)**	449.33(19)	618.92(18)	1078.54(1)
MaxCSP (dense loose)	20	697.74(1)	633.31(1)	**1162.49(2)**	0.00(0)
MaxCSP (dense tight)	20	209.22(18)	**199.18(18)**	202.71(18)	350.83(15)
MaxCSP (sparse loose)	20	296.48(16)	**272.89(16)**	408.06(15)	853.86(7)
MaxCSP (sparse tight)	20	235.98(19)	**216.19(19)**	230.31(19)	563.63(12)
WCSP (w-queens)	7	54.00(6)	230.25(7)	**228.06(7)**	258.10(7)

As a conclusion, we could say that variable saturation is an effective preprocessing technique that may produce substantial speed-ups, as well as increase the number of solved instances. As future work we plan to incorporate additional simplification techniques into our preprocessor, and explore the application of variable saturation to a limited number of nodes of the search space because its application at each node is too costly.

Table 6. Weighted Partial Max-SAT benchmarks with SR(w)

Instance set	#	SR-W	SR-W(6)	SR-W(10)	SR-W(14)
Auctions (paths)	88	**173.42(77)**	161.15(76)	169.32(72)	353.90(66)
Auctions (regions)	84	146.54(82)	136.45(82)	126.93(82)	**119.52(82)**
Auctions (scheduling)	84	276.91(56)	240.61(56)	**270.71(57)**	239.26(56)
Pseudo (factor)	186	0.00(0)	2.86(37)	**520.50(185)**	1091.88(1)
Pseudo (miplib)	16	**2.62(5)**	3.04(4)	216.89(4)	4.12(1)
QCP	25	715.58(5)	**572.40(5)**	675.34(5)	674.26(5)
WCSP (planning)	71	**379.57(57)**	371.42(53)	286.88(46)	285.58(25)
WCSP (spot5 dir)	21	2.95(6)	**1.90(6)**	9.27(4)	61.92(3)
WCSP (spot5 log)	21	14.56(6)	**11.53(6)**	25.30(5)	10.83(4)

References

1. Alsinet, T., Manyà, F., Planes, J.: Improved exact solver for weighted Max-SAT. In: SAT-2005, pp. 371–377 (2005)
2. Argelich, J., Manyà, F.: Exact Max-SAT solvers for over-constrained problems. Journal of Heuristics 12(4–5), 375–392 (2006)
3. Argelich, J., Manyà, F.: Partial Max-SAT solvers with clause learning. In: SAT-2007, pp. 28–40 (2007)
4. Bonet, M.L., Levy, J., Manyà, F.: A complete calculus for Max-SAT. In: SAT-2006, pp. 240–251 (2006)
5. Bonet, M.L., Levy, J., Manyà, F.: Resolution for Max-SAT. Artificial Intelligence 171(8–9), 240–251 (2007)
6. Fu, Z., Malik, S.: On solving the partial MAX-SAT problem. In: SAT-2006, pp. 252–265 (2006)
7. Heras, F., Larrosa, J.: New inference rules for efficient Max-SAT solving. In: AAAI-2006, pp. 68–73 (2006)
8. Heras, F., Larrosa, J., Oliveras, A.: MiniMaxSat: A new weighted Max-SAT solver. In: SAT-2007 (2007)
9. Larrosa, J., Heras, F.: Resolution in Max-SAT and its relation to local consistency in weighted CSPs. In: IJCAI-2005, pp. 193–198 (2005)
10. Li, C.M., Manyà, F., Planes, J.: Exploiting unit propagation to compute lower bounds in branch and bound Max-SAT solvers. In: CP-2005, pp. 403–414 (2005)
11. Li, C.M., Manyà, F., Planes, J.: Detecting disjoint inconsistent subformulas for computing lower bounds for Max-SAT. In: AAAI-2006, pp. 86–91 (2006)
12. Li, C.M., Manyà, F., Planes, J.: New inference rules for Max-SAT. Journal of Artificial Intelligence Research 30, 321–359 (2007)
13. Marques-Silva, J., Planes, J.: Algorithms for Maximum Satisfiability using Unsatisfiable Cores. In: DATE-2008 (2008)
14. Pipatsrisawat, K., Darwiche, A.: Clone: Solving weighted max-sat in a reduced search space. In: 20th Australian Joint Conf. on AI, AI-2007, pp. 223–233 (2007)
15. Ramírez, M., Geffner, H.: Structural relaxations by variable renaming and their compilation for solving MinCostSAT. In: CP-2007, pp. 605–619 (2007)

A Generalized Framework for Conflict Analysis

G. Audemard[1], L. Bordeaux[2], Y. Hamadi[2], S. Jabbour[1], and L. Sais[1]

[1] CRIL - CNRS UMR 8188, Artois, France
{audemard,jabbour,sais}@cril.fr
[2] Microsoft Research Cambridge, UK
{lucasb,youssefh}@microsoft.com

Abstract. This paper presents an extension of Conflict Driven Clauses Learning (CDCL). It relies on an extended notion of implication graph containing additional arcs, called inverse arcs. These are obtained by taking into account the satisfied clauses of the formula, which are usually ignored by conflict analysis. This extension captures more conveniently the whole propagation process, and opens new perspectives for CDCL-based approaches. Among other benefits, our extension leads to a new conflict analysis scheme that exploits the additional arcs to back-jump to higher levels. Experimental results show that the integration of our generalized conflict analysis scheme within two state-of-the-art solvers improves their performance.

1 Introduction

This paper extends *Conflict-Driven Clause-Learning (CDCL)*, which is one of the key components of modern SAT solvers [7,5]. In the CDCL approach a central data-structure is the *implication graph*, which records the partial assignment that is under construction together with its implications. This data-structure enables conflict analysis, which, in turn, is used for intelligent backtracking, clause learning, for the adjustment of the variable selection heuristic. An important observation is that the implication graph built in the traditional way is "incomplete" in that it only gives a partial view of the actual implications between literals. A solver only keeps track of the first explanation that is encountered for the deduced literal. This strategy is obviously very much dependent on the particular order in which clauses are propagated. We present here an extended notion of implication graph in which a deduced literal can have several explanations. An extended version of our work can be found in [1]. The paper is organized as follows, definitions and notations are presented in the next section. Section three describes classical conflict analysis. Section four presents our extension, and finally before the conclusion, section five presents some experimental results.

2 Preliminary Definitions and Notations

A *CNF formula* $\mathcal{F}$ is a set (interpreted as a conjunction) of *clauses*, where a clause is a set (interpreted as a disjunction) of *literals*. A literal is a positive (x) or negated ($\neg x$) propositional variable. The two literals x and $\neg x$ are called *complementary*. We note $\bar{l}$

H. Kleine Büning and X. Zhao (Eds.): SAT 2008, LNCS 4996, pp. 21–27, 2008.

the complementary literal of l. For a set of literals L, $\bar{L}$ is defined as $\{\bar{l} \mid l \in L\}$. A *unit clause* is a clause with only one literal (called *unit literal*). An *empty clause*, noted $\bot$, is interpreted as false, while an *empty CNF formula*, noted $\top$, is interpreted as true.

The set of variables occurring in $\mathcal{F}$ is noted $V_{\mathcal{F}}$. A set of literals is *complete* if it contains one literal for each variable in $V_{\mathcal{F}}$, and *fundamental* if it does not contain complementary literals. An *interpretation* ρ of a boolean formula $\mathcal{F}$ associates a value $\rho(x)$ to some of the variables $x \in \mathcal{F}$. An interpretation is alternatively represented by a complete and fundamental set of literals, in the obvious way. A *model* of a formula $\mathcal{F}$ is an interpretation ρ that satisfies the formula; noted $\rho \vDash \Sigma$.

The following notations will be heavily used throughout the paper:

- $\eta[x, c_i, c_j]$ denotes the *resolvent* between a clause c_i containing the literal x and c_j a clause containing the literal $\neg x$. In other words $\eta[x, c_i, c_j] = c_i \cup c_j \setminus \{x, \neg x\}$.
- $\mathcal{F}|_x$ will denote the formula obtained from $\mathcal{F}$ by assigning x the truth-value *true*. Formally $\mathcal{F}|_x = \{c \mid c \in \mathcal{F}, \{x, \neg x\} \cap c = \emptyset\} \cup \{c \setminus \{\neg x\} \mid c \in \mathcal{F}, \neg x \in c\}$. This notation is extended to interpretations: given an interpretation $\rho = \{x_1, \ldots, x_n\}$, we define $\mathcal{F}|_\rho = (\ldots ((\mathcal{F}|_{x_1})|_{x_2}) \ldots |_{x_n})$.
- $\mathcal{F}^*$ denotes the formula $\mathcal{F}$ closed under unit propagation, defined recursively as follows: (1) $\mathcal{F}^* = \mathcal{F}$ if $\mathcal{F}$ does not contain any unit clause, (2) $\mathcal{F}^* = \bot$ if $\mathcal{F}$ contains two unit-clauses $\{x\}$ and $\{\neg x\}$, (3) otherwise, $\mathcal{F}^* = (\mathcal{F}|_x)^*$ where x is the literal appearing in a unit clause of $\mathcal{F}$.

Let us now introduce some notations and terminology on SAT solvers based on the DPLL backtrack search procedure [3]. At each node the assigned literals (decision literal and the propagated ones) are labeled with the same *decision level* starting from 1 and increased at each branching. The current decision level is the highest decision level in the assignment stack. After backtracking, some variables are unassigned, and the current decision level is decreased accordingly. At level i, the current partial assignment ρ can be represented as a sequence of decision-propagations of the form $\langle (x_k^i), x_{k_1}^i, x_{k_2}^i, \ldots, x_{k_{n_k}}^i \rangle$ where the first literal x_k^i corresponds to the decision literal x_k assigned at level i and each $x_{k_j}^i$ for $1 \leq j \leq n_k$ represents a propagated (unit) literal at level i. Let $x \in \rho$, we note $l(x)$ the assignment level of x, $d(\rho, i) = x$ if x is the decision literal assigned at level i. For a given level i, we define ρ^i as the projection of ρ to literals assigned at a level $\leq i$.

3 Conflict Analysis Using Implication Graphs

Implication graphs capture the variable assignments ρ made during the search, both by branching and by propagation. This representation is a convenient way to analyze conflicts. In classical SAT solvers, whenever a literal y is propagated, we keep a reference to the clause at the origin of the propagation of y, which we note $\overrightarrow{cla}(y)$. The clause $\overrightarrow{cla}(y)$ is in this case of the form $(x_1 \vee \cdots \vee x_n \vee y)$ where every literal x_i is false under the current partial assignment ($\rho(x_i) = $ *false*, $\forall i \in 1..n$), while $\rho(y) = $ *true*. When a literal y is not obtained by propagation but comes from a decision, $\overrightarrow{cla}(y)$ is undefined, which we note for convenience $\overrightarrow{cla}(y) = \bot$.

When $\overrightarrow{cla}(y) \neq \bot$, we denote by $exp(y)$ the set $\{\overline{x} \mid x \in \overrightarrow{cla}(y) \setminus \{y\}\}$, called set of *explanations* of y. In other words if $\overrightarrow{cla}(y) = (x_1 \vee \cdots \vee x_n \vee y)$, then the explanations are the literals $\overline{x_i}$ with which $\overrightarrow{cla}(y)$ becomes the unit clause $\{y\}$. Note that for all i we have $l(\overline{x_i}) \leq l(y)$, i.e., all the explanations of the deduction come from a level at most as high. When $\overrightarrow{cla}(y)$ is undefined we define $exp(y)$ as the empty set. The explanations can alternatively be seen as an implication graph, in which the set of predecessors of a node corresponds to the set of explanations of the corresponding literal:

Definition 1 (Implication Graph). *Let $\mathcal{F}$ be a CNF formula, ρ a partial ordered interpretation, and let exp denote a choice of explanations for the deduced literals in ρ. The implication graph associated to $\mathcal{F}$, ρ and exp is $(\mathcal{N}, \mathcal{E})$ where:*

- $\mathcal{N} = \rho$, *i.e. there is exactly one node for every literal, decision or implied;*
- $\mathcal{E} = \{(x, y) \mid x \in \rho, y \in \rho, x \in exp(y)\}$

Example 1. $\mathcal{G}_{\mathcal{F}}^{\rho}$, shown in Figure 1 and restricted to plain arcs is an implication graph for the formula $\mathcal{F}$ and the partial assignment ρ given below : $\mathcal{F} \supseteq \{c_1, \ldots, c_9\}$

(c_1) $x_6 \vee \neg x_{11} \vee \neg x_{12}$ (c_2) $\neg x_{11} \vee x_{13} \vee x_{16}$ (c_3) $x_{12} \vee \neg x_{16} \vee \neg x_2$
(c_4) $\neg x_4 \vee x_2 \vee \neg x_{10}$ (c_5) $\neg x_8 \vee x_{10} \vee x_1$ (c_6) $x_{10} \vee x_3$
(c_7) $x_{10} \vee \neg x_5$ (c_8) $x_{17} \vee \neg x_1 \vee \neg x_3 \vee x_5 \vee x_{18}$ (c_9) $\neg x_3 \vee \neg x_{19} \vee \neg x_{18}$

$\rho = \{\langle \ldots \neg x_6^1 \ldots \neg x_{17}^1 \rangle \langle (x_8^2) \ldots \neg x_{13}^2 \ldots \rangle \langle (x_4^3) \ldots x_{19}^3 \ldots \rangle \ldots \langle (x_{11}^5) \ldots \rangle\}$. The current decision level is 5.

We consider ρ, a partial assignment such that $(\mathcal{F}|_\rho)* = \bot$ and $\mathcal{G}_{\mathcal{F}}^{\rho} = (\mathcal{N}, \mathcal{E})$ the associated implication graph. Assume that the current decision level is m. As a conflict is reached, then $\exists x \in$ st. $\{x, \neg x\} \subset \mathcal{N}$ and $l(x) = m$ or $l(\neg x) = m$. Conflict analysis is based on applying resolution from the top to the bottom of the implication graph using the different clauses of the form $(exp(y) \vee y)$ implicitly encoded at each node $y \in \mathcal{N}$. We call this process a conflict resolution proof. More formally,

Definition 2 (Asserting clause). *A clause c of the form $(\alpha \vee x)$ is called an asserting clause iff $\rho(c) = false$, $l(x) = m$ and $\forall y \in \alpha, l(y) < l(x)$. x is called asserting literal, which we note in short $\mathcal{A}(c)$. We can define $jump(c) = max\{l(\neg y) \mid y \in \alpha\}$.*

Definition 3 (Conflict resolution proof). *A conflict resolution proof π is a sequence of clauses $\langle \sigma_1, \sigma_2, \ldots \sigma_k \rangle$ satisfying the following conditions :*

1. $\sigma_1 = \eta[x, \overrightarrow{cla}(x), \overrightarrow{cla}(\neg x)]$, *where $\{x, \neg x\}$ is the conflict.*
2. σ_i, *for $i \in 2..k$, is built by selecting a literal $y \in \sigma_{i-1}$ for which $\overrightarrow{cla}(\overline{y})$ is defined. We then have $y \in \sigma_{i-1}$ and $\overline{y} \in \overrightarrow{cla}(\overline{y})$: the two clauses resolve. The clause σ_i is defined as $\eta[y, \sigma_{i-1}, \overrightarrow{cla}(\overline{y})]$;*
3. σ_k *is, moreover an asserting clause.*

It is called elementary iff $\nexists i < k$ s.t. $\langle \sigma_1, \sigma_2, \ldots \sigma_i \rangle$ is also a conflict resolution proof.

4   Extended Implication Graph

In modern SAT solvers, clauses containing a literal x that is implied at the current level are essentially ignored by the propagation. More precisely, because the solver does not maintain the information whether a given clause is satisfied or not, a clause containing x may occasionally be considered by the propagation, but only when another literal y of the clause becomes false. When this happens the solver typically skips the clause. However, in cases where x is true *and all the other literals are false*, an "arc" was revealed for free that could as well be used to extend the graph. Such arcs are those we propose to use in our extension.

To explain our idea let us consider, again, the formula $\mathcal{F}$ and the partial assignments given in the example 1. We define a new formula $\mathcal{F}'$ as follow : $\mathcal{F}' \supseteq \{c_1, \ldots, c_9\} \cup \{c_{10}, c_{11}, c_{12}\}$ where $c_{10} = (\neg x_{19} \vee x_8), c_{11} = (x_{19} \vee x_{10})$ and $c_{12} = (\neg x_{17} \vee x_{10})$.

The three added clauses are satisfied under the instantiation ρ. c_{10} is satisfied by x_8 assigned at level 2, c_{11} is satisfied by x_{19} at level 3, and c_{12} is satisfied by $\neg x_{17}$ at level 1. This is shown in the extended implication graph (see Figure 1) by the doted edges. Let us now illustrate the usefulness of our proposed extension. Let us consider again the the asserting clause Δ_1 corresponding to the classical first UIP. We can generate the following strong asserting clause: $c_{13} = \eta[x_8, \Delta_1, c_{10}] = (x_{17}^1 \vee \neg x_{19}^3 \vee x_{10}^5)$, $c_{14} = \eta[x_{19}, c_{13}, c_{11}] = (x_{17}^1 \vee x_{10}^5)$ and $\Delta_1^s = \eta[x_{17}, c_{14}, c_{12}] = x_{10}^5$. In this case we backtrack to the level 0 and we assign x_{10} to *true*. Indeed $\mathcal{F}' \models x_{10}$.

As we can see Δ_1^s subsumes Δ_1. If we continue the process we also obtain other strong asserting clauses $\Delta_2^s = (\neg x_4^3 \vee x_2^5)$ and $\Delta_3^s = (\neg x_4^3 \vee x_{13}^2 \vee x_6^1 \vee \neg x_{11}^5)$ which subsume respectively Δ_2 and Δ_3. This first illustration gives us a new way to minimize the size of the asserting clauses.

If we take a look to the clauses used in the implication graph $\mathcal{G}_\mathcal{F}^\rho$ (plain edges) all have the following properties: (1) $\forall x \in \mathcal{N}$ the clause $c = (\overline{exp(x)} \vee x)$ is satisfied by only one literal i.e. $\rho(x) = $ *true* and $\forall y \in exp(x)$, we have $\rho(y) = $ *true* and (2) $\forall y \in exp(x), l(\neg y) \leq l(x)$. Now in the extended implication graph, the added clauses satisfy property (1) and, in addition, the property (2') $\exists y \in exp(x)$ st. $l(\neg y) > l(x)$.

Let us now explain briefly how the extra arcs can be computed. Usually unit propagation does not keep track of implications from the satisfiable sub-formula. In this extension the new implications (deductions) are considered. For instance in the previous

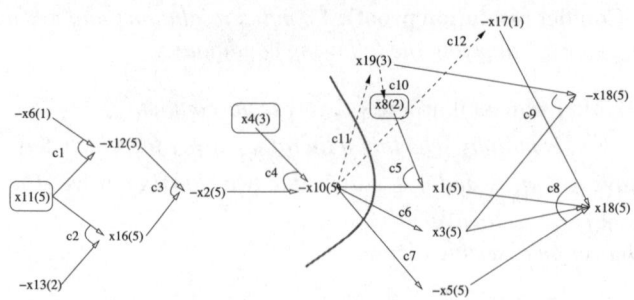

Fig. 1. Implication graph / extended implication graph

example, when we deduce x_{19} at level 3, we "rediscover" the deduction x_8 (which was a choice (decision literal) at level 2). Our proposal keeps track of these re-discoveries.

Before introducing the formal definition of our extended Implication Graph, we introduce the concept of inverse implication (inverse edge).

We maintain additionally to the classical clause $\overrightarrow{cla}(x)$ a new clause $\overleftarrow{cla}(x)$ of the form $(x \vee y_1 \vee \cdots \vee y_n)$. This clause is selected so that $\rho(y_i) = \mathit{false}$ for $i \in 1..n$; $\rho(x) = \mathit{true}$; and $\exists i. \, l(y_i) > l(x)$. This clause can be undefined in some cases (which we note $\overleftarrow{cla}(x) = \perp$). Several clauses of this form can be found for each literal, in which case one is selected arbitrarily: one can choose to consider the first one in the ordering. (It is easy to define a variant where we would take into account all of them, in which case $\overleftarrow{cla}(x)$ is a set of clauses; but we won't develop this variant).

We denote by $\overleftarrow{exp}(x)$ the set $\{\overline{y} \mid y \in \overleftarrow{cla}(x) \setminus \{x\}\}$, and, for clarity, by $\overrightarrow{exp}(x)$ the set that was previously noted exp. An extended implication graph is defined as follows (note that this graph is now not acyclic in general):

Definition 4 (Extended Implication Graph). *Let $\mathcal{F}$ be a CNF formula and ρ an ordered partial interpretation. We define the extended implication Graph associated to $\mathcal{F}$ and ρ as $\mathcal{G}s_{\mathcal{F}}^{\rho} = (\mathcal{N}, \mathcal{E} \cup \mathcal{E}')$ where, $\mathcal{N} = \rho$, $\mathcal{E} = \{(x,y) \mid x \in \rho, y \in \rho, x \in \overrightarrow{exp}(y)\}$ and $\mathcal{E}' = \{(x,y) \mid x \in \rho, y \in \rho, x \in \overleftarrow{exp}(y)\}$*

4.1 Learning to Back-Jump : A First Extension

In this section, we describe a first possible extension of CDCL approach using extended implication graph. Our approach makes an original use of inverses arcs to back-jump farther, i.e. to improve the back-jumping level of the classical asserting clauses.

Let us illustrate the main idea behind our proposed extension. Our approach works in three steps. In the first step (1) : an asserting clause, say $\sigma_1 = (\neg x^1 \vee \neg y^3 \vee \neg z^7 \vee \neg a^9)$ is learnt using the usual learning scheme where 9 is the current decision level. As $\rho(\sigma_1) = \mathit{false}$, usually we backtrack to level $jump(\sigma_1) = 7$. In the second step (2): our approach aims to eliminate the literal $\neg z^7$ from σ_1 using the new arcs of the extended graph. Let us explain this second and new processing. Let $c = (z^7 \vee \neg u^2 \vee \neg v^9)$ such that $\rho(z) = \mathit{true}$, $\rho(u) = \mathit{true}$ and $\rho(v) = \mathit{true}$. The clause c is an inverse arc i.e. the literal z assigned at level 7 is implied by the two literals u and v respectively assigned at level 2 and 9. From c and σ_1, a new clause $\sigma_2 = \eta[z, c, \sigma_1] = (\neg x^1 \vee \neg u^2 \vee \neg y^3 \vee \neg v^9 \vee \neg a^9)$ is generated. We can remark that the new clause σ_2 contains two literals from the current decision level 9. In the third step (3), using classical learning, one can search from σ_2 for another asserting clause σ_3 with only one literal from the current decision level. Let us note that the new asserting clause σ_3 might be worse in terms of back-jumping level. To avoid this main drawback, the inverse arc c is chosen if the two following conditions are satisfied : i) the literals of c assigned at the current level (v^9) has been already visited during the first step and ii) all the other literals of c are assigned before the level 7 i.e. level of z. In this case, we guaranty that the new asserting clause satisfies the following property : $jump(\sigma_3) \leq jump(\sigma_1)$. Moreover, the asserting literal of σ_3 is $\neg a$.

One can iterate the previous process on the new asserting clause σ_3 to eliminate the literals of σ_3 assigned at level $jump(\sigma_3)$ (for more details see [1]).

5   Experiments

Our extended learning scheme can be crafted to any CDCL based solver. See [1] for details. The experimental results reported in this section are obtained on a Xeon 3.2 GHz (2 GB RAM) and performed on a large panel of SAT instances (286) coming from SAT RACE2006 and SAT07 (industrial). All instances are simplified by the satellite preprocessor [4]. Time limit is set to 1800 seconds and results are reported in seconds. We implement our proposed extension to Minisat [5] and Rsat [8] and make a comparison between original solvers and extended ones (called Minisat^E and Rsat^E). Figure 2 shows the *time* (t) (figure on the left-hand side) and the *cumulated time* (ct) (figure on the right hand side) needed to solve a given number of instances (*nb instances*). t and ct represent respectively the number of instances with running time less than t seconds and the number of solved instances if we consider that all the instances are run sequentially within a time limit of t seconds. This global view clearly shows that as t or ct increase the extended versions solve more instances than the original ones.

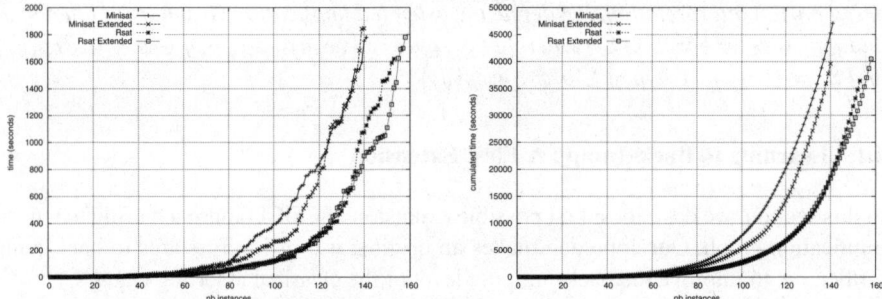

Fig. 2. Time and cumulated time for solving a given number of instances

6   Conclusion

In this paper, we have proposed a generalized framework for conflict analysis. This generalization is obtained by an original extension of the classical implication graph. This extension is obtained by considering clauses that come from the satisfiable part of the formula. Several learning schemes can be defined from this extension. The first extension of learning that improves the classical asserting clauses in term of back-jumping level shows the great potential of the new framework. Despite the different restrictions, our approach achieves interesting improvements of the SAT solvers (Rsat and MiniSat).

References

1. Audemard, G., Bordeaux, L., Hamadi, Y., Jabbour, S., Sais, L.: A Generalized Framework for Conflict Analysis. Microsoft Research, MSR-TR-2008-34 (2008)
2. Beame, P., Kautz, H., Sabharwal, A.: Towards understanding and harnessing the potential of clause learning. JAIR 22, 319–351 (2004)

3. Davis, M., Logemann, G., Loveland, D.W.: A machine program for theorem-proving. Communications of the ACM 5(7), 394–397 (1962)
4. Eén, N., Biere, A.: Effective preprocessing in SAT through variable and clause elimination. In: Bacchus, F., Walsh, T. (eds.) SAT 2005. LNCS, vol. 3569, pp. 61–75. Springer, Heidelberg (2005)
5. Eén, N., Sörensson, N.: An extensible sat-solver. In: Giunchiglia, E., Tacchella, A. (eds.) SAT 2003. LNCS, vol. 2919, Springer, Heidelberg (2004)
6. Marques-Silva, J.: Personal communication
7. Moskewicz, M.W., Madigan, C.F., Zhao, Y., Zhang, L., Malik, S.: Chaff: Engineering an efficient SAT solver. In: Proceedings of (DAC 2001), pp. 530–535 (2001)
8. Pipatsrisawat, K., Darwiche, A.: Rsat 2.0: Sat solver description. Technical Report D–153, Automated Reasoning Group, Computer Science Department, UCLA (2007)

Adaptive Restart Strategies
for Conflict Driven SAT Solvers

Armin Biere

Johannes Kepler University, Linz, Austria

Abstract. As the SAT competition has shown, frequent restarts improve the speed of SAT solvers tremendously, particularly on satisfiable industrial instances. This paper presents a novel adaptive technique that measures the agility of the search process dynamically, which in turn is used to control the restart frequency. Experiments demonstrate, that this new dynamic restart strategy improves speed of our SAT solver PicoSAT on crafted instances considerably and on industrial instances slightly.

1 Introduction

SAT solvers may benefit from restarts [3]. Particularly on satisfiable industrial examples frequent restarts improved the performance of our SAT solver PicoSAT [1] tremendously. Even though PicoSAT is a winner of the SAT competition 2007 in the category of satisfiable industrial instances, an analysis of PicoSAT's performance on unsatisfiable instances in general and on crafted instances in particular reveals, that frequent restarts can also be harmful.

In this short paper we address this issue and present a novel adaptive technique that measures the "agility" of the SAT solver as it traverses the search space, based on the rate of recently flipped assignments. The level of agility dynamically determines the restart frequency. Low agility enforces frequent restarts, high agility prohibits restarts. Our experiments demonstrate, that this new dynamic restart strategy improves the speed of PicoSAT on crafted instances considerably and on industrial instances slightly.

As has been argued in [3] combinatorial search has heavy-tail behavior. Even if an instance is easy to satisfy (or to refute), the search may get stuck in a complex part of the search space. As a solution to this problem, the authors suggest to use randomization, and in particular *restarts*. To *restart* means to stop the current search after a certain time has passed and start over again.

Our focus is on industrial and crafted instances. For random benchmarks randomized algorithms are more successful. There has been work on dynamic restart algorithms for randomized search, see for instance [4,6]. This work is not applicable to our setting. We want to improve the performance of conflict driven SAT solvers with learning, such as RSAT [9] and PicoSAT [1]. Additionally these solvers always pick the last assignment for a decision variable. Enforcing these heuristics without learning makes restarts useless. Furthermore, statistics, such as the number of satisfied clauses, which are crucial in adaptive restart scheduling for local search [4], are not available in the solvers we want to improve.

H. Kleine Büning and X. Zhao (Eds.): SAT 2008, LNCS 4996, pp. 28–33, 2008.

Techniques, as implemented in the SAT solver TiniSAT [5] inspired by [7] and further improved in RSAT [9] and PicoSAT [1], show, that frequent restarts in combination with saving and reusing the previous phase can speed up SAT solvers on industrial instances tremendously, particularly on satisfiable ones. In this category PicoSAT was a clear winner of the SAT'07 Competition.

Beside fast low level data structures [1], the major improvement in version 535 of PicoSAT as submitted to the SAT'07 Competition, is an aggressive restart schedule in combination with saving and reusing phases of assigned variables: The first restart occurs after 100 conflicts. Then this restart interval is increased by 10%, which means the next restart happens after another 110 conflicts, then after another 121 conflicts etc. However, this sequence of longer and longer inner restart intervals is reset to its initial value of 100 conflicts after the end of an outer restart interval is reached. Then the outer restart interval is also increased by 10%. This results in "bursts" of restarts. The restart frequency in one burst sequence slows down at the end and its length, the burst duration, slowly increases over time. More details can be found in [1].

RSAT [9] follows TiniSAT [5] with respect to restarts. Both have a less aggressive restart strategy than PicoSAT. They also do use the same kind of preprocessing [2] as MiniSAT. As a result RSAT, TiniSAT and MiniSAT turned out to be faster than PicoSAT on unsatisfiable industrial instances. On unsatisfiable *crafted* instances the situation is even worse. PicoSAT and in this case also RSAT can solve far less benchmarks than MiniSAT.

After this analysis it seems a valid conjecture, that frequent restarts may also be harmful, particularly on unsatisfiable crafted instances. The question then is, how to measure the effectiveness of frequent restarts, or better, to determine criteria, when to disable restarts.

2   Measuring Agility

In all our recent SAT solvers we monitor the average decision height and print it as a kind of progress report. The average decision height is calculated by summing up the decision levels at decision points and dividing the result by the number of decisions. If the average decision height is going up, we are "close" to a satisfying assignment. If the average decision height goes down[1], the solver will eventually resolve the empty clause, or at least some new unit clauses, and its getting "closer" to a refutation. Intuitively the solver is stuck if the average height is not changing much, and it may be a good idea to restart. On the other hand restarts should not happen if the average decision height is changing fast.

Our first *failed* attempt to dynamically control restarts was based on this observation. Restarts are disabled if the derivative of the average decision height becomes small. However, we were not able to get any positive results. In particularly, it seems to be impossible to come up with good "magic constants". The absolute values of the derivative of the average decision height varies considerably from instance to instance.

---

[1] This only applies to a conflict driven SAT solver with learning.

2.1   Flips

As pioneered by RSAT [9], PicoSAT always picks the last phase resp. direction to which a variable was assigned when assigning a decision variable. For instance if a decision variable was assigned to *true*, the last time it was assigned, then again it is assigned to *true*. If a variable is picked as decision variable and was not assigned before, then the phase is picked depending on the number of positive resp. negative occurrences.

Therefore, whenever a variable becomes assigned to a certain value, in particular if the assignment is forced by some other decision, PicoSAT and RSAT have to remember this value. During backtracking the variable is unassigned again, but the old value is *saved*.

This apparatus easily allows to determine when a new forced[2] assignment to a variable *flips* the old value of the variable. Flipping the value of a variable means, that it is assigned to the opposite value, as it was assigned the last time.

Clearly, if the frequency of flips is small, then the SAT solver literally does not move much, using for instance hamming distance in the boolean space as metric. This may be a good time to restart. On the other hand if many flips have occurred recently then there is no point in restarting, it may be even counterproductive.

2.2   A Fresh Look at VSIDS

In order to obtain a robust metric for measuring agility, we follow a reformulation of the seminal work on VSIDS [8]. The basic idea of VSIDS is to concentrate on those variables that recently were involved in conflicts: a variable v is *involved* in a conflict, if v is resolved in the conflict analysis to produce the learned clause or is contained in the learned clause.

Every variable has a counter, called the VSIDS score, which counts how often this variable was used in deriving a learned clause. This counter essentially sums up all these involvements. However, and this is the intriguing idea of VSIDS, it is much better to slowly forget past involvement. Variables with higher VSIDS score are picked as decision variable, which increases the focus of the search. Explaining the effectiveness of VSIDS is out of the scope of this paper.

One way to implement this scheme, is to multiply the VSIDS counters of all variables *not* involved in the current conflict by a constant factor[3] $0 < f < 1$, but not change the counters of involved variables. However, this does not quite work, because the counters will never increase. The solution is to first punish *all* variables by multiplying their score with f, including variables involved in the conflict, and only then additionally increment the score of the latter by $1 - f$.

$$s, f \leq 1, \quad \text{then} \quad s' \leq \underbrace{s \cdot \overbrace{f}^{\text{decay in any case}} + 1 - f}_{\text{increment if involved}} \leq f + 1 - f = 1$$

---

[2] An assignment for a decision variable will always use the old value according to the direction resp. phase saving and reusing heuristics.

[3] MiniSAT, RSAT: $f = 95\% \approx 1/1.05$, PicoSAT: $f = 1/1.1 \approx 91\%$.

This reformulation of VSIDS [8] has the benefit that it produces a rational number between 0 and 1, and can be interpreted as the percentage of the number of times a variable was involved in a conflict "recently". Unfortunately we do not have a more precise definition for "recently" at this moment.

The details are as follows. Let δ_n denote the normalized n^{th} increment of a variable v in the nth conflict. It is either 0 if v is *not* involved in the nth conflict, or 1 if v is involved, and we have $i_n = (1 - f) \cdot \delta_n$ for the actual increment i_n. Then the nth score s_n of v after conflict n can be calculated as

$$s_n = (\ldots (i_1 \cdot f + i_2) \cdot f + i_3) \cdot f \cdots) \cdot f + i_n = \sum_{k=1}^{n} i_k \cdot f^{n-k} = (1 - f) \cdot \sum_{k=1}^{n} \delta_k \cdot f^{n-k}$$

which we call *normalized* VSIDS (NVSIDS).

In practice it is too costly to update the VSIDS resp. NVSIDS score of all variables at every conflict, in particular for industrial examples. In the original Chaff implementation, this overhead is avoided, by accumulating and delaying punishment: variables are only punished after 256 conflicts have passed, by multiplying their score by 0.5. Meanwhile involvements increment the score by 1.

MiniSAT 1.13 has shown that it is also possible, much more accurate, more efficient and *more effective* to just update the scores of variables involved in the conflict. The same scheme is used in PicoSAT and in the following we explain and relate this optimized score calculation to our NVSIDS.

In MiniSAT's new *exponential* VSIDS scheme (EVSIDS) variables are *not* punished, but the EVSIDS score S_n has to be interpreted as $s_n \cdot f^{-n}/(1 - f)$, where n is the number of conflicts and s_n is the NVSIDS score. The increment becomes f^n at the nth conflict and with $I_k = \delta_k \cdot f^{-k}$ we get

$$s_n = (1 - f) \cdot f^n \cdot \sum_{k=1}^{n} \delta_k \cdot f^{-k} = (1 - f) \cdot f^n \cdot \sum_{k=1}^{n} I_k = (1 - f) \cdot f^n \cdot S_n$$

As the equation shows the EVSIDS score is linearly related to NVSIDS and thus can be used instead of NVSIDS to compare activity of variables. Moreover, it can be kept up-to-date by just adding f^{-k} to the score of those variables involved in the kth conflict. The EVSIDS scores of other variables, which are usually many more, do not have to be touched.

2.3   Average Number of Recently Flipped Assignments (ANRFA)

To obtain a concrete metric for the agility a we follow the same idea as our NVSIDS reformulation of VSIDS. The global variable a is initialized to zero and intuitively measures the average number of recently flipped assignments.

Whenever a variable v is forced to be assigned, a is updated. First a is multiplied by $0 < g < 1$. If the assignment is a *flip*, e.g. it assigns the opposite value as in the previous assignment to v, then we increment a by $1 - g$. Assignments of decision variables and variables not assigned before do not increment a. As discussed for NVSIDS this enforces $0 \le a \le 1$, if we start with $a = 0$:

$$a, g \le 1, \quad \text{then} \quad a' \le a \cdot \underbrace{g + 1 - g}_{\text{increment if flipped}} \le g + 1 - g = 1$$

$\overbrace{}^{\text{decay in any case}}$

Also note that we do not need an "exponential" reformulation of EVSIDS as for VSIDS, because there is only one single global agility counter.

A value of $g = 0.9999 = 1 - 1/10000$ was effective in our experiments. Slightly different values did not change the result much (in contrast to f in VSIDS). Note, that there are orders of magnitude more assignments than conflicts in a SAT run and therefore g naturally has to be much closer to 1 than f.

We logged a over industrial and crafted benchmarks on which the old version of PicoSAT performed much worse than competitors. It turned out that in those cases, where we conjectured that restarts should be slowed down, the agility a varied between 15% and 40%. For many industrial benchmarks a was way below 20%. Therefore we picked 20% as the limit at which a scheduled inner restart is disabled. Outer restarts are only disabled if the agility reaches 25% and more. Slightly different values do not change experimental results much.

The restart schedule controls the garbage collection limit for learned clauses, as in MiniSAT. Thus the restart schedule per se should *not* change. If a scheduled restart is disabled resp. skipped the solver simply does not backtrack and continues at the same decision level.

Table 1. Number of solved instances: "adaptive = *no*" is without dynamic restart control, "adaptive = **yes**" uses the ANRFA agility a to disable backtracking. Columns *sat*, *unsat*, and *solved* denote the number of solved satisfiable instances, then the number of unsatisfiable instances, and the sum of these two numbers. Time out is only 900 seconds which matches the one used in the SAT Race'06, but is much less than the time limit in the SAT Competition'07. The three rows with AAS-RSAT, show the number of solved instances for a modified version of RSAT, which is more similar to PicoSAT. The percentages "25%" and "30%" are the two values on the limit of the ANRFA agility a. Above this limit AAS-RSAT does not backtrack if a restart is scheduled.

		SAT Race'06			SAT Competition'07					
					industrial			crafted		
	adaptive	sat	unsat	solved	sat	unsat	solved	sat	unsat	solved
MiniSAT 2.0	no	32	38	70	37	57	94	22	46	68
orig. RSAT 2.0	no	38	36	74	41	51	92	10	20	30
AAS-RSAT	no	33	33	66	45	48	93	11	21	32
AAS-RSAT 25%	yes	34	32	66	44	49	93	11	24	35
AAS-RSAT 30%	yes	36	33	69	48	48	96	12	23	35
PicoSAT 741	no	35	39	74	43	54	97	14	24	38
PicoSAT 741	yes	36	39	75	44	57	101	16	36	52

3 Experiments

We added calculating ANRFA and the adaptive restart strategy to PicoSAT and measured its effect on the SAT Race'06 instances and the SAT'07 Competition benchmarks with a time out of 900 seconds and a memory limit of 1.5 GB on Linux PCs with 3 GHz Pentium IV. As Tab. 1 shows we slightly improved on industrial examples. PicoSAT with the adaptive restart schedule can solve 36% more crafted instances. This is mainly due to the improvement on unsatisfiable instances, where 50% more instances are solved.

We also implemented the suggested adaptive technique in RSAT 2.0, the version submitted to the SAT'07 Competition. Before we changed the basic restart interval from 512 to 100 as in PicoSAT and always enforced saving and reusing phases to match PicoSAT more closely. This results in an "aggressive always saving" RSAT, called AAS-RSAT, with and without adaptive restart control. Using adaptive control for restarts in RSAT is not as impressive as for PicoSAT, but we did not spend much time to optimize magic constants either.

4 Conclusion and Future Work

We presented a new adaptive restart strategy, which slows down restarts if the agility of the SAT solver is high. The key insight is to apply the same filtering technique to the number of flipped assignments as in a new reformulation of VSIDS. For PicoSAT considerable performance improvements have been achieved. In future work we want to apply similar ideas to dynamically control the number of garbage collected clauses resp. the limit on the number of conflicts.

References

1. Biere, A.: PicoSAT essentials. Journal on Satisfiability, Boolean Modeling and Computation (submitted, 2008)
2. Eén, N., Biere, A.: Effective preprocessing in SAT through variable and clause elimination. In: Bacchus, F., Walsh, T. (eds.) SAT 2005. LNCS, vol. 3569, Springer, Heidelberg (2005)
3. Gomes, C., Selman, B., Kautz, H.: Boosting combinatorial search through randomization. In: Proc. AAAI 1998 (1998)
4. Hoos, H.: An adaptive noise mechanism for WalkSAT. In: Proc. AAAI 2002 (2002)
5. Huang, J.: The effect of restarts on the eff. of clause learning. In: Proc. IJCAI 2007 (2007)
6. Kautz, H., Horvitz, E., Ruan, Y., Selman, B., Gomes, C.: Dynamic restart policies. In: Proc. AAAI 2002 (2002)
7. Luby, M., Sinclair, A., Zuckerman, D.: Optimal speedup of Las Vegas algorithms. Information Processing Letters 47 (1993)
8. Moskewicz, M., Madigan, C., Zhao, Y., Zhang, L., Malik, S.: Chaff: Engineering an efficient SAT solver. In: Proc. DAC 2001 (2001)
9. Pipatsrisawat, K., Darwiche, A.: RSat 2.0: SAT solver description. Technical Report D–153, Automated Reasoning Group, Comp. Science Dept., UCLA (2007)

New Results on the Phase Transition for Random Quantified Boolean Formulas[*]

Nadia Creignou[1], Hervé Daudé[2], Uwe Egly[3], and Raphaël Rossignol[4]

[1] Université d'Aix-Marseille, Laboratoire d'Informatique Fondamentale,
Luminy, F-13288 Marseille, France
[2] Université d'Aix-Marseille, Laboratoire d'Analyse, Topologie et Probabilités,
Chateau Gombert F-13453 Marseille, France
[3] Institut für Informationsysteme 184/3, Technische Universität Wien
Favoritenstrasse 9-11, A-1040 Wien, Austria
[4] Université de Paris 11, Département de Mathématiques, Bâtiment 425,
F-91405 Orsay Cedex, France

Abstract. The QSAT problem is the quantified version of the satisfiability problem SAT. We study the phase transition associated with
random QSAT instances. We focus on a certain subclass of closed quantified Boolean formulas that can be seen as quantified extended 2-CNF
formulas. The evaluation problem for this class is coNP-complete. We
carry out an advanced practical and theoretical study, which illuminates
the influence of the different parameters used to define random quantified
instances.

1 Introduction

Recently there has been a growth of interest in a powerful generalization of the
Boolean satisfiability, namely the satisfiability of quantified Boolean formulas,
QBFs. Compared to the well-known propositional formulas, QBFs permit both
universal and existential quantifiers over Boolean variables. Thus QBFs allow for
the modeling of problems having higher complexity than SAT, ranging in the
polynomial hierarchy up to PSPACE. These problems include problems from the
areas of verification, knowledge representation and logic. The numerous applications of QBFs have stimulated the development of practically efficient QBF
solvers.

A significant tool for SAT research has been the study of random instances.
It has stimulated fruitful interactions among the areas of artificial intelligence,
theoretical computer science, mathematics and statistical physics. Encouraged
by the widespread embrace of the random SAT model, random instances of
QBF have started to attract some attention (see [8,2,11]). Models for generating random instances of QBF have been initiated in [8]. Experimental studies
have revealed that QBFs in prenex conjunctive normal form show a sharp transition from satisfiability to unsatisfiability, similar to the one observed for SAT.

---

[*] This work has been supported by EGIDE 10632SE, ÖAD Amadée 2/2006 and ACI
NIM 202.

H. Kleine Büning and X. Zhao (Eds.): SAT 2008, LNCS 4996, pp. 34–47, 2008.

Chen and Interian [2] proposed a mathematically tangible model for generating random instances of QBF. The parameter space of the model offers a richer framework for exploring random instances and their complexity than the SAT model. Our work takes place in this framework. Our goal is to illuminate the role of the different parameters. We focus on particular problems for which we can combine practical experiments with theoretical studies. A first step in this line of research was made in [5], where we studied the QXOR-SAT problem. This problem deals with quantified CNF formulas in which the usual "or" is replaced by the "exclusive or". It has the property of being polynomial time solvable, and thus is a natural candidate to carry out both practical and theoretical studies. Thus, we got new insight on the parameters that influence the nature of the transition from satisfiability to unsatisfiability for XOR-CNF formulas. Here, we continue in this line of research in studying another subclass of formulas, but this time, the evaluation problem is coNP-complete. We focus on a certain subclass of closed quantified Boolean formulas that can be seen as quantified extended 2-CNF formulas. This feature provides instances that are still in the reach of the current QBF solvers and also induces some good combinatorial properties that are of use to derive theoretical results.

More precisely, we are interested in closed formulas in conjunctive normal form having two quantifier blocks, namely in formulas of the type $\forall X \exists Y \varphi(X, Y)$, where X and Y denote distinct sets of variables, and $\varphi(X, Y)$ is a conjunction of 3-clauses, each of which contains exactly one universal literal and two existential ones. It is worth noticing that the evaluation problem for this subclass of formulas is coNP-complete. Moreover it provides a fixed-length-clause class that smoothly "interpolates" in between P and coNP-complete (see Section 2.1).

In order to generate random instances we have to introduce several parameters. The first one is the pair (m, n) that specifies the number of variables in each quantifier block, i.e., in X and Y. The second one is $L = cn$, the number of clauses. To sum up the generated formulas are of the form $\forall X \exists Y \varphi(X, Y)$, where X has m variables, Y has n variables, each clause in φ has one literal from X and two from Y and there is a total number of cn clauses in φ. We are interested in the probability that a formula drawn at random uniformly out of this set of formulas evaluates to true as n tends to infinity. We will denote by $\mathbb{P}_{m,c}$ this probability. We are thus interested in

$$\lim_{n \to +\infty} \mathbb{P}_{m,c}(n).$$

We prove that the transition between satisfiability and unsatisfiability for such a random formula occurs when c is in between 1 and 2. Moreover we show that the parameter that controls the location of the transition is m the number of universal variables. For m big enough (as a function of n), there is a *critical value* (or a *threshold*) of c, $c = 1$, above which the likelihood of a random formula being satisfiable vanishes as n tends to infinity, and below which it goes to 1. For m small enough, the critical value is at $c = 2$. An intermediate regime is obtained when m is of logarithmic order compared to n. Our main result is

Theorem 1. *Let* $m = \lceil \alpha \ln n \rceil$ *where* $\alpha > 0$. *There exist two decreasing functions* a *and* b *with* $1 < a(\alpha) \leq b(\alpha) \leq 2$ *such that the following holds:*

- *if* $c < a(\alpha)$, *then* $\mathbb{P}_{m,c}(n) \xrightarrow[n \to +\infty]{} 1$,

- *if* $c > b(\alpha)$, *then* $\mathbb{P}_{m,c}(n) \xrightarrow[n \to +\infty]{} 0$.

According to the following partition in three intervals for α *we have:*

1. *if* $\alpha \leq \dfrac{1}{\ln 2}$, *then* $a(\alpha) = b(\alpha) = 2$,

2. *if* $\dfrac{1}{\ln 2} < \alpha \leq \dfrac{2}{\ln 2 - 1/2}$, *then* $a(\alpha) < b(\alpha) = 2$ *and* a *is strictly decreasing,*

3. *if* $\alpha > \dfrac{2}{\ln 2 - 1/2}$, *then* $a(\alpha) < b(\alpha) < 2$, a *and* b *are strictly decreasing*
    and $\lim\limits_{\alpha \to +\infty} a(\alpha) = \lim\limits_{\alpha \to +\infty} b(\alpha) = 1$.

The following figure gives a synthetic picture of the evolution of both lower and upper bounds $a(\alpha)$, $b(\alpha)$ mentioned in Theorem 1 and explicitly defined in Section 4.

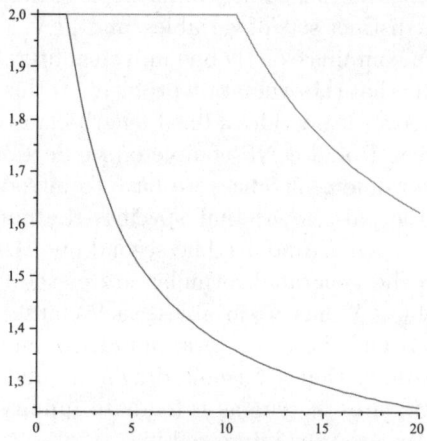

Fig. 1. $a(\alpha)$ and $b(\alpha)$

The paper is organized as follows. First in Section 2 we precisely define the problem we are interested in. We discuss its complexity and finally present the random model. In Section 3, we report some experiments and we show how they have lead to first informations on the phase transition from satisfiability to unsatisfiability. We also illustrate in this section the limits of the experiments. The proof of our main result is inspired by the investigation done by Chvátal, Reed and Goerdt [3,10] in establishing a sharp threshold phenomenon for random 2-SAT (the associated critical ratio being $c = 1$). It is based on a digraph representation of our formulas presented in Section 4. Then, first and second moment methods used on specific structures on these graphs give lower and upper bounds for the location of the phase transition. The main steps of the final analytical analysis are given in Section 5.

2   Definition of Our Problem

2.1   The Problem (1,2)-QSAT and Its Complexity

A *literal* is a propositional variable or its negation. The *atom* or the propositional variable of a literal l, denoted by $|l|$, is l itself, if l is of the form p, and p if l is of the form $\bar{p}$. A *clause* is a finite disjunction of literals. A formula is in *conjunctive normal form* (CNF) if it is a conjunction of clauses. A formula is in k-CNF, if any clause consists of exactly k literals.

We assume familiarity with the syntax and semantics of quantified Boolean formulas (QBFs). We only consider *closed* QBFs, i.e., QBFs without free variables. A *universal* (*existential*) *literal* is a literal whose atom is universally (existentially) quantified.

Here we are interested in formulas of the form

$$F = \forall X \exists Y \varphi(X, Y)$$

where $X = \{x_1, \ldots, x_m\}$, and $Y = \{y_1, \ldots, y_n\}$, and $\varphi(X, Y)$ is a 3-CNF formula, with exactly one universal and two existential literals in each clause. We will call such formulas (1,2)-QCNFs. These formulas can be considered as quantified extended 2-CNF formulas, because deleting the only universal literal in each clause and removing the then superfluous $\forall$-quantifiers results in an existentially quantified set of binary clauses. In the following, the 2-CNF formula so obtained will be denoted by $F_Y = \exists Y \varphi(Y)$.

A (1,2)-QCNF formula is *true* (or *satisfiable*) if for every assignment to the variables X, there exists an assignment to the variables Y such that φ is true.

Let us give some information about the complexity of the evaluation (true or false) of such formulas. The exhaustive algorithm which consists in deciding whether for all assignment to the variables X, there exists an assignment to the variables Y such that φ is true provides a first upper bound for the worst case complexity. Indeed, since the satisfiability of a 2-CNF formula can be decided in linear time [1], the evaluation of the formula $\forall X \exists Y \varphi(X, Y)$ can be performed in time $O(2^m \cdot |\varphi|)$, where m is the number of universal variables. Observe that if m is bounded by a constant, then it provides a linear time algorithm, and if m is of the order of $\log n$, then it provides a polynomial time algorithm. If m has the same order as n, then the above algorithm runs in exponential time. Moreover this problem is in coNP: to prove that such a formula is unsatisfiable, guess a vector of truth values $v_1, \ldots, v_m$ corresponding to $x_1, \ldots, x_m$. Replace in $\exists Y \varphi(X, Y)$ all free occurrences of any x_i by v_i, remove $\bot$ from the clauses and delete clauses with $\top$. The resulting formula is a usual 2-CNF formula, whose unsatisfiability can be checked in polynomial time. It is also hard for this class as shown in [7].

Theorem 2

- *For every fixed α, when restricted to formulas having m universal variables and n existential variables with $m = \lceil \alpha \ln n \rceil$, the evaluation problem for (1,2)-QCNF formulas is decidable in polynomial time.*
- *In its full generality, this evaluation problem is* coNP-*complete.*

It is interesting to note that the same functional dependency between the number of universal variables and the number of existential one, namely $m = \lceil \alpha \ln n \rceil$, appears in Theorem 1 and in Theorem 2, thus controlling the location of the transition as well as the complexity of the evaluation problem.

2.2   Random Instances

Let us now describe our model, which is a model suggested in [8] and systematically defined in [2]. The model has several parameters. The first parameter is a pair (m, n) specifying the number of variables in each quantifier block, respectively in X and Y. The second parameter is L, the number of clauses. To sum up the generated formulas are of the form $\forall X \exists Y \varphi(X, Y)$, where X has m variables, Y has n variables, each clause in φ has one variable from X and two from Y and there is a total number of L clauses in φ.

Throughout the paper, we reserve m for the number of universal variables, n for the number of existential variables. Note that there are

$$N = m \cdot \binom{n}{2} \cdot 2^3 = 4 \cdot m \cdot n(n-1) \tag{1}$$

clauses. We consider random formulas $\forall X \exists Y \varphi(X, Y)$ obtained by choosing uniformly independently and with replacement L clauses from all the possible N clauses. We will always consider the parameter m as a function of n, i.e., $m = m(n)$ and L as a fraction of n, i.e., $L = cn$. Thus, we are interested in the probability that a formula drawn at random uniformly out of this set of formulas is true as n tends to infinity. It is well-known that equivalently, we can consider a formula drawn at random in choosing independently each possible clause with probability p, where $N \cdot p = c \cdot n$, that is

$$p \sim \frac{c}{4nm}.$$

We will denote by $\mathbb{P}_{m,c}(n)$ the probability that such a random formula is true. For fixed n and m, $\mathbb{P}_{m,c}(n)$ is a decreasing function of $c = L/n$, which is a control parameter for the transition from satisfiability to unsatisfiability. We will be interested in studying $\lim_{n \to +\infty} \mathbb{P}_{m,c}(n)$ as a function of the parameters m and c. Any value of c such that $\mathbb{P}_{m,c}(n) \to 1$ (resp. s. t. $\mathbb{P}_{m,c}(n) \to 0$) gives a lower (resp. upper) bound for the threshold effect associated to the phase transition.

3   Experimental Results and a First Estimate for the Location of the Threshold

Before we start discussing the empirical results, let us first describe how we performed the experiments. All experiments have been conducted according to the same scheme, which is described with the help of Fig.2. One experiment consisted in generating at random (in drawing uniformly and independently) (1,2)-QCNF formulas over given values of m universal variables and n existential variables, with a ratio "number of clauses/number of existential variables" varying from 0.85 to 1.2 in steps of 0.05. In Fig. 2, $m = n$ and the values are 5000, 10000, 20000 and 40000. For each of the chosen values of ratio, a sample of 1000 formulas have been studied using the QBF solver QuBE [9], thus computing the truth value of each formula. The proportion of true (or satisfiable) instances for each considered value of ratio has been plotted in Fig. 2.

The experimental results shown in Fig. 2 suggest that, if $m = n$, then the transition between satisfiability and unsatisfiability occurs when the ratio of number of clauses to number of existential variables, c, is equal to 1. Fig. 3 shows that if m is constant, $m = 2$, then the transition occurs at $c = 2$. Moreover, the experiments reported in Fig. 4 indicate that an intermediate regime, with a transition occurring in between 1 and 2, can also be observed.

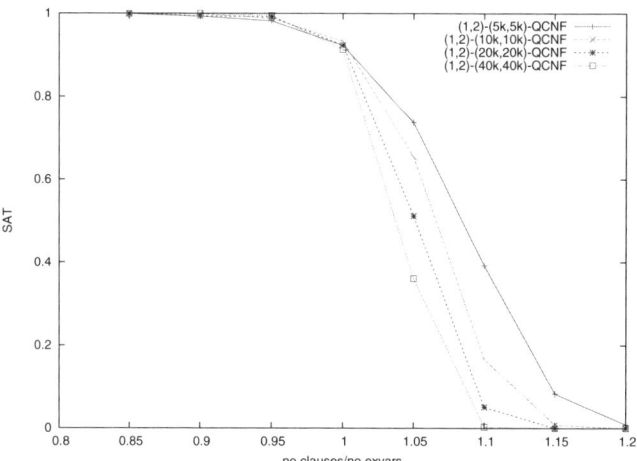

Fig. 2. $\mathbb{P}_{m,c}$ when $m(n) = n$. The threshold occurs at $c = 1$.

These first experiments indicate that the phase transition from satisfiability to unsatisfiability for (1,2)-QCNF formulas occurs when $1 \leq c \leq 2$. The following easy result confirms this observation.

Proposition 1. *Let $m = m(n)$ be any sequence of integers.*

- *If $c < 1$ then $\mathbb{P}_{m,c}(n) \xrightarrow[n \to \infty]{} 1$.*

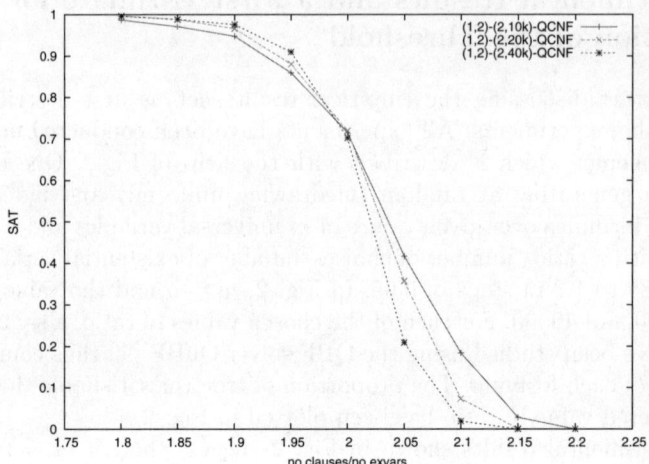

Fig. 3. $\mathbb{P}_{m,c}$ when $m(n) = 2$. The threshold occurs at $c = 2$.

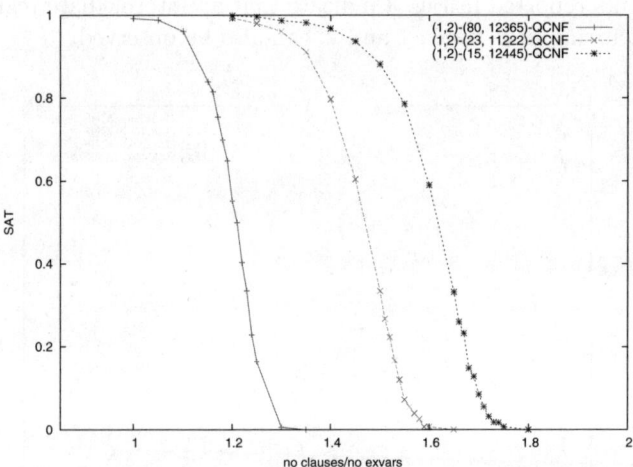

Fig. 4. $\mathbb{P}_{m,c}$ when $m(n)$ is varying

– If $c > 2$ then $\mathbb{P}_{m,c}(n) \xrightarrow[n \to \infty]{} 0$.

Proof. Let F_t be the 2-CNF formula obtained from F by setting all the variables $x_1, \ldots, x_m$ to *true* and omitting all quantifiers. If F is satisfiable, then so is F_t. Notice that F_t can be obtained by picking independently each possible 2-clause with probability $q(n) = 1 - (1 - p(n))^m = \frac{c}{4n} + O\left(\frac{1}{n^2}\right)$. Thus the average number of clauses in F_t is equal to $4\binom{n(n-1)}{2} \cdot q \sim c/2 \cdot n$. It follows from the threshold of 2-SAT [3,10] that F_t is unsatisfiable with probability tending to 1 if $c > 2$. Thus, the same holds for F.

Now, we look at the existential part of the formula, F_Y. Observe that if F_Y is satisfiable, then so is F. In F_Y, each of the $4\binom{n}{2}$ 2-clauses appear independently with probability $q'(n) = 1 - (1 - p(n))^{2m} = \frac{c}{2n} + O\left(\frac{1}{n^2}\right)$. Therefore, the threshold of 2-SAT tells us that when $c < 1$, the formula F_Y is satisfiable with probability tending to one.

For m constant the critical value seems to be at 2, for $m = n$ it seems to be at 1. Then a natural question arises: at what speed should m vary so that the critical value is strictly in between 1 and 2? The curves shown in Fig. 4 suggest that a good candidate to look at is when m is of logarithmic order compared to n. Indeed, each of the curves in this figure corresponds to $m = \lceil \alpha \ln n \rceil$ for some value α, respectively for $\alpha = 9/8$, $3/2$ and $15/8$. The following proposition confirms that the logarithmic scale is indeed a good candidate.

Proposition 2. *Let $m = m(n)$ be a sequence of integers such that $m \leq \ln n / \ln 2$. If $c < 2$ then $\mathbb{P}_{m,c}(n) \xrightarrow[n \to \infty]{} 1$.*

Observe that this result together with Proposition 1 shows a threshold at $c = 2$ when m is small enough, that is when $m \leq \ln n / \ln 2$. In Theorem 1, this corresponds to the first interval, namely $\alpha \leq \dfrac{1}{\ln(2)}$.

To take a step further, a question is whether we can continue to use experiments in order to make precise the critical value when $m = \lceil \alpha \ln n \rceil$. Are the solvers, and the machines, powerful enough to provide experiments at a scale big enough?

Figures 5 and 6 show that the critical value of the threshold is very difficult to estimate from the experiments. The experimental results reported in Figure 5 could suggest that all the curves pivot about a single point, thus indicating a critical ratio at $c \sim 1.8$. However, as evidenced in Figure 6, which consists

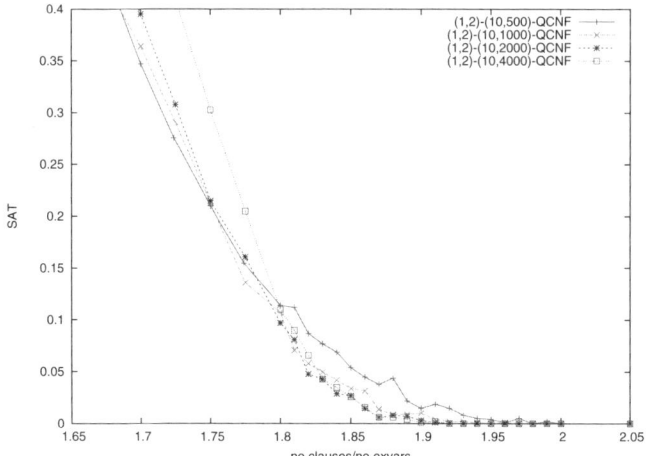

Fig. 5. $\mathbb{P}_{m,c}$ when $m(n) = 10$. Is the threshold at $c = 1.8$?

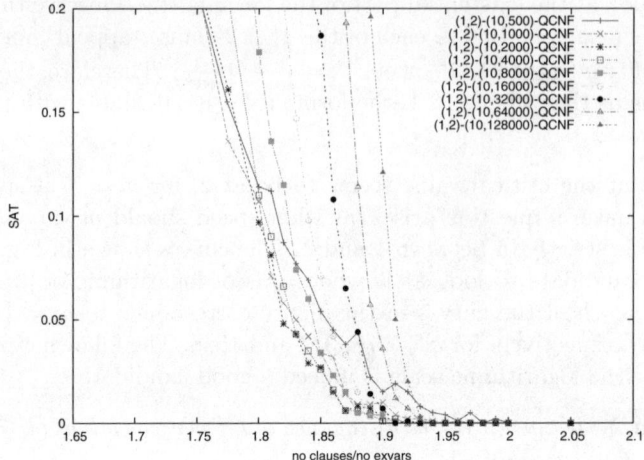

Fig. 6. $\mathbb{P}_{m,c}$ when $m(n) = 10$. The critical value is difficult to estimate.

in experiments on a finer scale for bigger values of n, one can have successive crossings of pairs of curves for increasing values of n, which provide only a rough estimate of a possible critical ratio. Moreover, the asymptotical behavior (here according to Proposition 2, we have a critical ratio at 2) is still not reached for very big values of n, e.g., for $n = 128000$.

For this reason, when looking at the case $m = \lceil \alpha \ln n \rceil$ (for which the complexity is higher than in the case $m = 10$) one cannot hope that the experiments furnish a reliable estimate on the relationship between the location of the threshold and α.

4   Main Result and Its Relation to 2-SAT

Our main result, which is stated in Theorem 1, shows that the transition occurs for c strictly in between 1 and 2 when the number of universal variables is of a sufficiently large logarithmic order compared to the number of existential variables. Two functions $a(\alpha)$ and $b(\alpha)$, which give respectively a lower and an upper bound for the threshold, are announced in Theorem 1 and shown in Fig. 1. Our probabilistic analysis shows that they are implicitly defined as follows:

$a(\alpha)$ is the unique solution of $H(c) = \dfrac{1}{\alpha}$ for $c \in]1, 2[$ where

$$H(c) = \ln(c) + \left(\frac{2}{c} - 1\right) \ln(2 - c),$$

$b(\alpha)$ is the unique solution of $K(c) = \dfrac{1}{\alpha}$ for $c \in]1, 2[$ where

$$K(c) = \frac{1}{2}\left(\ln c + \frac{1}{c} - 1)\right).$$

We have $\lim\limits_{\alpha \to +\infty} a(\alpha) = \lim\limits_{\alpha \to +\infty} b(\alpha) = 1$. Thus, when $m/\ln n \xrightarrow[n \to +\infty]{} +\infty$, Theorem 1 together with Proposition 1 establish a sharp threshold for the satisfiability of (1,2)-QCNF formulas with a critical ratio at $c = 1$. Since it is easy to derive from [7] that the evaluation problem of (1,2)-QCNF formulas is coNP-complete when restricted to the case $m = n$, this proves a sharp threshold for a quantified satisfiability problem which is coNP-complete.

In order to prove our main result we will use the relation of our problem to random 2-SAT. Chvátal and Reed introduced specific substructures (bicycles and snakes) on digraphs associated to 2-CNF formulas. Below we will show that their analysis can be adapted to study (1,2)-QCNF random formulas in considering *labeled* digraphs, *pure* bicycles and *simple* snakes. Although the digraph structures associated to 2-CNF and (1,2)-QCNF formulas are very similar, we will need a more involved analysis to describe the probabilistic behavior of *pure* bicycles and *simple* snakes associated to our quantified formulas.

4.1   Representation of (1,2)-QCNF Formulas as Labeled Digraphs

Any (1,2)-QCNF-formula can be represented as a digraph with labeled arcs. For constructing the digraph, we construct the implication digraph [1] associated with the existential 2-CNF formula, and we put the universal literal as a label of the two arcs derived from each clause. Two labels are *dual* if one is x and the other $\bar{x}$ for some universal variable x. We say that a subgraph of a labeled digraph is *pure* if its set of labels does not contain two dual labels. The maximal pure subgraphs correspond to the implication graphs of the 2-CNF formulas obtained after instantiating the universal variables in the original formula and deleting the quantifiers. Therefore the quantified formula is satisfiable if and only if all the 2-CNF formulas corresponding to the maximal pure subgraphs are satisfiable.

Let $\phi\colon \forall x_1 \exists y_1 y_2 ((x_1 \vee y_1 \vee y_2) \wedge (\overline{x_1} \vee y_1 \vee \overline{y_2}))$. The labeled digraph of ϕ is shown on the left in Fig. 7 together with its two maximal pure subgraphs. The first one corresponds to the instantiation $x_1 = 1$, whereas the second one corresponds to the instantiation $x_1 = 0$.

In order to get lower and upper bounds for the location of the phase transition the idea is to identify specific structures in these graphs that guarantee a formula to be satisfiable (respectively unsatisfiable).

By a *bicycle* of length $s+1 \geq 3$, we mean a set of $s+1$ clauses $C_0, \ldots, C_s$ that have the following structure: there are s distinct existential literals $w_1, \ldots, w_s$

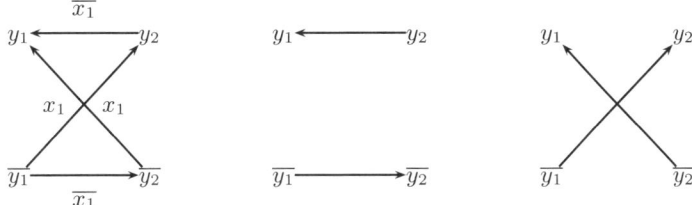

Fig. 7. The digraph for ϕ together with its maximal pure subgraphs

such that no w_i is the complement of another, there is a sequence $v_0, \ldots, v_s$ of $s + 1$ universal literals (or labels), each C_r with $0 < r < s$ is $(v_r \vee \overline{w_r} \vee w_{r+1})$, and $C_0 = (v_0 \vee u \vee w_1)$, $C_s = (v_s \vee \overline{w_s} \vee v)$ with literals u, v chosen from $w_1, \ldots, w_s, \overline{w_1}, \ldots, \overline{w_s}$ with $(u, v) \neq (w_s, w_1)$. We consider *pure bicycles*, which are bicycles such that no label is the complement of another.

Claim. Every unsatisfiable (1,2)-QCNF formula contains a pure bicycle.

Let B be the number of pure bicycles in a (1,2)-QCNF formula. In our probabilistic model, we deduce from the above claim and the Markov inequality

$$1 - \mathbb{P}_{m,c}(n) \leq \Pr(B \geq 1) \leq \mathbb{E}(B). \tag{2}$$

By a *snake* of length $s+1$, we mean a set of $s+1$ clauses $C_0, \ldots, C_s$, that have the following structure: there are s distinct existential literals $w_1, \ldots, w_s$ with $s = 2t - 1$ such that no w_i is the complement of another, there is a sequence $v_0, \ldots, v_s$ of $s + 1$ universal literals (or labels), each C_r with $0 \leq r \leq s$ is $(v_r \vee \overline{w_r} \vee w_{r+1})$ with $w_0 = w_{s+1} = \overline{w_t}$. We consider *simple snakes*, which are snakes such that no label is the same as or the complement of another. Note that a simple snake is pure. Simple snakes are easier to enumerate than pure ones and will be sufficient for our purpose. Observe that $\forall X \exists Y C_0 \wedge \ldots \wedge C_s$ is unsatisfiable.

Claim. Every (1,2)-QCNF formula that contains some simple snake is unsatisfiable.

Let X be the number of simple snakes of size $s + 1 = 2t$ in a (1,2)-QCNF formula. In our probabilistic model, we deduce from the above claim and the Cauchy-Schwarz inequality:

$$1 - \mathbb{P}_{m,c}(n) \geq \Pr(X \geq 1) \geq \frac{(\mathbb{E}(X))^2}{\mathbb{E}(X^2)} \tag{3}$$

4.2   The First Moment of B and the Second Moment of X

In our probabilistic model, where $p = \dfrac{c}{4m(n - 1)} \sim \dfrac{c}{4nm}$, the following result will be the starting point to get lower bounds for the location of the phase transition.

Proposition 3. *The mean of the number B of pure bicycles in a random (1,2)-QCNF formula is given by*

$$\mathbb{E}(B) = \sum_{s=2}^{n} (n)_s 2^s [(2s)^2 - 1] c(m, s + 1) p^{s+1}, \tag{4}$$

where

$$c(m, s + 1) = \sum_{k=1}^{min(m, s+1)} \binom{m}{k} \cdot 2^k \cdot S(s + 1, k) \cdot k! \tag{5}$$

with $S(m, k)$ denoting the Stirling number of the second kind.

Proof. To count B, choose s, the s distinct literals $w_1, \ldots, w_s$ such that no w_i is the complement of another, choose u and v, and choose the pure sequence of $s+1$ labels $v_0, \ldots, v_s$ (they are not necessarily distinct but no literal can be the complement of another).

Let $c(m, s+1)$ be the number of pure sequences of literals of length $s+1$, having a set of m variables from which the literals can be built. Let us recall that $S(m, k) \cdot k!$ is the number of applications from a set of m elements onto a set of k elements. A pure sequence of literals of length $s+1$ is obtained by exactly one sequence of choices of the following choosing process.

1. Choose the number k of different variables occurring in the sequence.
2. Choose the k variables
3. For each such variable, choose whether it occurs positively or negatively.
4. Choose their places in the sequence.

As in [3], the following observation will be the starting point to get upper bounds for the location of the phase transition:

Proposition 4. *Let X be the number of simple snakes of size $s + 1 = 2t$ in a (1,2)-QCNF formula. Then*

$$\frac{(\mathbb{E}(X))^2}{\mathbb{E}(X^2)} = \frac{1}{q_0(m, n) + \sum_{i=1}^{2t} q_i(m, n) \cdot p^{-i}} \tag{6}$$

where

$$q_i(m, n) = \frac{\#\{simple\ snakes\ B\ such\ that\ |A_0 \cap B| = i\}}{\#\{simple\ snakes\}} \tag{7}$$

for any fixed simple snake A_0, with $|A_0 \cap B|$ denoting the number of clauses A_0 and B share.

5 Proofs

5.1 Proof of Proposition 2

Coming back to the first moment of B, we get from equation (4):

$$\mathbb{E}(B) \leq \frac{c}{nm} \sum_{s=2}^{n} s^2 \left(\frac{c}{2m}\right)^s c(m, s+1) . \tag{8}$$

Notice that $c(m, s+1)$ is bounded from above by $2^{\min\{m, s+1\}}$ times the number of applications from $\{1, \ldots, s+1\}$ to $\{1, \ldots, m\}$:

$$c(m, s+1) \leq 2^{\min\{m, s+1\}} m^{s+1} . \tag{9}$$

When $x \in]0, 1[$ and $r \geq 1$, standard computations show that:

$$\sum_{s=r}^{\infty} s^2 x^s \leq r^2 \frac{x^r}{(1-x)^3} . \tag{10}$$

Thus, when $1 < c < 2$, then $\mathbb{E}(B) \leq \dfrac{c}{n} \displaystyle\sum_{s=2}^{m-1} s^2 c^s + \dfrac{c2^m}{n} \displaystyle\sum_{s=m}^{\infty} s^2 \left(\dfrac{c}{2}\right)^s = O\left(m^2 \dfrac{c^m}{n}\right),$

which goes to zero as n goes to infinity when $m \leq \ln n / \ln 2$. Proposition 2 is proved.

5.2   Proof of the Lower Bound in Theorem 1

By using precise results for the behavior of Stirling numbers of the second kind [12] (already used in [6] and [5]), a finer analysis of the expected number of pure bicycles as expressed in (4) gives the following result.

Theorem 3. *When $1 < c < 2$, and $m = \lceil \alpha \ln n \rceil$ with $\alpha > \dfrac{1}{\ln(2)}$, the average number of pure bicycles satisfies*

$$\mathbb{E}(B) \leq C(\ln n)^{9/2} \cdot n^{\alpha H(c) - 1} + o(1),$$

where C is a constant depending only on α and c, and $H(c) = \ln(c) + \left(\dfrac{2}{c} - 1\right) \ln(2 - c)$.

Let $a(\alpha)$ be the solution of the equation $\alpha \cdot H(c) = 1$, then for $c < a(\alpha)$ the above result shows that $\mathbb{E}(B) = o(1)$. Thus, with (2) we deduce the lower bound stated in Theorem 1.

5.3   Proof of the Upper Bound in Theorem 1

When considering simple snakes and making a similar estimation as in equations (8) and (9) in [3], we get the following result.

Theorem 4. *When $1 < c < 2$, $m = \lceil \alpha \ln n \rceil$ and for $t = \left\lceil \dfrac{\alpha}{2}\left(1 - \dfrac{1}{c}\right) \ln(n) \right\rceil$ we have*

$$\sum_{i=1}^{2t} q_i(m,n) p^{-i} = O\left(\max\left(\ln(n) \cdot n^{1-\alpha K(c)}, \dfrac{(\ln n)^{10}}{n}\right)\right)$$

where $K(c) = \dfrac{1}{2}\left(\ln c + \dfrac{1}{c} - 1\right)$.

Let $b(\alpha)$ be the solution of the equation $\alpha \cdot K(c) = 1$ then $b(\alpha) < 2$ when $\alpha > \dfrac{2}{\ln(2) - 1/2}$. For $c > b(\alpha)$ the above result shows that $\sum_{i=1}^{2t} q_i(m,n) p^{-i} = o(1)$.

Observe that $\displaystyle\sum_{i=0}^{2t} q_i(m,n) = 1$, then with (3) and (6) we get the upper bound stated in Theorem 1.

6   Conclusion

We have made an extensive study of a natural and expressive quantified problem. The obtained results have several interesting features. They highlight the role of different parameters and their influence on the transition. These results are based on experiments that make use of a current QBF solver. These experiments are carried out at a scale large enough in order to give a useful intuition on the asymptotical behavior of random instances. We have shown that functional dependencies other than $m = \rho n$ can be important. Indeed, we have demonstrated that $m = \lceil \alpha \ln n \rceil$ is the scale which is crucial, both for the complexity and the behavior of random instances (see Theorems 1 and 2). Moreover, we give the precise location of the sharp phase transition (namely at $c = 1$) for a natural quantified problem (namely when $m = n$) which is coNP-complete.

References

1. Aspvall, B., Plass, M.F., Tarjan, R.E.: A linear-time algorithm for testing the truth of certain quantified Boolean formulas. Information Processing Letters 8(3), 121–123 (1979)
2. Chen, H., Interian, Y.: A model for generating random quantified Boolean formulas. In: Proceedings of the 19th International joint Conference on Artificial Intelligence, IJCAI 2005, pp. 66–71 (2005)
3. Chvátal, V., Reed, B.: Mick gets some (the odds are on his side). In: Proceedings of the 33rd Annual Symposium on Foundations of Computer Science, FOCS 1992, pp. 620–627 (1992)
4. Creignou, N., Daudé, H., Dubois, O.: Expected number of locally maximal solutions for random Boolean CSPs. In: Proceedings of the13th International Conference on Analysis of Algorithms, AofA 2007, Antibes, June 2007. DMTCS, pp. 507–516 (2007)
5. Creignou, N., Daudé, H., Egly, U.: Phase transition for random quantified XOR-formulas. Journal of Artificial Intelligence Research 19, 1–18 (2007)
6. Dubois, O., Boufkhad, Y.: A general upper bound for the satisfiability threshold of random r-SAT formulae. Journal of Algorithms 24(2), 395–420 (1997)
7. Flögel, A., Karpinski, M., Kleine Büning, H.: Subclasses of quantified Boolean formulas. In: Schönfeld, W., Börger, E., Kleine Büning, H., Richter, M.M. (eds.) CSL 1990. LNCS, vol. 533, pp. 145–155. Springer, Heidelberg (1991)
8. Gent, I.P., Walsh, T.: Beyond NP: the QSAT phase transition. In: Proceedings of AAAI 1999 (1999)
9. Giunchiglia, E., Narizzano, M., Tacchella, A.: QuBE: A System for Deciding Quantified Boolean Formulas Satisfiability. In: Goré, R.P., Leitsch, A., Nipkow, T. (eds.) IJCAR 2001. LNCS (LNAI), vol. 2083, pp. 364–369. Springer, Heidelberg (2001)
10. Goerdt, A.: A threshold for unsatisfiability. Journal of of Computer and System Sciences 53(3), 469–486 (1996)
11. Interian, Y., Corvera, G., Selman, B., Williams, R.: Finding small unsatisfiable cores to prove unsatisfiability of QBFs. In: Proceedings of the 9th International Symposium on Artificial Intelligence and Mathematics (2006)
12. Temme, N.M.: Asymptotic estimates of Stirling numbers. Stud. appl. Math. 89, 223–243 (1993)

Designing an Efficient Hardware Implication Accelerator for SAT Solving

John D. Davis[1], Zhangxi Tan[2,*], Fang Yu[1], and Lintao Zhang[1]

[1] Microsoft Research Silicon Valley Lab
{joda,fangyu,intaoz}@microsoft.com
[2] UC Berkeley
xtan@cs.berkley.edu

Abstract. This paper discusses the design of a hardware accelerator for Boolean Constraint Propagation (BCP) using Field Programmable Gate Arrays (FPGA). In particular, we describe the detailed implementation of the inference engine, a key component of the accelerator that performs implications. Unlike previous efforts in FPGA assisted SAT solving, our design uses Block RAM (BRAM) to store instance information. This novel design not only facilitates fast lookup and update, but also avoids synthesizing overhead for each SAT instance. We demonstrate that SAT instances can be easily partitioned into multiple groups that can be processed by multiple inference engines in parallel. By exploiting parallelism in hardware, the BCP accelerator can infer implications in 6 to 17 clock cycles for a new variable assignment. In addition, our design supports dynamic insertion and deletion of learned clauses. Cycle accurate simulation shows that our BCP accelerator is 5~16 times faster than the conventional software based approach for BCP.

1 Introduction

Hardware-assisted SAT solving has attracted much research in recent years. Designs based on Field Programmable Gate Arrays (FPGAs) have been described in [2][3][4][5][6][7][8], and were compared in a survey [9]. Unfortunately, most of these accelerators were designed before the prevalence of the so called "chaff-like" modern SAT solvers [10][1]. Due to the tremendous improvements of modern SAT solvers and the stringent requirements from industrial applications, many of the existing hardware solvers are either obsolete in some cases and/or severely constrained in others. For example, compared with modern software solvers, the hardware accelerators from previous work were usually slow and capacity limited, and they are unable to accommodate important features in software SAT solvers such as learning.

Our main goal in this research is to leverage hardware acceleration to build a practical SAT solver. In a previous paper [11], we described the overall architecture of the FPGA-based SAT accelerator and compared it with previous works. Our design concentrates on accelerating the Boolean Constraint Propagation (BCP) part of the SAT solving process in hardware. We target the acceleration of the BCP phase because it is

---

* This work was done during the author's internship at Microsoft Research.

H. Kleine Büning and X. Zhao (Eds.): SAT 2008, LNCS 4996, pp. 48–62, 2008.

a stable component in all SAT solvers and accounts for 80~90% of total runtime in a modern software SAT solver. The rest of the work such as branching, restarting and conflict analysis is left to the software on the host computer. Our accelerator is application specific instead of instance specific. It does not require time consuming re-synthesizing the logic in the FPGA for each new CNF instance and can load a SAT instance into the FPGA in sub-second.

In this paper, we describe the key component of the accelerator, namely the inference engines. The inference engine is responsible for inferring new implications from new variable assignments. In our design, each inference engine is in charge of a number of clauses. The information of these clauses such as the literals and their corresponding values is stored locally in FPGA Block RAM (BRAM). Multiple inference engines can operate in parallel to perform inferences on the same newly assigned variable. In our design, clauses need to be partitioned and the groups of clauses are distributed across multiple inference engines. Finding the optimal partitioning that uses the least amount of memory is itself an NP hard problem. In this paper, we present a simple and efficient clause partitioning algorithm that generates high quality partitions in practice.

In our previous work [11], the design lacked the ability to accommodate learned clauses, which is a critical feature of modern SAT solvers. One of the main contributions of this paper is the improved inference engine design with the ability to dynamically insert and delete learned clauses.

2   An Overview of the Hardware SAT Accelerator

For the completeness of the paper, we briefly review the design of the accelerator. We refer the readers to [11] for a detailed description of the architecture and the rationales behind the design. The overall system architecture of the accelerator is shown in Figure 1. The shaded blocks are modifications to accommodate learning (as described in section 3.4). It is composed of the following major components:

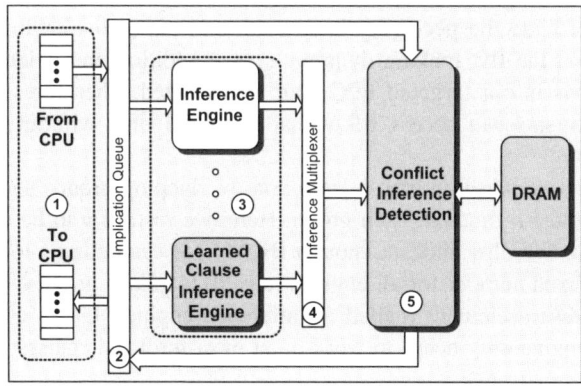

Fig. 1. FPGA Boolean Constraint Propagation Accelerator Architecture

1. **CPU Communication Module:** This module receives branch decisions from and returns inference results back to the CPU.
2. **Implication Queue:** Decisions from the CPU and implications derived from the inference engines are queued in a FIFO and sent to multiple implication inference engines. This module also puts the implications performed in a buffer to be sent back to the CPU.
3. **Parallel Inference Engines:** Clauses of the SAT formula are partitioned and stored in multiple inference engines. We present a more detailed description of the inference engines in Section 3.
4. **Inference Multiplexer:** This module serializes the data communications between the parallel inference engines and the sequential conflict inference detection stage.
5. **Conflict Inference Detection:** This module stores the global variable values and detects conflict inference results generated by the inference engines.

3 Inference Engines

The inference engine is the key component of any SAT solver, regardless of whether it is hardware or software based. Given a new variable assignment, the implication machinery in a SAT solver needs to infer the implications caused by the new assignment and current variable assignments. To accomplish this, it must store the clause information. In software SAT solvers, the clauses are stored in main memory as arrays or lists of literals. Previous hardware SAT solvers such as [3] synthesize clauses into gates using Lookup Tables or LUTs in the FPGA. This design is inflexible because each new SAT instance requires the FPGA to be re-synthesized, which is time consuming[1]. Moreover, most current FPGAs cannot be dynamically re-configured or the process is very cumbersome, thus making dynamic clause addition/removal difficult. In our novel design, we leverage the fact that modern FPGAs have many banks of Block RAM (BRAM), which are distributed around the FPGA with the configurable logics (LUTs). We use BRAM to store clause information, thus avoiding re-synthesizing the logic in the FPGA. Multiple BRAM blocks can be accessed at the same time to provide the necessary bandwidth and parallelism. Moreover, BRAMs can be loaded on the fly, making dynamic clause addition and deletion for learning possible. BRAMs in our targeted FPGA are dual ported. Therefore, implication and learning mechanisms can access BRAM at the same time without disrupting each other's operation.

In our design, we partition clauses into non-overlapping groups so that each literal only occurs at most p times in each group. Here, we restrict p to be a small number, e.g., one or two. We also allocate enough BRAM for each engine to store c clauses, with c being a fixed number for all engines (e.g. 1024). Each group of clauses is processed by a hardware element (called an inference engine). Given a newly assigned variable, each engine only needs to work on at most p related clauses -- a process that

---

[1] [8] uses memory to store clauses, but the solver stores a full matrix of the clause data, instead of the traditional sparse matrix representation as in software SAT solvers.

takes a fixed number of cycles. Thus, by limiting p, multiple inference engines process literal assignments in parallel rather than in serial as in software solvers. By partitioning clauses into groups, the number of engines can be significantly smaller than the number of clauses, more efficiently utilizing FPGA resources. We allow p to be larger than one because slightly larger p can help reduce the number of engines required. This is especially helpful for long clauses such as the learned clauses because they share variables with many other clauses. Although it is outside of the scope of this paper, we can adjust p and c to optimize the number of inference engines and the memory utilization within the inference engine. In Section 3.1, we present the partitioning algorithm and study the effect of p in the simulation section.

We will describe the high-level inference engine functionality, followed by the design considerations and challenges. Inference engines use a two step operation to process a new variable assignment and produce possible implications, as shown in Figure 2. In the first step, the inference engine needs to find out whether the assigned variable actually is related to any clauses stored in the engine, and if so, identify these clauses. In the second step, the engine examines these clauses to see whether they actually imply a new variable. The algorithm and data structure for the first step is described in Section 3.2, and the second step is described in Section 3.3. In Section 3.4, we describe how we dynamically add clauses into and remove clauses from inference engines to enable learning.

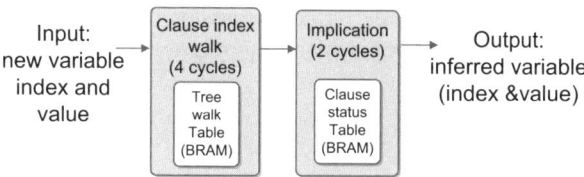

Fig. 2. Inference engine overview

3.1 Clause Partition for Inference Engines

As mentioned previously, our design restricts the number of clauses associated with any inference engine to be at most c clauses, and restricts the max number of occurrences of any variable in an inference engine to be p. In this section, we discuss the algorithm for partitioning a SAT instance into multiple groups that satisfy these restrictions.

If we restrict each literal to be associated with at most one clause ($p=1$) in each group, and allow unlimited group size ($c=\infty$), the problem is essentially a graph coloring problem. Each vertex in the graph represents a clause. An edge between two vertices denotes that these two clauses share a common literal. The graph coloring process ensures that no two adjacent vertices have the same color. This process is equivalent to dividing the clauses into groups with each color denoting a group and no two clauses in a group share any literal. Therefore, we can use the graph coloring algorithms to solve a relaxed partitioning problem ($c=\infty$ and $p=1$).

The graph coloring problem is a well know NP Complete problem [13] and has been extensively studied. DSATUR, TABU, MAXIS, to name a few, are well known graph coloring algorithms [14]. These algorithms are not directly applicable because our problem restricts the group size c and allows $p>1$ (p adjacent edges have the same color). Since an optimal solution for the partition is not required, we use a simple yet effective greedy algorithm to partition the clauses.

The goal of the partitioning algorithm is to ensure that any literal is associated with at most p clauses per group. Meanwhile, we want to evenly distribute clauses across groups and minimize the number of groups (inference engines) required. Our greedy algorithm is shown in Figure 3. It begins with zero groups and loops through all the clauses that have not been assigned a group and inserts the clause into the first group G_i that can accommodate it. The accommodation criteria is checked in lines 5-8 of the pseudo-code -- for each variable in clause C_i, there should be no more than p-1 related clauses in group G_i. If there exists a group G_i that can accommodate this clause, we insert it into the group. Otherwise, the algorithm creates a new group (Line 12) and adds the clause to the new group (Line 13). After the algorithm finishes processing all the clauses, it returns all groups in G. This algorithm is obviously polynomial with respect to the size of the input and in Section 4, we will compare it with graph-coloring based algorithms. Note that the greedy algorithm we listed here is just one possible algorithm. In general, the group selection criteria can be very flexible (e.g. instead of selecting the first group, choose the group that contains the least number of clauses). In Section 4.1, we provide further discussion of the group size, distribution, and techniques like introducing new variables to improve the clause partitioning.

```
Algorithm: Partition clauses into multiple engines
Input:      Clauses list C, the maximum number of clauses associated
            with one variable is p
Output:     Groups of clauses, each group fits into one engine

1 Begin
2 Groups G = Ø
3 For each clause C_i that has not been assigned a group yet
4      For each group G_i in G
5          For each variable V_j in C_i
6              If V_j has p related clauses in group G_i already
7                      pass to next group G_i+1 (Goto line 4);
8          End for
9          Assign C_i to the group G_i;
10         pass to next clause (Goto line 3);
11     End for
12     Create a new group G_new and add it to G;
13     Add clause C_i to group G_new;
14 End for
15 Return all groups in G
16 End
```

Fig. 3. Clause partitioning algorithm

3.2 Literal Occurrence Lookup

Given a newly assigned variable as input, the inference engine first needs to efficiently locate the clauses associated with the variable. In a software SAT solver, this can be implemented by associating each variable with an array of its occurrences (the occurrence list). A more efficient implementation may only store the watched clauses in each array (the watched list [10]). This scheme reduces the number of clauses to be examined, but does not reduce the total number of arrays, which is proportional to the number of variables. In our design, given an inference engine, each variable has at most p occurrences and most variables have no occurrence at all. Storing an entry for each variable in every inference engine is inefficient space-wise since SAT benchmarks often contain thousands of variables. A possible solution is to use a Content Addressable Memory (CAM), the hardware equivalent of a hash table. Unfortunately, most FPGAs do not contain CAMs as hard blocks and implementing CAM in an FPGA is expensive [15]. Instead, we implemented a novel tree walk algorithm for this purpose.

We organize the literal occurrences in an inference engine in a trie. A leaf node stores the clause ID where the literal occurs, as well as the literal index in the clause that corresponds to the literal. An internal node contains an offset pointer (the base index) to help locate the leaf nodes that share the same variable ID prefix. The tree is stored in the *tree walk table* in an on-chip BRAM block local to the inference engine module. Suppose the variable index has a width of k (so that the accelerator can handle 2^k variables) and every non-leaf tree node has 2^m child nodes, then the tree will be k/m deep. Here both k and m are configurable. Given a non-leaf node, the address of its leftmost child in the tree walk table is called the *base index* of this tree node. The rest of the children are stored sequentially in the table following the leftmost child. Therefore, to locate the ith child, the address can be calculated by adding i to the base index. If a leaf node is not associated with any clauses, we store a no-match (-1) tag in the entry. For an internal node, if all of its 2^m children have no-match, we do not expand its sub tree and just store a no-match tag in the node itself.

Figure 4 provides a simple example with the literal index size $k =4$ and the tree branch width $m=2$. There are two clauses, $(x_1 \lor x_{14})$ and $(x_{12} \lor x_{13})$, where variable x_1's index is 0001, x_{12}'s index is 1100, x_{13}'s index is 1101, and x_{14}'s index is 1110. Suppose the newly assigned variable is 1101. The thick arrows in Figure 4 represent the two memory lookups needed to locate the clauses associated with the decision variable 1101 (x_{13}). The base index of the root node is 0000 and the first two bits of the input are 11. The table index is the sum of the two: 0000+11= 0011. Using this table index, the first memory lookup is conducted by checking the 0011 entry of the

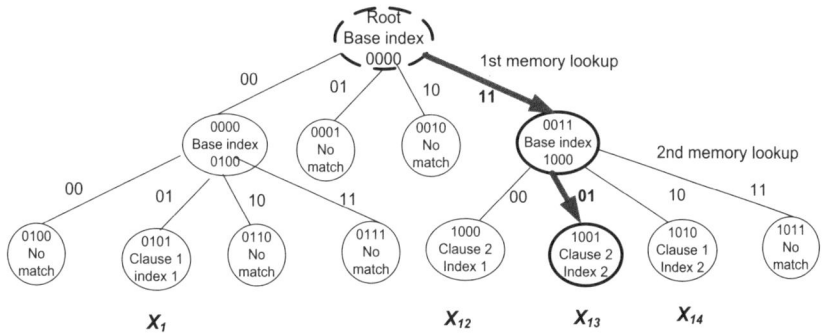

Fig. 4. Clause index tree walk in the inference engine

table. This entry shows that the next lookup is an internal tree node with the base index 1000. Following this base index, adding it to the next two bits of the input 01, we reach the leaf node 1000+01 = 1001. This leaf node stores the literal information; in this case, the literal is the second literal of clause two.

Figure 5 illustrates the tree structure mapping in memory. Note the last m bits of the base index are always zeros because each internal node has exactly 2^m children. Even if a child is not associated with any related clauses, we still store the child's index, using a no-match tag. With this design choice, we do not need addition to calculate addresses. We can use the top k-m bit of the base index and concatenate it with m bits in the input to obtain the address, thus eliminating the need for a hardware adder and saving one clock cycle.

For a leaf node, the table stores the related clause information as shown in Figure 6. It contains the Clause ID (CID), the position in the clause (PID), and its sign (whether it's a positive or negative literal in the clause). These three pieces of information will be used by the literal value inference module for generating new inferences. Note that the CID does not need to be a globally unique among all inference engines. A locally unique CID is sufficient to distinguish different clauses associated with one inference engine. The mapping between a local CID to a global CID is stored in DRAM and maintained by the conflict inference detection engine, shown as the rightmost block of Figure 1.

If $p>1$, one variable can be associated with multiple clauses per inference engine. They can be stored sequentially at the leaf nodes and processed sequentially with one implication module. If hardware resource permits, it is also possible to process them in parallel because they are associated with different clauses.

Table index	base index
0000	0100
0001	-1 (No-match)
0010	-1 (No-match)
0011	1000
0100-1011	leaf nodes

Fig. 5. Clause index walk table for internal tree nodes

Table index	Info stored at leaf nodes
0100	-1
0101	CID 1, PID 1, positive
0110	-1
0111	-1
1000	CID 2, PID 1, negative
1001	CID 2, PID 2, positive
1010	CID 1, PID 2, positive
1011	-1

Fig. 6. Clause index walk table for leaf tree nodes

In our implementation, we set k to be 16 and m to be 4. Therefore, our current design can accommodate 64K variables and it takes 4 tree walk steps to locate a literal. To store the tree in on-chip memory, there are two options. The first is to put the entire tree into BRAM. An inference engine requires four cycles to identify the related clause in the BRAM. Using a single port of the BRAM, inference engines can only service a new lookup every four cycles. The second design uses distributed RAM to store the first two levels of the tree. Similar to BRAM, distributed RAM is also dynamically readable and writable, but with much smaller total capacity. Since the top two levels of the tree are very small, we can easily fit them into distributed RAM. The rest of the tree is stored in BRAM. By doing this, we break the 4-cycle pipeline stage into two pipeline stages with two cycles each, thus improving inference engine throughput by enabling lookups every two cycles.

3.3 Inference Generation

After finding a clause to examine, the second step of the inference process is to examine the clause that contains the newly assigned variable to see whether it infers any new implications. This is shown in the second block of Figure 2. The literals' values in each clause are stored in a separate BRAM called a clause status lookup table. The implication inference module takes the output of the previous stage as inputs, which includes the CID, PID in addition to the variable's newly assigned value. With this information, it examines the clause status table, updates its status and outputs possible implications in two cycles. This step is described in more detail in [11].

3.4 Dynamic Learned Clauses Insertion and Deletion

Learning is one of the most important features in modern SAT solvers. Learned clauses are generated during conflict analysis and added to the clause database to prune the search space. These learned clauses can be long. Our inference engine design has a fixed maximum length for clauses (usually 9, 18, or a multiple of the size of a BRAM word). Clauses longer than the length cannot be added to the engines directly. There are two solutions to this problem. The first method is to break a longer clause into multiple shorter clauses by introducing new variables. For example, clause $(x_1 \vee x_2 \ldots \vee y_1 \vee y_2 \ldots)$ is equi-satisfiable to $(z \vee x_1 \vee x_2 \ldots) \wedge (\neg z \vee y_1 \vee y_2 \ldots)$ where z is a new variable. The benefit of performing this translation is that the transformed formula is logically equivalent (modulo existentially quantified bridging variables) to the original one. The drawbacks are that the number of literals is increased, which takes hardware resources. Extra implications are needed to pass through the bridging variable, which slows down the solver. We use this scheme in the evaluation section.

Another method is to abbreviate the learned clauses. When a learned clause is generated from conflict analysis, it is an *asserting clause* [1] and may contain many false literals assigned at lower decision levels. At higher decision levels, these literals can be omitted because their values do not change. We can exploit this by throwing away lower decision level literals and marking the clause valid only after a certain decision level. To maintain the correctness of the solver, the clause needs to be invalidated when the solver backtracks to an earlier decision level and as a result, must be garbage collected. The advantage of this scheme is that it only stores a smaller number of literals for each clause. The drawback of this scheme is that the system needs to be able to invalidate clauses dynamically, thus complicating the solver logic. Moreover, most learned clauses will be deleted after deep backtracks and restarts, thus reducing the possibility of future pruning of the search space. We decided against this scheme, mainly because this scheme affects the heuristics taken by the software SAT solver, while the previous scheme is transparent.

In our FPGA design, we partition the number of inference engines dedicated to original clauses and learned clauses. The inference engines for original clauses contain static content for a given SAT instance. The learned inference engines have dynamic content. We overload the inference engine programming port to handle the dynamic content operations. There are three types of operations: clause insertion, clause deletion (by invalidation), and garbage collection. In order to insert and delete clauses in the learned clause inference engines, several components must be modified, as shown in the shaded areas of Figure 1. The modified modules include the tree walk table, clause status table, global status table, and global-to-local translation table. In the rest of the section, we describe these operations in more details and describe the modifications on the modules accordingly.

3.4.1 Clause Insertion

After the conflict analysis process derives a new learned clause, we need to find an inference engine that can accommodate the clause and insert the clause into the engine. It would be time consuming to use software to examine hundreds of inference engines in the accelerator to find out the ones that can accommodate the clause. Instead, we leverage the parallelism and our tree-walk algorithm to find a potential inference engine in hardware.

First, the software sends the newly learned clause to all the learned clause inference engines. We use the second port of the BRAM to search the table for these literals sequentially. Suppose there are m literals in the clause, for each literal, we walk down the tree to see whether a no-match tag is found at a node in the tree, or if there is space in the tree leaf node for insertion. If so, the engine can accommodate this literal. If all m literals can be accommodated, this clause can be inserted into the engine. This checking process requires 4 cycles per literal to traverse the entire tree or $4m$ cycles for one clause. All learned-clause inference engines can perform the checking in parallel, and since the checking uses the second memory port, it can be performed without disrupting the implication process.

If no engine indicates that the clause can be inserted, the accelerator notifies the software to trigger garbage collection. On the other hand, multiple inference engines may signal that they can accommodate the clause. We use a priority encoder or simple round-robin logic to select the inference engine for clause insertion. The engine picked by the priority encoder receives an insertion enable signal and proceeds to insert the literals into the tree walk table. Each inference engine keeps a free-index pointer to indicate the starting point of un-used entries in the table. It inserts the literals sequentially by traversing the tree m times again. This requires a tree traversal and update to nodes at various levels in the tree. If a no-match tag is encountered, we need to create the subtree by accessing and updating the free-index pointer to insert new nodes. The final tree walk table step updates the leaf node with the CID, PID, and sign of the literal.

The clause status table can then be updated accordingly. This is followed by updating the global status table and local-to-global translation table in the conflict inference detection unit. All of these updates can be performed after the learned clause inference engine has been selected. It should be noted that these updates can be done in parallel with the actual insertion into the tree walk table because all the information is known at that point. Moreover, software is notified that the clause insertion was successful and the ID of the inference engine that stores the clause. This information is maintained in software for deletion and garbage collection purposes. The overall learned clause system architecture is shown in Figure 7 and is orthogonal to the normal BCP operation.

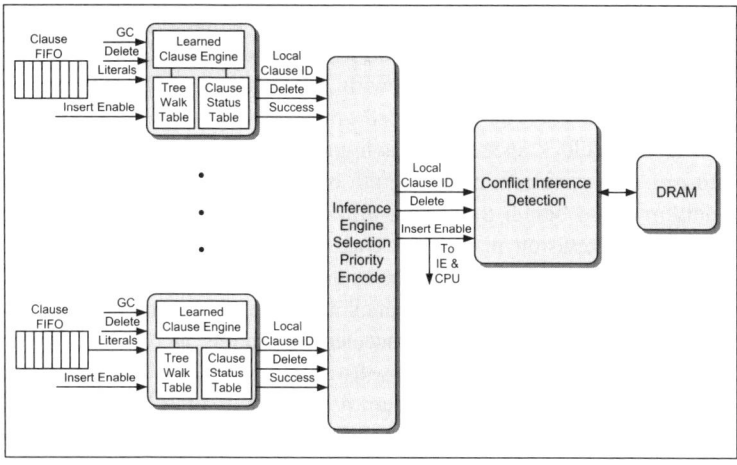

Fig. 7. Learned clause insertion, deletion, and garbage collection details

3.4.2 Clause Deletion

Clause deletion is very simple. Software knows the learned clause to inference engine mapping and simply sends a delete clause notice to the learned clause's inference engine. The learned clause's inference engine simply updates the clause status table and invalidates the clause entry by adding a tag to prevent future implications from being generated by this clause. Even though the clause information remains in the tree walk table, subsequent lookups in the tree walk table will result in no inferences.

3.4.3 Garbage Collection

Even though invalidated clauses will not generate implications, they still occupy spaces in the BRAM. Garbage collection is required to clean inference engines that contain a significant portion of invalidated clauses. Garbage collection is a software directed task that can be triggered by a threshold value of invalidated clauses or the inability to insert a new clause into the tree walk table. The garbage collection operation is controlled at the granularity of a single inference engine. Thus, implications from the rest of the engines can be generated while one or more engines are garbage collected.

We perform garbage collection by first reinitializing the inference engine and then adding the valid clauses back into the engine. For initialization, we write all entries in the BRAM to their initial value. Using both BRAM ports, the worst case number of writes can be reduced to half the table size. By targeting inference engines with only a few valid clauses, we minimize the re-insertion overhead.

4 Evaluation Results

In this section, we experimentally evaluate our approach. We first evaluate the effectiveness of the clause partitioning algorithm, and then report the speed of the accelerator and compare it with software-based solvers.

We obtained speedup numbers from a cycle accurate simulator that simulates events in the accelerator. We implemented the non-learning version of our accelerator design in 4000 lines of VHDL and synthesized with Xilinx ISE 9.2i SP3. On our target device, a Xilinx Virtex 5 LX110T with speed grade -3, we can fit 64 inference engines running at 200 MHz clock speed, and each engine has a capacity of 2048 tree walk entries with $p=1$ and maximum clause length of 9. The design is simulated and tested using ModelSim 6.3 to obtain the timing information in our C++ based event driven simulator. Since the revision of the inference engines for learning is minor without requiring additional BRAM resources (the limiting resource in our design), we expect to be able to achieve similar clock frequency and capacity for the learning version of the accelerator. We compare the FPGA accelerator running at 200 MHz with a state-of-the-art software-only SAT solver. The software SAT solver is a modified version of Zchaff [11]. It runs on a 3.6 GHz Pentium 4 with 2 GB of RAM. Since the FPGA accelerator only performs the BCP part of the solver, we compare it with the software BCP module. The same solver is used to drive the simulator with branching, conflict analysis and restarts.

4.1 Clause Partitioning Algorithm

We apply the greedy partitioning algorithm to a large number of real world SAT instances. Due to space limitation, Table 1 only lists representative SAT instances in each category. Here, we restrict each literal to be associated with at most one clause per group ($p=1$), and the group size can be unlimited ($c=\infty$). The fourth column of the table shows the total number of groups needed using the graph coloring algorithm DSATUR[14], and the fifth column is the results using our greedy algorithm. We can see that the results are very similar, but the DSATUR implementation we used is not able to handle large instances. Generally, the bigger the instance, the more groups are needed. However, there are a few exceptions. The crypto-md5_48 instance is the largest among all instance (66.9K variables and 279.3K clauses) and only needs 262 groups. A much smaller instance fvp-1.0-1dlx_c_mc_ex_bp_f requires a larger number of groups (280). The reason is that in crypto-md5_48, each literal is associated with at most 36 clauses, while in fvp-1.0-1dlx_c_mc_ex_bp_f, one literal can be associated with up to 141 clauses. Therefore, we need at least 141 groups to accommodate this instance because each variable can be associated with at most one clause per group. One possible optimization is to introduce equivalent variables for the top occurring variables. For example, if variable x appears in 1000 clauses, we can introduce a new variable a and use it to replace x in 500 clauses. Of course, we need to add clauses $(\neg a \vee x)(a \vee \neg x)$ to force x and a to the same value. We leave this to future work.

To understand the BRAM utilization, we measure the *Literal space ratio*, which is defined as the number of entries used in the tree walk table per literal. In software solvers, each variable has an array or list recording its occurrences. So every literal only consumes one pointer in memory. In our tree based design, some space is wasted due to the "no-match" entries. To quantify the BRAM resource cost of entries per literal, we report the literal space ratio in Table 1. This ratio varies from 2 to 7, showing that the number of wasted entries is relatively modest.

Table 1. Greedy partition algorithm results

	Vars	Cls	DSATUR groups	Greedy groups	Literal Space Ratio
miters-c3540	3451	9327	47	48	2.86
miters-c5315	5400	15025	48	48	2.66
miters-c880	958	2591	32	32	2.70
bmc-galileo-8	58075	294822	N/A	372	5.17
bmc-ibm-12	39599	194661	N/A	453	5.42
crypto-md4_wang5	53229	221185	N/A	201	2.97
crypto md5_48	66893	279265	N/A	262	2.94
fvp-1.0-1dlx_c_mc_ex_bp_f	777	3726	280	280	5.58

The above results are based on the assumption that one tree can have an unlimited number of entries. This assumption is unrealistic as each engine has limited resources. Next, we performed a sensitivity study that restricts the tree size, thereby changing the grouping results. We used three tree sizes c: 1024, 2048, and 4096. We recorded the number of groups needed in all three cases and compare the ratio to the unlimited tree size case (Table 1). Figure 8 plots the results. For SAT instances with a small number of variables, the ratio is close to one. This is because *collision*, defined as two clauses sharing the same variable, is high when the number of variables is small. When the number of variables becomes larger, the collision rate decreases, and consequently we can accommodate more clauses per group. This creates very large groups. If we limit the tree size, these large groups will need to be split into smaller groups. Therefore, large instances such as crypto-md4_wang5 need a large number of groups when tree walk table size is small, thus increasing the number of inference engines needed.

We also tested different p values. $p>1$ allows one variable to be associated with multiple clauses in one group. The experimental results (not presented here due to

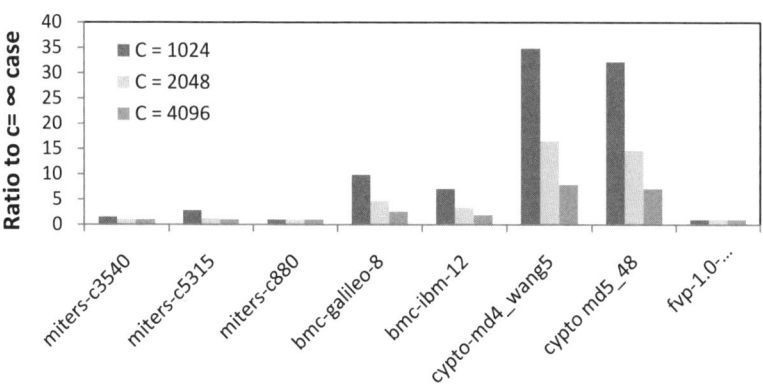

Fig. 8. Groups needed with limited tree size (normalized by $c=\infty$ case)

space limitations) show that when the tree size is unlimited, the number of groups decreases almost linearly to p because we can pack p trees into one tree. When the tree has limited size, the number of groups decreases close to linear for instances with small variables. But for larger instances, the number of groups decreases sub-linearly as most trees are almost close to their capacity limit.

4.2 Comparison of FPGA-Based BCP Accelerator to Software

We use the cycle accurate simulator to test the performance of our system using a set of benchmarks. We evaluate the two design points as described in Section 3.2. The first design places the entire tree in BRAM. The second design splits the tree, placing the top two levels of the tree in distributed RAM and the rest in BRAM. We terminate the simulator after 1 million implications if the solver cannot solve the instance within the given time limit. In the simulation, we assume the interconnect between CPU and the accelerator is a 16 lane 800 MHz HyperTransport [12].

Figure 9 presents the average number of FPGA cycles per implications. The number ranges from 6 to 17 cycles. The scheme using distributed RAM and BRAM is around 30% faster than the BRAM only scheme. By using both distributed RAM and BRAM, we are able to reduce the latency of this pipeline stage to two cycles, matching the latency of other pipeline stages in the system. This removes the performance bottleneck of a 4-cycle pipeline stage for the BRAM-only design.

Figure 9 also presents the learning overhead, which is the total number of cycles for inserting, deleting and garbage collecting learned clauses, amortized to each implication. It varies from 1% to 25% over different benchmarks. This cost is largely related to the number of learned clauses generated.

Finally, we compare our design to the Software-based BCP implication engines. Figure 10 presents the converted CPU cycles per implication. The speedup ratio of our FPGA-based BCP accelerator is 5 to 16 times, demonstrating the effectiveness of our system.

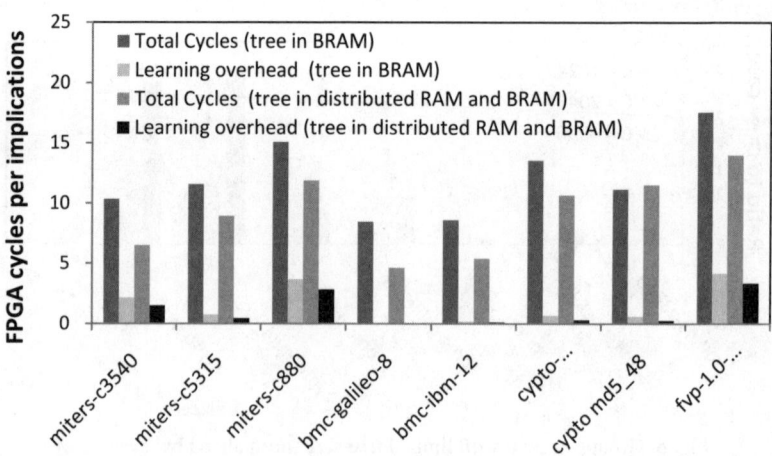

Fig. 9. FPGA cycles needed per implication

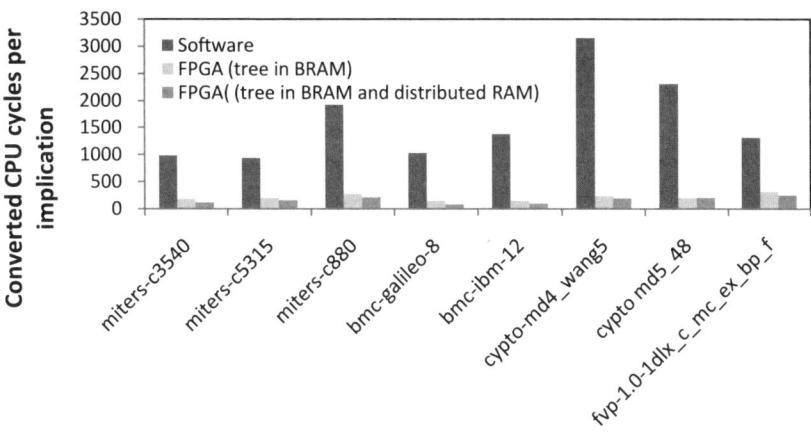

Fig. 10. Comparison of the FPGA-based and software-based methods

5 Conclusions

In this paper we describe the detailed design of the inference engines of a Boolean Constraint Propagation accelerator for SAT solvers. Our design leverages the Block RAM (BRAM) resources in modern FPGAs to store instance specific information. Compared with previous works on hardware-assisted SAT accelerators, our designs eliminates the need to re-synthesize each new SAT instance, and can accommodate dynamic learned clause insertion and deletion. Empirical evaluation on a large set of benchmarks demonstrates that our method of achieving parallelism by partitioning clauses into groups is both efficient and effective. Our cycle accurate simulator demonstrates speed-up over an optimized software-only BCP implementation by approximately 5 to 16 times. For future work, we intend to implement the new learned clause inference engine extensions in VHDL and map the system to hardware.

References

[1] Zhang, L., Malik, S.: The Quest for Efficient Boolean Satisfiability Solvers. In: Brinksma, E., Larsen, K.G. (eds.) CAV 2002. LNCS, vol. 2404, Springer, Heidelberg (2002)

[2] Suyama, T., Yokoo, M., Sawada, H., Nagoya, A.: Solving Satisfiability Problems Using Reconfigurable Computing. IEEE Trans. VLSI Systems 9(1), 109–116 (2001)

[3] Zhong, P., Martonosi, M., Ashar, P., Malik, S.: Using Configurable Computing to Accelerate Boolean Satisfiability. IEEE Trans. Computer-Aided Design of Integrated Circuits and Systems 18(6), 861–868 (1999)

[4] Zhong, P., Martonosi, M., Ashar, P., Malik, S.: Solving Boolean Satisfiability with Dynamic Hardware Configurations. In: Hartenstein, R.W., Keevallik, A. (eds.) FPL 1998. LNCS, vol. 1482, pp. 326–335. Springer, Heidelberg (1998)

[5] Abramovici, M., de Sousa, J.T.: A SAT Solver Using Reconfigurable Hardware and Virtual Logic. J. Automated Reasoning 24(1-2), 5–36 (2000)

[6] Dandalis, A., Prasanna, V.K.: A Parallel Pipelined SAT Solver for FPGA's. FPGAACM Trans. Design Automation of Electronic Systems 7(4), 547–562 (2002)

[7] de Sousa, J., Marques-Silva, J.P., Abramovici, M.: A Configware/Software Approach to SAT Solving. In: Proc. Ninth IEEE Int'l Symp. Field-Programmable Custom Computing Machines (2001)

[8] Skliarova, I., Ferrari, A.B.: A Software/Reconfigurable Hardware SAT Solver. IEEE Trans. Very Large Scale Integration (VLSI) Systems 12(4), 408–419 (2004)

[9] Skliarova, I., Ferrari, A.B.: Reconfigurable Hardware SAT Solvers: A Survey of Systems. IEEE Transactions on Computers 53(11), 1449–1461 (2004)

[10] Moskewicz, M., Madigan, C., Zhao, Y., Zhang, L., Malik, S.: Chaff: Engineering an Efficient SAT Solver. In: 38th Design Automation Conference, Las Vegas (June 2001)

[11] Davis, J.D., Tan, Z., Yu, F., Zhang, L.: A Practical Reconfigurable Hardware Accelerator for Boolean Satisfiability Solvers. In: 45th Design Automation Conference, Anaheim (June 2008)

[12] HyperTransport Technology I/O link, AMD (2001)

[13] Cormen, T.H., Leiserson, C.E., Rivest, R.L.: Introduction to Algorithms. McGraw-Hill, New York (1990)

[14] Culberson, J.: Graph Coloring Programs, available at http://www.cs.ualberta.ca/~joe/Coloring/Colorsrc/index.html

[15] Xilinx product specification, Content Addressable Memory V5.1, available at http://www.xilinx.com/ipcenter/catalog/logicore/docs/cam.pdf

Attacking Bivium Using SAT Solvers

Tobias Eibach, Enrico Pilz, and Gunnar Völkel

Ulm University, Institute of Theoretical Computer Science,
James-Franck-Ring 27, 89069 Ulm, Germany
{tobias.eibach,enrico.pilz,gunnar.voelkel}@uni-ulm.de

Abstract. In this paper we present experimental results of an application of SAT solvers in current cryptography. Trivium is a very promising stream cipher candidate in the final phase of the eSTREAM project. We use the fastest industrial SAT solvers to attack a reduced version of Trivium – called Bivium. Our experimental attack time using the SAT solver is the best attack time that we are aware of, it is faster than the following attacks: exhaustive search, a BDD based attack, a graph theoretic approach and an attack based on Gröbner bases. The attack recovers the internal state of the cipher by first setting up an equation system describing the internal state, then transforming it into CNF and then solving it. When one implements this attack, several questions have to be answered and several parameters have to be optimised.

Keywords: SAT Solver, Application, Cryptography, Stream Cipher, Rsat, eSTREAM, Bivium, Trivium, BDD, Gröbner Base.

1 Introduction

Stream ciphers are used in many applications like GSM, UMTS, RFID, Bluetooth and online encryption of big amounts of data in general. The eSTREAM project ([1]) was started in October 2004 to find a new stream cipher, after the NESSIE project ([2]) ended in 2003 without recommending one. Starting with a call for candidates the project is organised into several phases and in each phase weak candidates dropped out. The final stream cipher candidates should be fast and cryptographically secure. All eSTREAM candidates are divided into two categories: hardware-oriented and software-oriented ciphers. The eSTREAM project is now in the last phase with only few candidates left in each category. One of the hardware-oriented ciphers is Trivium, introduced in [3]. Until now no attacks have been successfully applied to Trivium – i.e. it has not been possible to prove a running time faster than exhaustive search. In [4] a reduced version of the cipher has been introduced: Bivium (initially called Bivium B). The intention is to find attacks on Bivium and then extend them to Trivium. In this paper we focus on Bivium, however this "algebraic attack" concept is generic and can also be applied to other stream ciphers.

Stream ciphers are symmetric cryptographic primitives – the communicating parties already share a secret key. Like most stream ciphers, Trivium can be

H. Kleine Büning and X. Zhao (Eds.): SAT 2008, LNCS 4996, pp. 63–76, 2008.

described as a finite automaton whose initial state is derived from the secret
key and a public known initialisation vector (IV) by filling the register with
the key and the IV and then making a few transitions to "shuffle" the internal
state. After this initialisation phase the cipher starts to produce output bits (the
"keystream"). Trivium and Bivium produce one keystream bit with every clock
(step) of the cipher. To encrypt a message m, one uses the XOR function to add
the message- and keystream-bits bitwise, e.g: $m_i \oplus z_i = c_i$ for $i = 1, 2, 3, \dots$. The
receiver produces the same keystream z and also adds it bitwise to decrypt the
message.

We use the common and realistic attack scenario, that we know a part of the
keystream z (from a known-plaintext attack) and try to reconstruct the internal
state of the cipher from it. If we are successful, we can clock the cipher backwards
to reconstruct the secret key. More importantly, we can clock the cipher forward
to produce the whole keystream and thereby decrypt the whole message.

In order to use a SAT solver to reconstruct the internal state, we first set up
an equation system, given by the Bivium definition and the observed keystream.
The solution of the equation system is the internal state of the cipher. Then we
transform the equation system into a CNF formula, by using the truth table and
Quine McCluskey algorithm. Finally we use the fastest complete SAT solvers
of the SAT competition 2007 ([5]) in the industrial category to solve the CNF.
Doing this, there are several parameters to be optimised and several questions
that have to be answered.

We are aware of several attack concepts that can be applied to Bivium as well.
We implemented 3 of them and we quote the results of the remaining ones to
have a complete comparison of the SAT solver speed to the other attack speeds.
According to our experiments, the SAT solver attack is by far the fastest attack
type. It is faster than our exhaustive search on the key/IV-setup, also faster than
our attack based on BDDs (binary decision diagrams) and faster than our attack
based on Gröbner bases. It is also faster than the attack times that we found
in other papers: an attack based on a graph-theoretic approach ([4]), a guess-
and-determine strategy ([6]) and an attack based on the birthday paradoxon
([7]).

Most attack concepts like the one based on SAT solvers, BDDs or Gröbner
bases use heuristical algorithms that are theoretically not well understood – at
least there is a significant gap between the proven bounds on the running times
and the actual running times. This is the reason why experiments are needed.
We run our experiments on a multi-processor system. The important process
is computed on a fast 2 GHz CPU with 2GB of memory to avoid interference
with other processes. We address several questions and optimisations and try to
answer them isolatedly.

1.1   About This Paper

In Section 2 we describe the stream cipher Bivium. In Section 3 we describe,
how we use SAT solvers to recover the internal state of Bivium. In Section 4
we show, how we optimised the attack and which results we found. In Section 5

we describe briefly our attack on Bivium using BDDs. In Section 6 we briefly describe our attack using Gröbner bases. Finally in Section 7 we discuss and compare our experimental results, we also compare them to results that have been published so far and give an outlook in Section 8.

2 Description of Bivium

We focus our attack on Bivium, a stream cipher with an internal state of 177 bits (that can be seen as 2 registers, of size 93 and 84 bits) and a key size of 80 bits. The internal state of Bivium is initialized with the secret key, the initialization vector (IV) and zeros. Then the cipher is clocked $4 \cdot 177$ times and then starts producing keystream bits, according to the scheme given in Figure 1. As one can see from the figure, Bivium is obviously a reduced version of Trivium, as it uses only 2 registers instead of 3. The update-functions of the 3 internal registers are non-linear, as each involves one AND gate. The output-function is linear as it just combines 4 (for Bivium) or 6 (for Trivium) bits of the internal state by a XOR gate. One also notices the very low amount of gates used. On the one hand this leads to a low power consumption of the cipher and a fast implementation, but on the other hand the equation system describing the cipher will be sparse and thereby can be rather efficiently converted to a CNF (conjunctive normal form) formula (see also [8]).

The size of the secret key (used to initialise the cipher) is just 80 bits. However, so far the most efficient way to attack the key directly is exhaustive search (see Section 7). For the SAT solver attack, we decided not to attack the key directly. We try to reconstruct the internal state (177 bit) from a part of the keystream. Of course it is a disadvantage to have a search space of 177 bits instead of the 80 bits. However the equation system in the 80 key-bits gets far too difficult

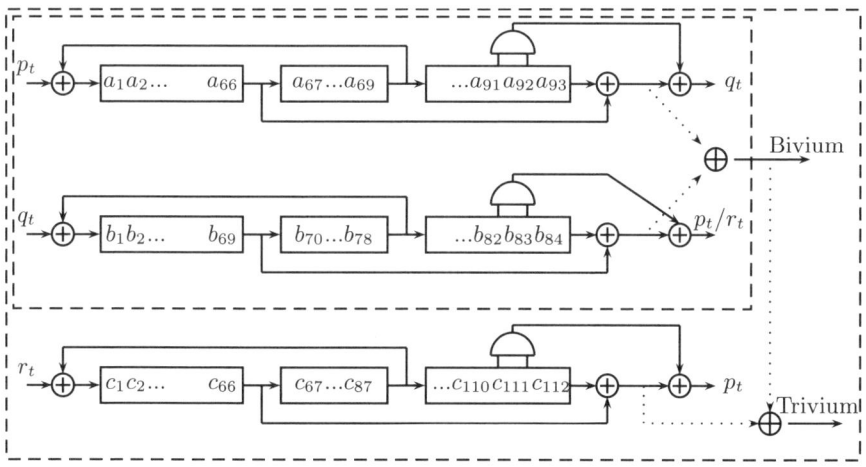

Fig. 1. Trivium and Bivium scheme

through the key/IV-setup phase. Below we give the pseudocode of Bivium. For a more detailed description please see [4] for the description of Bivium or [3] for Trivium.

Bivium pseudocode

```
1: for i = 1,2,3,... do
2:    t[1] := s[66] + s[93]
3:    t[2] := s[162] + s[177]
4:    z[i] := t[1] + t[2]
5:    t[1] := t[1] + s[91] * s[92] + s[171]
6:    t[2] := t[2] + s[175] * s[176] + s[69]
7:    (s[1],s[2],...,s[93]) := (t[2],s[1],...,s[92])
8:    (s[94],s[95],...,s[177]) := (t[1],s[94],...,s[176])
9: end do
```

$s[1]...s[177]$ denote the internal state of the cipher and $z[i]$ $(i = 1, 2, 3, ...)$ denotes the output of the cipher. $t[1]$ and $t[2]$ are temporary variables.

3   Describing the Attack

The concept of using SAT solvers for attacking stream ciphers has been proposed in [8] and [9]. Before this, there have already been other remarkable applications of SAT solvers in cryptography. The actual running time of a SAT solver can be hardly estimated. Here experiments are needed to determine the running time on Bivium instances.

First we generate an equation system describing the internal state of the cipher. By clocking the cipher we get one new equation in every step from line 4 of the pseudo code that connects the (known) output to the internal state. In line 5 and 6 we can decide whether we want to introduce 2 new variables for t[1] and t[2] or whether we use the given construction of t[1] and t[2] in line 7 and 8. The first option obviously increases the number of variables and equations but keeps the equations short and the degree low while the second increases the degree and length of the equations. It turned out that introducing two new variables for t[1] and t[2] has many advantages: The equation system produced has only 2 types of equations – one with 4 and one with 5 variables, which can be converted into CNF without producing too many clauses. Also we believe that this way the "structure" of the initial problem can be maintained for the SAT solver while algebraic operations on the equation system might reduce the link between the original structure and the structure in the CNF formula.

Having the equation system we now transform every single equation into a CNF formula that equals 1 in case the equation is fulfilled and 0 otherwise. By combining all formulas with AND we get an equivalent Boolean formula. Transforming an algebraic equation into CNF can be done by looking at the truth table to construct the clauses and then the formula can be minimized by using the Quine McCluskey algorithm. Transforming an equation with n

variables produces about 2^{n-1} clauses. Actually 2^{n-1} is the maximum, but the actual number is close to this, as the equation system consists mainly of XOR operations and has only few AND operations. The fact that the number of clauses grows exponentially in n means that at a certain point we have to introduce new variables that substitute several old variables, to keep the number of variables in each equation small. In [8] a "cutting number" of 6 variables per equation is suggested. This means that if an equation has more than 6 variables one should substitute half of them by a new variable, leading to 2 new equations with about half the size and fewer clauses in the final CNF. By introducing 2 new variables in every step we do not have to consider this cutting number rule, as our equations do not get bigger than 5 variables.

Having the Boolean formula we have to transform it into the DIMACS format that serves as input to the SAT solver program and is just a convention ([10]). Before we let the SAT solver solve the formula we have to reduce the complexity of the instance by guessing some variables. Here we have to decide how many variables we guess and which ones. If we decide to guess m variables the expected number of runs of the SAT solver will be approximately 2^{m-1} so the total expected running time will be 2^{m-1} times the average running time for one instance (if we guess the variables in a way so that we do not guess the same assignment twice).

We skipped one more consideration above: We have to decide how many equations we want to produce for our equation system. We need at least 177 equations – but more equations would make the problem instance overdefined and could thereby speedup the solving time. Our experiments showed that less than 180 used keystream bits are not enough (the attack was very slow or did not return the internal state that we were looking for). We did not see a difference in the range of 180 to 300 used keystream bits, just a slight increase in the running times above 250 bits used. So we decided to use 200 keystream bits to set up our equation system. Also as noted in [4] it is not necessary to add new variables to the equation system, if they do not get "connected" to the keystream (the last introduced 66 variables for the first register and the last introduced 69 variables for the second register do not get connected to the keystream). This way the number of variables can be reduced.

4   Experimental Results of the SAT Attack

In this section we present the experimental results of our implementation of the attack using SAT solvers. All times are given in seconds and are averaged over 100 instances. In Table 1 we compare several SAT solvers to find out which one would solve our kind of instances fastest. In the second column we guess 40 values of the internal state, in the next column 45 and then 50. We used the guessing strategy "Ending2" (see below). The fastest SAT solver is Rsat combined with the SatElite preprocessor (version 2.01), followed by MiniSAT (version 2/070721) – available at [11] and [12].

Table 1. Comparing SAT solvers

	guess 40	guess 45	guess 50
Rsat & SatElite	46.10	3.32	0.26
MiniSat	67.32	5.06	0.36
Picosat	103.96	5.78	0.42
Rsat	229.09	11.49	0.79
Zchaff	735.08	17.36	0.78

Rsat (with SatElite) and MiniSat were also the two fastest solvers in the SAT competition 2007 in the UNSAT industrial category (as we are guessing m bits to reduce the complexity, the outcome is "UNSAT" in all runs except one). Consequently the following experiments have been done using Rsat (with SatElite). In Table 2 we compare several guessing strategies. Inspired by [6] we used the strategy "ThreeFour" that is guessing 3 variables in a way to directly compute a fourth variable. This way it is possible to start the SAT solver with 64 variables guessed – at the same cost as guessing 48 independent variables. However it turned out that guessing the last 48 variables of the second register ("Ending2") helps more and leads to a faster average running time. We also tried "Ending1" that is guessing the end of the first register and "Ending-halved" that is guessing the endings of both registers. We did the same for the beginning positions of the registers. We also tried to guess 3 random sets of variables that show quite a variation. One possible explanation why the "Ending2" strategy gives the fastest running time might be, that many of the most frequent occurring variables are in the "Ending2" set.

Table 2. Comparing different guessing strategies

strategy	time
Beginning1	3246
Beginning2	21.27
Beginning-halved	2712
Ending1	3.94
Ending2	0.116
Ending-halved	0.718
ThreeFour	0.275
Random1	1.144
Random2	51.988
Random3	17.993

In Figure 2 we determine the optimal number of variables to guess. The x-axis shows the number of variables guessed using the "Ending2" strategy. The y-axis shows the expected running time (scaled by 10^{-10}) of the whole attack. The dark curve shows the case that we guess randomly (e.g. with high probability incorrect, leading to an UNSAT result of the SAT solver) and the dotted curve

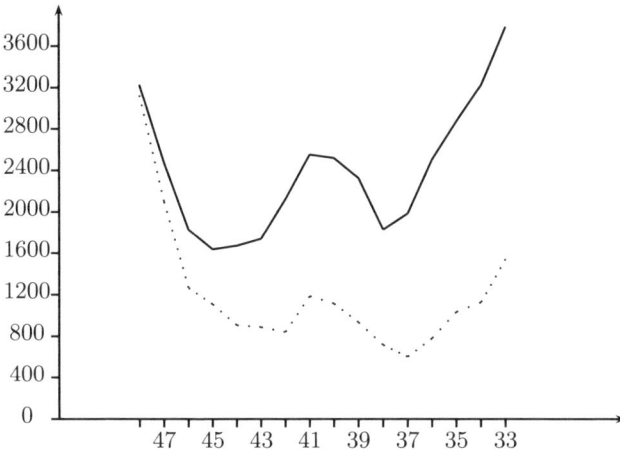

Fig. 2. Optimal guessing number

shows the running time if we guess the variables correctly. This however will happen just once in our attack scenario. One should expect SAT solvers to be better than guessing a variable randomly at the cost of multiplying the running time with 2. This is – at a certain point – not the case, so that we measure a minimum at guessing 45 variables with an expected running time of 1637E10 seconds.

In Figure 3 we determine the correlation between the Hamming-weight of the internal state and the time needed to solve the CNF. When we guess the 36 variables correctly there is a huge correlation (dotted curve). This however happens only once in the attack scenario and we do not see this correlation if we guess randomly (dark curve). The x-axis shows the Hamming weight of the internal state and the y-axis the time needed to solve one instance by guessing 36 variables with the "Ending2" strategy. We average the running time here over just 50 instances.

Following the idea of the attack described in [6], we made one experiment, whether certain parts of the keystream are easier to attack. This however was not the case. In all the experiments, we observed a huge variation in the running times – up to a factor of 20. This is why we averaged the running times over 100 instances.

One way to optimise the attack is to guess the m variables earlier than in the final CNF formula. This allows us to do some further simplifications in the equations. However this work has to be done for every guess and not just once. We did not see a big influence on the running times here, but it allows us to give a more realistic table on the CNF statistics. So in Table 3 we give the averaged numbers of the instance size that we get if we guess m bits in the equation system and do some fast simplifications there (roughly: expand equations, remove duplicates and remove duplicate monomials, where possible and substitute to 4-CNF).

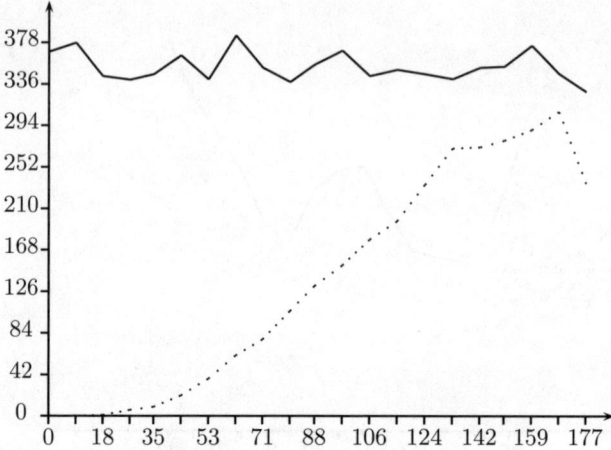

Fig. 3. Influence of the Hamming weight

Table 3. Number of guessed variables vs. number of clauses and variables

Nb. guessed	Nb. clauses	Nb. variables
0	9909.0	2677.00
31	5631.0	1543.50
35	5200.2	1427.70
39	4933.8	1361.50
43	4679.4	1293.50
45	**4560.4**	**1265.10**
47	4463.6	1236.80
51	4277.4	1189.45

In [8] it is further suggested to use Gaussian elimination to reduce the number of variables in each equation. This did not help in our experiments.

5   An Attack Based on BDDs

We also implemented an attack on Bivium that is based on BDDs. We explain the idea of this attack roughly, present our attack times and in Section 7 we compare them to the SAT solver attack times.

A BDD is a way to represent a Boolean formula as a directed graph, with the variables at the nodes, the values of the variables at the outgoing edges and the corresponding function values at the leaves. There are several fast operations on BDDs, especially if we delimit them to OBDDs. These are BDDs with an order on the variables, so every variable can only be read once and in every path the reading order is the same. In this case, the following operations are efficient: count the remaining number of paths leading to 1 ("sat-count"), minimize the BDD for the given order and combine two BDDs. For an introduction please see [13].

The idea of the BDD attack has been published in [14], more details are published in [15] and improvements in [16] and [17]. The BDD attack uses one BDD to efficiently represent all possible internal states of Bivium at a given time. The main BDD that we construct in step i represents the characteristical function of the set of all possible internal states after i observed keystream bits. To construct the main BDD we describe the 2 internal update-functions and the output-function of Bivium as 3 BDDs for every keystream bit. We combine all those BDDs to get the main BDD. We initialize the main BDD with the constant 1-BDD that represents the fact that all internal states are possible (as we have not read a single keystream bit). Then, for every keystream bit, we combine the 3 new BDDs with the main BDD using the AND operation. This reduces in every step the number of possible internal states. After the minimal amount of keystream bits the BDD represents only the one internal state that we wanted to recover.

Also in this attack there are several questions to be answered and parameters to be optimised. First we had to decide which BDD library to use. We only achieved a fast implementation using the CUDD library ([18]). CUDD uses a hash-table of initially fixed size to speed up the operations on the BDDs. The size of this hash-table is critical. In our experiments we got good running times with about 1GB of memory, less memory led to swapping and thereby to much worse running-times. The running time of the BDD operations depends mainly on the size of the BDD (number of nodes) and the width of the BDD (the size of the maximal level of the BDD). There is no particular order in which the single BDDs have to be combined. One can start by combining only the BDDs describing the output-function and then at a later point add the BDDs of the update-function. The optimisation target here is to keep the main BDD as small as possible.

The CUDD library does not support an explicit minimisation operation, but the resulting BDD of an AND operation is already minimised (for a fixed variable order). The reordering operation tries to further minimize the BDD by reordering the variables. This operation requires most of the time in the BDD attack and is performed every few steps (based on a heuristic). The reordering operation is also just a heuristic and does usually not find the best ordering of the variables. However it significantly reduces the size of the BDD (up to a factor of 10) and makes future operations faster.

5.1   Experimental Results of the BDD Attack

The variance in the running times is much lower for BDDs than for SAT solvers. So we averaged our experimental results over just 10 runs for every instance. After identifying the CUDD library as the best BDD library for our purpose we continued the same way as for the SAT solver attack. We tried several guessing strategies, to find out how the complexity of the problem can be reduced most efficiently. Again it was most useful to guess the bits close to the "end" of Bivium (the bits close to the output). With a small difference: it is best to guess half of the bits at the end of the first register and half of the bits at the end of the second register ("Ending-halved").

In Figure 4 we determine the optimal number of variables to guess in the BDD attack. The x-axis shows the number of variables that we guess (using the "Ending-halved" strategy). The y-axis shows the expected running time in seconds scaled by 10^{-17}. It is optimal to guess 55 variables, with an expected running-time of 4.22E17 seconds. As for the SAT attack there is a point from which on it is better to guess more variables at a cost of "factor 2" in the running time, than using the BDD construction.

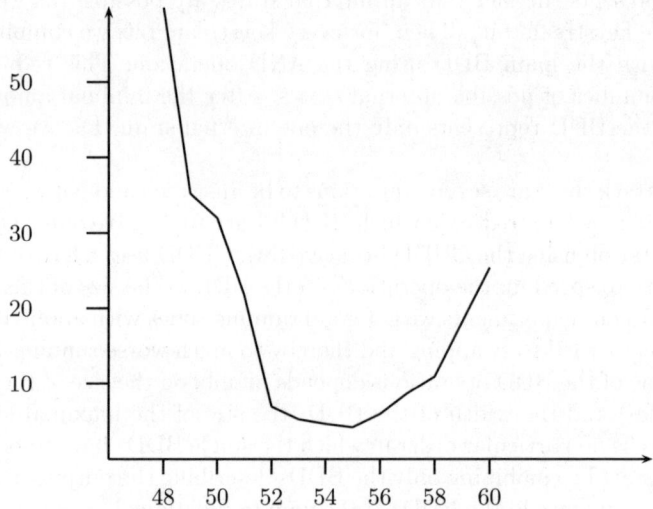

Fig. 4. Optimal guessing number for the BDD attack

6   An Attack Based on Gröbner Bases

Gröbner bases are the most common and usually fastest way to solve a system of (non-linear) equations. They are used in algebra software like Mathematica, Maple, MAGMA and Singular ([20]). We use the algebra software SAGE ([19]) that integrates the Singular software, as it is free, offers several algorithms to compute a Gröbner base and also Singular is well known for its fast and optimised Gröbner base algorithms.

A Gröbner base G for a given set of polynomials F by definition generates the same ideal as F and has certain nice properties that allow fast solutions to many problems that are hard for F. This is especially true for the solution of the equation system: the Gröbner base G has the same solution(s) as F but due to the "elimination property" the solution(s) can be derived easily. The "elimination property" implies that one of the polynomials is univariate – meaning that it depends on just one of the variables – and so this variable can be computed directly. When we substitute this variable in the other equations we get again an equation in just one variable and so on. This can be seen as a generalised form of the Gaussian elimination (please see [21] for an introduction).

First we set up the Bivium equation system – in contrast to the SAT solver attack, it is much better not to introduce any new variables, so we set up the equation system in just the 177 unknown internal bits. We set up 200 equations and then we guess several variables to reduce the complexity of one instance. We use different algorithms of the Singular packet to construct the Gröbner base, this step took most of the time. If we guess the variables incorrectly (most likely) the Gröbner base will be just the 1-polynomial and we guess again. If we guess correctly, we will be able to derive the solution (the internal state) very efficiently. In contrast to the SAT solver approach here the running time when guessing incorrectly is much faster than when guessing correctly.

Table 4. Comparing guessing numbers and Gröbner algorithms

Nb. guessed	std	slimgb
64	0.2852	0.1887
62	0.2308	0.2264
60	0.1824	**0.1823**
58	8.1856	82.019
56	267.35	1114.3

In Table 4 we give the averaged times for the 2 fastest Gröbner base algorithms in the Singular packet (out of 5 algorithms) for solving one instance. We used the "Ending-halved" guessing strategy for which it is optimal to guess 60 variables (randomly), resulting in an expected running time of 1.051E17 seconds. The "std" algorithm used from 100MB to 5 GB of memory – depending on the number of variables guessed. The "slimgb" algorithm used just up to 400 MB.

There are many different algorithms to compute a Gröbner base for a given equation system and a given ordering on the monomials. Two well known algorithms are the F4 and F5 algorithm by Faugère. In the Singular portfolio of Gröbner base algorithms the "slimgb" algorithm was slightly the fastest in our experiments. It is optimised to keep coefficients small and polynomials short on the computation and this pays off in computation time and memory usage ([22]). It is partly based on ideas of the F4 algorithm by Faugère. It also offers a direct access to adjust the algorithm to special problem classes through its weighted length computation. However we did not exploit this parameter.

A reduced Gröbner base is unique for any given ideal and monomial ordering. However there are huge differences when switching from one order to another. Usually the "fastest" ordering is "Graded Reverse Lex Order", meaning that a monomial has the higher order if the sum of its degrees is smaller (in case the degrees sum up to the same sum, one further distinguishes by a lexicographical order).

The expected running time of 1.051E17 seconds is surprising to us, as in [8] the authors say "if Magma or Singular do not crash, then they tend to be faster" (than SAT solvers). However in our experiments the SAT solvers are about 6400-times faster – and both implementations are stable.

7   Discussing and Comparing the Results

The expected running time of the SAT solver attack is 1.64E13 seconds, while the BDD attack takes 4.22E17 seconds and the attack using Gröbner bases 1.051E17 seconds. The other main differences are that on the one hand the SAT solver attack is faster, probably because it combines several sophisticated search heuristics, but at the cost that the SAT solver can almost only be used as a black box. While on the other hand the BDD attack uses heuristics only in the reordering algorithm. Also the BDD in construction, can already be interpreted in every step, as it represents all possible internal states of the cipher at this point. Also there are some theoretical bounds for the BDD attack, as published in [17]. We used the Gröbner bases attack also almost only as a black box, however the theory offers more insight and optimisation as sketched above.

Another resource to compare the 3 attack concepts on, is the amount of memory required: SAT solvers require almost no memory, while BDDs use up to 1 GB and Gröbner bases up to 400 MB. All three attacks can be easily parallelised due to the guessing loop. If we had several CPUs we would just divide the guessing-space among them, leading to parallelisation without overhead, as no communication between the processes is required. The amount of keystream needed is very low for all attacks (about 200 bits).

The influence of randomness is much higher in the SAT solver attack, while the BDD attack only uses randomness for guessing the bits in the beginning. This also supports the fact that the variance in the running times of SAT solvers is much higher (up to a factor of 20) while the variance for BDD running times is below a factor of 2. The variance for the Gröbner base approach is also rather low, especially when computing the same instance twice there is almost no variance in the running times.

For all three attack concepts we had to decide, which part of the internal state we wanted to guess. It turned out that for SAT solvers and BDDs it is most useful to guess the internal bits close to the end of the 2 registers. The difference is that it is optimal for the SAT solvers to only guess the end of the second register and for the BDDs, it is optimal to share the guessed bits between the ends of both registers. Looking at the equation system describing Bivium, we notice that the variables close to the output occure rather frequent (those of the second register a little more than those of the first). This might be one reason, why these guessing strategies helped most.

We also implemented an exhaustive search on the keysetup, this resulted in an expected running time of 1.5E17 seconds.

7.1   Comparing Against Other Attacks

We want to cite some more results that we are aware of, but have not implemented ourselves: In [4] Raddum proposed to solve the equation system by a graph-theoretic approach, resulting in a running time of about $2^{56} \approx 7.2\text{E}16$ seconds. The attack published in [6] gives a running time of about $c \cdot 2^{36.1}$, for

$c \approx 2^{14}$, leading to $2^{50.1} \approx 1.2E15$ seconds. In [9] the SAT solver attack has also been implemented – we were not able to reconstruct the results, so we just quote them: when guessing 34 variables the average running time using MiniSAT is given as $2^{8.7}$, leading to an expected running time of $2^{33} \cdot 2^{8.7} = 2^{41.7} \approx 3.57E12$ seconds. A classic time-memory trade-off technique based on the birthday paradoxon ([7]) gives a theoretical running time of $O(\sqrt{2^{177}})$ (also for the amount of memory needed).

8   Outlook

The attack concepts based on SAT solvers, BDDs and Gröbner bases are generic, so one could run many more experiments to get cryptoanalytic results also on other stream ciphers, not just for Bivium. Of course one open question is how to extend the attack to Trivium. Also we believe that there is much potential for optimisation left. For example one could try to guess the variables not just independently with probability one-half 0 or 1. We expect further speedups for the SAT attack, by just applying the latest SAT solvers that will lead to faster running times.

While SAT solvers are much faster and need almost no memory the huge disadvantage is that they can almost only be used as a black box. This might be a starting-point for further research. Also we hope to show by this paper that there is a very interesting benchmark class for industrial SAT solvers. An expected running time of about 533,000 years for the attack of course does not break Bivium and especially not the harder cipher Trivium. However parallelisation is possible without overhead and combined with further improvements this attack concept could become practicable.

References

1. eSTREAM: eSTREAM – The ECRYPT Stream Cipher Project.
   http://www.ecrypt.eu.org/stream/
2. NESSIE: NESSIE – New European Schemes for Signatures, Integrity and Encryption. https://www.cosic.esat.kuleuven.be/nessie/
3. De Cannière, C., Preneel, B.: TRIVIUM – a stream cipher construction inspired by block cipher design principles. eSTREAM, ECRYPT Stream Cipher Project, Report 2005/030 (2005), http://www.ecrypt.eu.org/stream/trivium.html
4. Raddum, H.: Cryptanalytic results on TRIVIUM. eSTREAM, ECRYPT Stream Cipher Project, Report 2006/039 (2006), http://www.ecrypt.eu.org/stream
5. Le Berre, D., Simon, L.: Special Volume on the SAT 2005 competitions and evaluations. Journal of Satisfiability (JSAT) (March 2006),
   http://www.satcompetition.org/
6. Maximov, A., Biryukov, A.: Two Trivial Attacks on Trivium. In: Selected Areas in Cryptography 2007, pp. 36–55 (2007)
7. Biryukov, A., Shamir, A.: Cryptanalytic time/memory/data tradeoffs for stream ciphers. In: Okamoto, T. (ed.) ASIACRYPT 2000. LNCS, vol. 1976, pp. 1–13. Springer, Heidelberg (2000)

8. Bard, G., Courtois, N., Jefferson, C.: Efficient Methods for Conversion and Solution of Sparse Systems of Low-Degree Multivariate Polynomials over GF(2) via SAT-Solvers. Cryptology ePrint Archiv, Report 2007/024 (2007)
9. McDonald, C., Charnes, C., Pieprzyk, J.: Attacking Bivium with MiniSat. Cryptology ePrint Archive, Report 2007/040 (2007)
10. DIMACS specification: http://www.satlib.org/Benchmarks/SAT/satformat.ps
11. Pipatsrisawat, K., Darwiche, A.: RSat 2.0: SAT Solver Description. Technical report D153. Automated Reasoning Group, Computer Science Department, University of California, Los Angeles (2007), http://reasoning.cs.ucla.edu/rsat/
12. Een, N., Sorensson, N.: MiniSat – A SAT Solver with Conflict-Clause Minimization. In: Bacchus, F., Walsh, T. (eds.) SAT 2005. LNCS, vol. 3569, Springer, Heidelberg (2005), http://www.cs.chalmers.se/Cs/Research/FormalMethods/MiniSat/MiniSat.html
13. Wegener, I.: Branching Programs and Binary Decision Diagrams. SIAM Monographs on Discrete Mathematics and Applications. SIAM, Philadelphia (2000)
14. Krause, M.: BDD-Based Cryptanalysis of Keystream Generators. In: Knudsen, L.R. (ed.) EUROCRYPT 2002. LNCS, vol. 2332, pp. 237–239. Springer, Heidelberg (2002)
15. Krause, M.: OBDD-Based Cryptanalysis of Oblivious Keystream Generators. Theory of Computing Systems 40(1), 101–121 (2007)
16. Krause, M., Stegemann, D.: Reducing the space complexity of BDD-based attacks on keystream generators. In: Robshaw, M.J.B. (ed.) FSE 2006. LNCS, vol. 4047, pp. 163–178. Springer, Heidelberg (2006)
17. Stegemann, D.: Extended BDD-based Cryptanalysis of Keystream Generators. In: Proceedings of SAC 2007. LNCS, vol. 4876, pp. 17–35 (2007)
18. Somenzi, F.: CUDD, version 2.4.1, University of Colorado, http://vlsi.colorado.edu/~fabio/CUDD/
19. Stein, W.: Sage Mathematics Software (Version 2.9.2) The SAGE Group (2007), http://www.sagemath.org.
20. Greuel, G.-M., Pfister, G., Schönemann, H.: Singular 3.0.4. A Computer Algebra System for Polynomial Computations. Centre for Computer Algebra, University of Kaiserslautern (2007), http://www.singular.uni-kl.de/
21. Buchberger, B.: Gröbner Bases: A Short Introduction for System Theorists. In: Moreno-Díaz Jr., R., Buchberger, B., Freire, J.-L. (eds.) EUROCAST 2001. LNCS, vol. 2178, pp. 1–14. Springer, Heidelberg (2001)
22. Brickenstein, M.: Slimgb: Gröbner Bases with Slim Polynomials. Reports on Computer Algebra 35, ZCA, University of Kaiserslautern (2005)

SAT Modulo the Theory of Linear Arithmetic: Exact, Inexact and Commercial Solvers

Germain Faure, Robert Nieuwenhuis, Albert Oliveras,
and Enric Rodríguez-Carbonell[*]

Abstract. Many highly sophisticated tools exist for solving linear arithmetic optimization and feasibility problems. Here we analyze why it is difficult to use these tools inside systems for SAT Modulo Theories (SMT) for linear arithmetic: one needs support for disequalities, strict inequalities and, more importantly, for dealing with incorrect results due to the internal use of imprecise floating-point arithmetic. We explain how these problems can be overcome by means of result checking and error recovery policies.

Second, by means of carefully designed experiments with, among other tools, the newest version of ILOG CPLEX and our own new Barcelogic T-solver for arithmetic, we show that, interestingly, the cost of result checking is only a small fraction of the total T-solver time.

Third, we report on extensive experiments running exactly the same SMT search using CPLEX and Barcelogic as T-solvers, where CPLEX tends to be slower than Barcelogic. We analyze these at first sight surprising results, explaining why tools such as CPLEX are not very adequate (nor designed) for this kind of relatively small incremental problems.

Finally, we show how our result checking techniques can still be very useful in combination with inexact floating-point-based T-solvers designed for incremental SMT problems.

1 Introduction

The applicability of current SAT solvers to many areas in and outside computer science is nowadays well known. However, some practical problems are more naturally described and more efficiently solved in logics that are more expressive than propositional logic. For example, for reasoning about timed automata or about intervals in scheduling problems, a good choice is *difference logic*, where formulas contain atoms of the form $a - b \leq k$. Similarly, the conditions arising from program verification usually involve arrays, lists and other data structures, so it becomes very natural to consider satisfiability problems *modulo* the theory T of these data structures. In such applications, problems may consist of thousands of clauses like

$$p \quad \vee \quad \neg q \quad \vee \quad a = b - c \quad \vee \quad read(s, b - c) = d \quad \vee \quad a - c \leq 7$$

containing purely propositional atoms as well as atoms over (combined) theories. This is known as the *Satisfiability Modulo Theories* (SMT) problem for a theory

---

[*] Tech. Univ. of Catalonia, Barcelona. All authors partially supported by Spanish Min. of Educ. and Science through the LogicTools-2 project, TIN2007-68093-C02-01.

H. Kleine Büning and X. Zhao (Eds.): SAT 2008, LNCS 4996, pp. 77–90, 2008.

T: given a formula F, determine whether F is T-satisfiable, i.e., whether there exists a model of T that is also a model of F. SMT has become an extremely active area of research and many SMT systems have been developed [DdM06a, dMB07, BBC+05, BT07, NO05a], as well as a library of benchmarks for SMT, called SMT-LIB [TR05].

The DPLL(T) approach to SMT couples a general DPLL(X) engine, in charge of enumerating propositional models of the formula, with a theory solver $Solver_T$, responsible for checking the consistency of these models over the theory T (e.g., if T is difference logic and the current boolean assignment contains $x - y \leq 0$, $y - z \leq 0$, and $x - z \geq 1$, then $Solver_T$ has to detect its T-inconsistency).

Here we consider SAT modulo the theories of Linear (Real or Integer) Arithmetic (LRA or LIA). So far, in SMT systems not much of the wide body of technology developed in the field of OR has been exploited. The reason for this is that the main application area of SMT is verification, which has some requirements that are not considered essential in OR: one needs to handle disequalities and strict inequalities, and, in order to guarantee correctness, employ arbitrary-precision arithmetic instead of floating-point arithmetic. The only work the authors are aware of the application of OR tools to SMT is [YM06], where nevertheless the issue of how incorrect answers from the solver should be handled was not addressed. Still, OR solvers *may* give wrong answers: for instance CPLEX 11 [ILO07], the newest version of ILOG CPLEX, returns that the following set of constraints (obtained from the industrial benchmark `clocksynchro_2clocks.main_invar.induct` from the SMT-LIB) is satisfiable:

$$-x - y + u \leq 0 \qquad -11z + v + 11t \leq 0$$
$$-u + z \leq 0 \qquad 11x - v \leq -10^{-5}$$
$$-t + y \leq 0 \qquad x \geq 10^{-5}$$

However, it is unsatisfiable, as the reader can easily check by multiplying the first three constraints by 11 and adding up all constraints but the last bound (which is not in the conflict but is needed to get a wrong answer from CPLEX).

In this paper we further study the applicability of OR tools for developing theory solvers for LA. We show how imprecise floating-point-based simplex solvers can be used in combination with result checking and error recovery policies for handling solver failures.

Furthermore, we report on a large number of carefully designed experiments with commercial and non-commercial OR solvers, including CPLEX 11, and with several versions of our own new Barcelogic $Solver_T$ for LRA and LIA. These experiments show, among several other interesting results, that (i) result checking takes only a small fraction of the total OR solver time and (ii) OR solvers are not designed for the incremental feasibility problems that occur in SMT and are often outperformed in this context by our specialized exact T-solver.

This closes some research directions and opens other new ones. In particular, it seems that a good approach may be to combine result checking with floating-point implementations of our current SMT-style incremental solvers. Following this idea we have implemented a prototype using floating-point arithmetic, which we have compared experimentally with CPLEX obtaining promising results.

This paper is structured as follows. We first give some background on SMT and DPLL(T) in Section 2. Section 3 studies which functionalities are offered and missing in OR solvers in order to be used as theory solvers. Then, Section 4 concentrates on how to use inexact solvers like CPLEX in DPLL(T). Next, Section 5 analyzes the performance of OR solvers when used as theory solvers. Finally, Section 6 presents preliminary results on the development of inexact solvers specifically designed for SMT, and we conclude in Section 7.

2   Background on SMT and DPLL(T)

In this section we give a quick overview of SMT and DPLL(T). We refer to [NOT06] for further details, extensions and references. The SMT problem consists of, given a ground first-order formula F and a theory T, deciding whether F is T-satisfiable (or T-consistent), i.e., whether there exists a model of T that is also a model of F. For that purpose, most state-of-the-art SMT solvers combine a boolean engine DPLL(X), very similar in nature to a SAT solver, with a theory solver $Solver_T$, thus producing a DPLL(T) system.

In the simplest version of such systems, the boolean engine initially considers each atom as a distinct propositional symbol. If the formula turns out to be propositionally unsatisfiable, it is T-unsatisfiable as well. Otherwise, DPLL(X) returns a propositional model M. This model, seen as a conjunction of literals, is then checked for T-consistency by $Solver_T$. If M is T-consistent then F is T-satisfiable; otherwise, in order to prevent M from later consideration, one can conjunct the negation of M (a disjunction of literals) to F and repeat the process until DPLL(X) finds a T-consistent model or returns unsatisfiable.

Example 1. Let F be $x \leq 2 \quad \wedge \quad (\neg(x + y = 1) \vee x \geq 3) \quad \wedge \quad x + y = 1$. In this case, DPLL($X$) will return the model $M = \{x \leq 2,\ x \geq 3,\ x + y = 1\}$ which will be detected T-inconsistent by $Solver_T$. After adding to F the clause $\neg(x \leq 2) \vee \neg(x \geq 3) \vee \neg(x + y = 1)$, the boolean engine will report the unsatisfiability of the formula.

In this simple setting all one needs from $Solver_T$ is the capability of checking the T-consistency of a conjunction of literals. However, for this approach to be efficient in practice several improvements need to be made. Here we list some of them, making special emphasis on the requirements they pose on $Solver_T$:

- The T-consistency of the assignment stored by DPLL(X) can be checked while it is being built, without delaying the check until a propositional model has been found (i.e., we are at a *leaf* of the search tree). This saves a large amount of useless work but requires $Solver_T$ to be incremental, that is, being faster in processing the addition of a single literal to a set of literals already found T-consistent than in reprocessing the whole set from scratch.

- When an assignment M is found T-inconsistent by $Solver_T$, one can ask DPLL(X) to backtrack to some point where the assignment was still T-consistent instead of restarting the search from scratch. This obviously forces $Solver_T$ to be able to support backtracking. Moreover, DPLL(X) needs to start its conflict analysis mechanism with an *inconsistency explanation* given by $Solver_T$, that is, a small subset of M that is also T-inconsistent (e.g, in Example 1, an inconsistency explanation is $\{x \leq 2,\ x \geq 3\}$).
- As a further optional refinement, if we want $Solver_T$ to play an active role in the search, instead of being used only to validate the search a posteriori, we can ask $Solver_T$ to detect unassigned input literals that are T-entailed by the current assignment M; that is, literals l such that $M \wedge T \models l$. This refinement, called *theory propagation*, allows DPLL(X) to assign them a truth value instead of having to guess an arbitrary value for them.

These improvements have allowed SMT solvers to be successfully used in a variety of applications. Many of them involve reasoning over the theory of *linear arithmetic* (LA), where atoms are of the form $a_1 x_1 + \ldots + a_n x_n \bowtie b$, being the a_i's rational numbers, the x_i's integer or rational variables and $\bowtie$ one of the operators $=,\ \leq,\ <,\ >,\ \geq$ or $\neq$. An interesting fragment of linear arithmetic is the one of difference logic (DL), where atoms are of the form $x_1 - x_2 \bowtie b$. In SMT benchmarks most LA constraints are indeed DL and their consistency can be checked very efficiently by means of negative-cycle-detection algorithms. Hence, when checking the T-consistency of a set of LA constraints it is not uncommon to first apply a specialized DL solver to filter out the inconsistencies that arise only taking into account DL atoms. On the other hand, for dealing with general LA constraints all state-of-the-art theory solvers in SMT tools are based on the *simplex method*. For further reading see, e.g., [Sch87].

3   Using OR Solvers as Theory Solvers for LA

In this section we summarize what OR solvers provide and miss so as to be applied to DPLL(LA).

All linear programming (LP) packages developed in OR allow the user to test the satisfiability of a conjunction of linear equations and non-strict inequations. Very often there is no specific facility for this purpose, since all that needs to be done is to optimize the null function over the system of constraints of interest: all models of the formula are optimal with respect to this objective function.

Moreover, most of these systems implement the so-called *bounded* simplex method [Mar86], which handles bounds on variables in a more efficient way than in the textbook version [Sch87]. This is important in the SMT context, since typically a significant amount of the literals in a problem are bounds: on average over 30% in the SMT-LIB, and in some benchmarks beyond 50%.

Also important as regards efficiency, the majority of these packages provide an API that avoids expensive communication through files and system calls.

Another issue that is paramount for the application to DPLL(LA) is incrementality: fortunately, most often the interfaces of these tools provide facilities for adding and removing constraints and modifying bounds, among others.

However, when one is faced with an unsatisfiable conjunction of constraints, as far as the authors know only commercial LP tools (or demo versions with limited capabilities of these) provide a means for computing an irredundant explanation for the inconsistency. Moreover some of these, such as CPLEX 9.1, produce explanations for LRA but not for LIA; besides, for some pathological instances, the explanations given by CPLEX 9.1 are redundant, though they should not be according to the documentation. For example, for the following system of constraints:

$$x + y \leq 2 \ \wedge \ x \leq 1 \ \wedge \ x \geq 1 \ \wedge \ y \leq 2 \ \wedge \ y \geq 2$$

CPLEX 9.1 considers the conjunctions $x \leq 1 \wedge x \geq 1$ and $y \leq 2 \wedge y \geq 2$ as the equations $x = 1$ and $y = 2$ respectively, and returns $E = \{x + y \leq 2, x = 1, y = 2\}$ as an irredundant explanation, whereas $E' = \{x + y \leq 2, x \geq 1, y \geq 2\}$ is a proper subset of E that is also inconsistent. Fortunately CPLEX 11 fixes these problems and does produce truly irredundant explanations for both LRA and LIA.

On the other hand, to the knowledge of the authors what all LP packages lack is support for handling disequalities and strict inequalities. Basically this is due to two facts: (1) optimization problems with these constraints may not have optimal solutions, and (2) in LP data are not usually absolutely precise, e.g., because they are subject to measurement errors.

Another feature that most LP tools lack is precise arithmetic. For the sake of efficiency, typically an OR solver works with floating-point arithmetic, instead of arbitrary-precision rationals as done in SMT solvers. This is the reason why, as shown in Section 1, an OR solver may give a wrong answer, i.e., return "SAT" for an unsatisfiable problem or "UNSAT" for a satisfiable one, or also compute a wrong explanation of inconsistency. Moreover, the use of floating-point arithmetic entangles the risk of a sudden unexpected failure; this is one of the reasons why optimization routines in LP libraries return a status value that indicates whether an internal error has occurred.

For instance, all of the versions of CPLEX we have experimented with just support floating-point arithmetic. On the other hand, the non-commercial OR solver GLPK [Mak07] additionally provides the user with exact arbitrary-precision arithmetic. See Section 5 for the results of our experiments with this feature.

Finally, no LP package supports theory propagation. This is natural, since in the context of OR this notion does not make any sense. Although the importance of theory propagation has been acknowledged elsewhere for LA and other theories [NO05b, DdM06b], one of the initial hypotheses of this research was that, given the huge amount of work done in the area of OR over the years, the performance of LP tools would be so outstanding that this limitation would be compensated for. Further, as seen in Section 2, theory propagation is not a necessary part of the core interface with DPLL(X), but an optimization on this interface.

4   How to Deal with Inexact Solvers in DPLL(T)

As discussed in Section 3, there are two issues that must be addressed so as to employ an inexact OR solver as a LA-solver:

(1) In the SMT context, constraints may be not only equalities and non-strict inequalities but also the negation of these, i.e., disequalities and strict inequalities; the OR solver must be able to handle them all.
(2) Due to imprecise arithmetic, the answers given by the OR solver may be wrong, and thus must be checked; moreover, there must be a policy for recovering from the possible errors and resuming the search.

In this section it is shown how this gap can be filled. As far as (1) is concerned, the problem of handling disequalities can be reduced to that of strict inequalities, since one can preprocess the input formula by splitting equalities into conjunctions of non-strict inequalities and disequalities into disjunctions of strict inequalities, which works very well in practice [DdM06b]. Now, given that the issue of correctness of the OR solver needs to be addressed anyway, a possibility is to strengthen strict inequalities by subtracting a small value ϵ; i.e., a constraint of the form $c^T x < d$ is transformed into $c^T x \le d - \epsilon$ (for instance, in our experiments we used $\epsilon = 10^{-5}$). Thus, the problem (1) of handling strict constraints has been reduced to (2), that of correctness of the inexact solver.

Now, regarding (2), a general solution for using inexact T-solvers in DPLL(T) (not necessarily OR solvers when T is LA) is to check results by means of an exact T-solver only when it is strictly necessary to ensure correctness. That is, (i) whenever the inexact solver returns "UNSAT", checking that the explanation for the conflict is indeed inconsistent; and (ii) whenever the inexact solver returns "SAT" (or an internal error occurs) *at a leaf*, checking that the assignment is indeed consistent with the theory. A corresponding error recovery policy can be easily described: in case (i), if the explanation is wrong, the exact solver is called again over the partial assignment, and the search is resumed using the result of this exact consistency check; similarly, in case (ii) the result of the check with the exact solver is employed to continue the search. Both result checking and error recovery policies are summarized in Algorithm 1.

Notice that the most expensive calls to the exact solver are those where the consistency of the whole partial assignment is checked. Under the hypotheses that the inexact solver will produce almost no wrong explanations of inconsistency and that internal errors will be infrequent too, these calls will be basically due to the uncommon event of the boolean search getting to a leaf of the tree. So it is reasonable to imagine that the cost of these calls will not be noticeable.

As regards the calls to the exact solver with inconsistency explanations, these will be much more frequent, since typically every few decisions a conflict arises.

Algorithm 1. Consistency check and error recovery policies

---

if *in a leaf* **then**
    if *there is an internal error or inexact solver returns "SAT"* **then**
        check consistency of partial assignment with exact solver;
        resume search using the result given by exact solver;
    else `//inexact solver returns ''UNSAT''`
        check consistency of inconsistency explanation with exact solver;
        if *exact solver returns "SAT"* **then**
            check consistency of partial assignment with exact solver;
            resume search using the result given by exact solver;
        else `//exact solver returns ''UNSAT''`
            resume search using inconsistency explanation for conflict analysis;
else `//in an internal node`
    if *there is an internal error or inexact solver returns "SAT"* **then**
        continue search as if partial assignment were theory consistent;
    else `//inexact solver returns ''UNSAT''`
        check consistency of inconsistency explanation with exact solver;
        if *exact solver returns "SAT"* **then**
            continue search as if partial assignment were theory consistent;
        else `//exact solver returns ''UNSAT''`
            resume search using inconsistency explanation for conflict analysis;

Fortunately, in general the number of literals in an explanation is below a few tens, and therefore these calls are often cheap.

In order to empirically assess the cost of result checking, we have carried out the following experiment: for all benchmarks in the LRA and LIA divisions (501 and 203 problems, respectively) of the SMT-LIB [TR05], we have run our SMT tool using CPLEX 11 as a LA-solver and implementing the result checking and error recovery policies presented in Algorithm 1 with our exact Barcelogic LA-solver. In order to avoid noise, no difference-logic pre-filtering has been applied. In this and in the rest of experiments in this paper, the machine used was a PC with an Intel(R) Xeon(TM) CPU 3.80GHz processor running Linux Debian 4.1.1. The timeout was set to 15 minutes.

In the graph in Figure 1 each dot represents an SMT instance of LRA. The horizontal axis shows the time spent in CPLEX (consistency checking, inconsistency explanation generation and backtracking); the vertical axis represents the time taken by result checking. Besides, the line $y = x/10$ is drawn as a reference.

As can be seen from the graph, for most problems in LRA the time taken by result checking is in general at most 10% of the time spent in CPLEX. In those instances for which result checking is significantly more expensive than that, this is due to either (1) the length of the inconsistency explanations (for some examples in the TM family, several hundreds of literals) or (2) the amount of errors produced by CPLEX. However, in more than 75% of the benchmarks, errors occur in at most 2% of the consistency checks.

The graph in Figure 2 is similar to that in Figure 1, but for benchmarks from LIA. In this case it is also clear that, in general, the cost of result checking is at most 10% of the time spent by CPLEX.

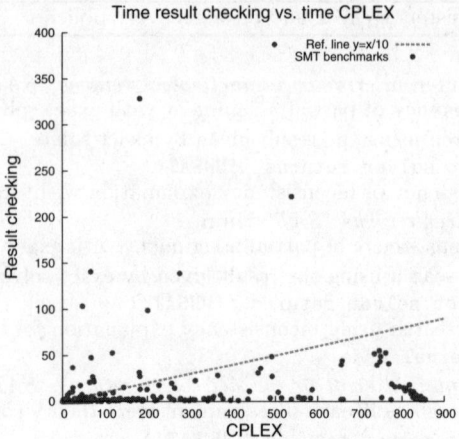

Fig. 1. Result checking evaluation for LRA

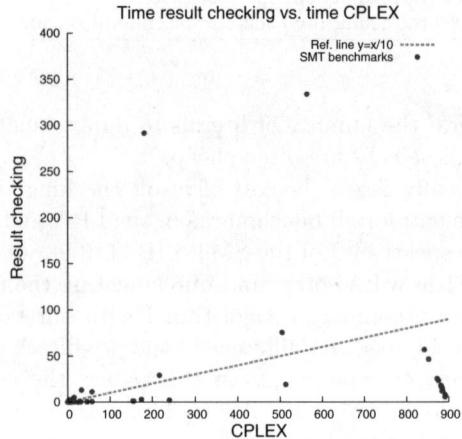

Fig. 2. Result checking evaluation for LIA

5   Performance of OR Solvers as T-Solvers

In this section we experimentally evaluate the performance of inexact OR solvers against that of specialized exact LA-solvers designed for SMT.

To this end, we have carried out the following experiment: guiding the search with our exact Barcelogic LA-solver, we have also run in parallel CPLEX 11 and compared the timings of the two tools, counting consistency checks [1], incon-

---

[1] There is a difference in the way consistency checks were performed with each tool. For our LA-solver, pending constraints were asserted one at a time; for CPLEX, all pending constraints were asserted at the same time. The reason for this is that, if CPLEX was asked to deal with constraints one at a time, it was much slower.

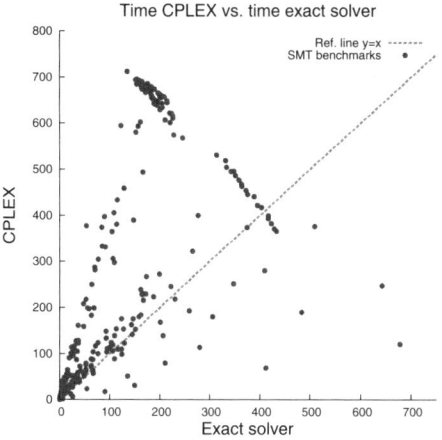

Fig. 3. Comparison between our exact solver and CPLEX in LRA

sistency explanation generation and backtracking. To make a fair comparison, apart from exploring the search space in the same way, neither difference-logic pre-filtering nor theory propagation have been applied.

The graph in Figure 3 shows the results of this experiment for LRA. Each dot represents an SMT instance. The horizontal axis is the time taken by our own exact LA-solver; the vertical axis is the time spent by CPLEX. Besides, the line $y = x$ is drawn as a reference.

Contrary to our initially expected results, when used as a theory solver in the DPLL(T) framework, CPLEX 11 tends to perform worse or not significantly better than our LRA solver. This is mainly due to consistency checks, and also inconsistency explanation generation, which are usually more expensive with CPLEX 11. The same experiments have also been carried out with other inexact OR solvers, namely CPLEX 9.1 and GLPK 4.25, the newest version of the GNU Linear Programming Kit, with similar outcome (although CPLEX 11 performs better than CPLEX 9.1, which is in turn better than GLPK). We have worked on several hypotheses in order to explain these results:

The default CPLEX parameter values are not adequate for SMT. The experiments above have been carried out using the default values for the parameters of CPLEX, so one could argue that these values are not the most appropriate for SMT problems. For this reason we have experimented changing several of the parameters that, according to CPLEX documentation, have most impact on the performance: simplex method (primal, dual, barrier), pricing strategy (standard, steepest edge, devex, ...), refactorization frequency, and several preprocessing options. No significant improvements have been achieved on the results obtained with the default values of the parameters.

The basis is refactored at each constraint addition/deletion. CPLEX allows writing a log file with information about the progress of the computation. From these log files it can be seen that refactorizations are not performed systematically each time constraints are added or removed, but more spacedly. Also, as mentioned above, we did not significantly enhance the results by modifying the refactorization frequency.

CPLEX is using a Phase I procedure that adds many new auxiliary variables and/or rows to the problem at each consistency check. As far as the authors could infer from the documentation, the Phase I primal algorithm implemented in CPLEX is based on [Mar86], where no extra rows or variables are added to the problem. Moreover, if the dual simplex method is employed, since the objective function is null any basis is trivially feasible, and thus all work is done in Phase II, where no auxiliary rows or variables are added either; still, we did not improve timings by using the dual simplex method, as said above.

CPLEX is not designed for being used as a *Solver*$_T$ **in DPLL(***T***), nor for the kind of problems that arise in SMT.** This is the most plausible explanation for the results obtained in this experiment, since the way CPLEX is commonly used in OR is remarkably different from that in this paper for SMT.

First of all, CPLEX is aimed at linear programs with up to millions of variables and constraints, whereas consistency checks from SMT involve few thousands of constraints over few hundreds of variables. Thus, using CPLEX for solving these problems may be an overkill.

Secondly, when in OR a linear program is solved, typically the user carries out some sensitivity analysis; in order to reuse computations in further reoptimizations, CPLEX provides the facilities not only for adding/removing constraints and changing bounds, but also changing coefficients of the objective function and the whole constraint matrix. However, efficiency in adding/removing constraints and changing bounds is not as determinant as in DPLL(*T*), where thousands of problems need to be solved incrementally for a single benchmark. As a result of this, CPLEX does not outperform our exact LA-solver in an incremental setting, whereas when solving large static problems it is better by orders of magnitude.

As regards inconsistency explanations, the typical scenario in OR is the following one: when dealing with big linear programs it may be tedious to detect errors in the data, for instance when the problem turns out to be infeasible whereas it should not; CPLEX offers functionalities for computing conflicting sets of constraints in order to help the user to diagnose where the error could be. Thus, in the context for which CPLEX has been designed, the computation of explanations of inconsistency is not critical, unlike in DPLL(*T*). In fact, while we were experimenting with a previous version of CPLEX, CPLEX 9.1, the bottleneck for many problems in LRA (namely, the sc and TM families) was precisely the generation of these explanations. CPLEX 11 is more efficient than its predecessor when computing inconsistency explanations, but there are still instances for which it does not perform very well.

Finally, CPLEX provides the user with finely tuned technology for optimizing hard problems, among others sophisticate pricing strategies, several optimization algorithms, advanced basis methods, etc. On the other hand, from the optimization point of view, the linear programs arising from SMT problems are easy and can be usually solved with few iterations of the simplex algorithm. Again, using CPLEX in this context may be excessive.

In order to look further into the cost of consistency checks in OR solvers, we experimented with the open-source OR solver GLPK 4.25 [2], which supports both floating-point and arbitrary-precision arithmetic. As regards inexact arithmetic, as mentioned above the results of the experiments were similar to those with CPLEX 11, although the performance of GLPK was worse than that of CPLEX. The execution profiles showed that about half of the time in GLPK was spent on factorizing the basis and the rest on (re)initializing data structures and simplex iterations, but did not reveal any deeper insights. Regarding exact arithmetic, GLPK performed two orders of magnitude worse than our exact LA-solver.

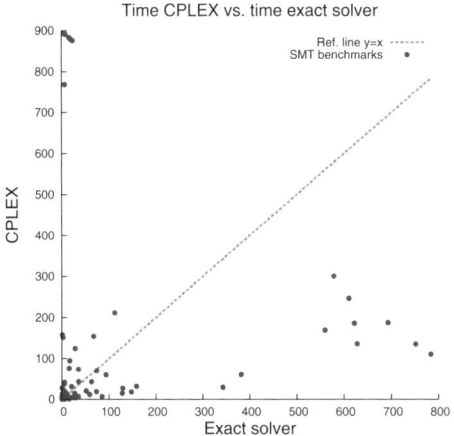

Fig. 4. Comparison between our exact solver and CPLEX in LIA

Finally, in Figure 4 we show the results of the experiment with CPLEX 11 and our exact solver on all LIA benchmarks from the SMT-LIB. As can be seen from the graph, CPLEX does perform better in general than our LIA solver (notice, however, that in some instances CPLEX is much slower; this is because for these particular LIA benchmarks it spends a huge amount of time computing explanations of inconsistency). This outcome is explained by the simplicity of our LIA solver, whose heuristics for branch & bound and cut generation have not been finely tuned. Nevertheless, given that the underlying engine for solving integer problems is a solver for reals, and given the above results for LRA, it seems reasonable to think that this difference between CPLEX and our LIA solver can be reduced if our search mechanism for integer solutions is improved.

---

[2] GLPK does not provide facilities for computing irredundant explanations of inconsistency, and so it was just used to check the consistency of partial assignments.

6   New Prospects: An Inexact Solver Designed for DPLL(T)

In this section we report on work in progress towards the use of inexact LA-solvers specifically designed for SMT as opposed to solvers developed in OR, based on the results obtained in the previous section.

Namely, in Section 5 our experiments in LRA have revealed that state-of-the-art OR solvers such as CPLEX 11 and GLPK 4.25, when applied in the DPLL(T) framework, are not competitive with specialized tools. Though in principle this is a negative result, in fact it suggests a new line for research: to combine result checking techniques with implementations of our current SMT incremental solvers using floating-point instead of arbitrary-precision arithmetic.

To assess the viability of this idea, we have run in parallel CPLEX 11 and an implementation of our LA-solver with floating-point numbers, using result checking and the error recovery policies described in Section 4. Our inexact LA-solver is currently a first-stage prototype that has been implemented basically by replacing exact rational variables by `double` variables, but without fine tuning for handling precision errors. As expected, using floating-point numbers in code designed for exact arithmetic may sometimes cause invariant violation and thus runtime errors and non-terminating behavior in the solver. For this reason, the experiment described here does not include all benchmarks from the LRA division of the SMT-LIB, but just those for which these errors did not occur. Interestingly enough, just 15% of the benchmarks were discarded; these are mainly the most difficult ones in the `clock_synchro`, `sc` and `tta_startup` families.

The outcome of this experiment is shown in the graph in Figure 5. Again, each dot represents an SMT benchmark. The horizontal axis is the time taken by our inexact LA-solver prototype; the vertical axis is the time spent by CPLEX. Besides, the lines $y = x$ and $y = 5x$ are drawn as a reference.

As can be seen from the graph, the results are promising. For most instances, CPLEX 11 spends at least five times as much time as our inexact solver. Taking into account the results obtained in the previous section, there is thus a potential gain in employing inexact LA-solvers implemented with floating-point arithmetic over exact LA-solvers implemented with arbitrary-precision numbers. [3]

However, two problems need to be addressed. First, although result checking was not an issue when using CPLEX, the situation is different with our prototype, for which the cost of ensuring correctness starts to become significant over the total time spent in theory reasoning. Therefore, result checking becomes eligible for optimization. A possibility in this direction is as follows. In all of the experiments reported here, result checking is implemented by having an auxiliary checking solver that each time it is called asserts all required constraints, and once the answer is returned it is emptied; this could be enhanced, for instance,

---

[3] A more precise experiment would have been to compare our inexact solver with the exact one. However this was not possible: our implementation employs static objects, which prevents us from having two solvers running simultaneously without resorting to system calls.

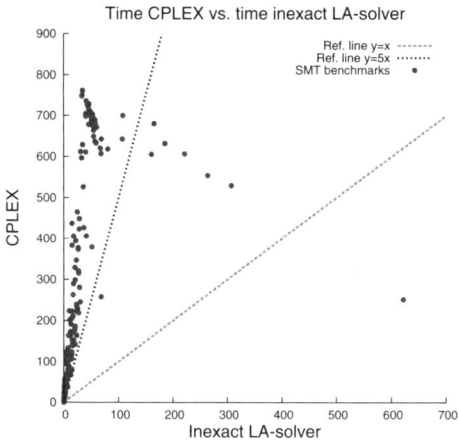

Fig. 5. Comparison between our inexact solver and CPLEX in LRA

by having two auxiliary solvers, one for checking explanations of inconsistency and another one for checking consistency of partial assignments, and working incrementally with the second one. Moreover, so far the inexact solver does not communicate any internal information to the auxiliary checking solver: still, the latter could use some data from the former to speed up the consistency check, for example which is the feasible basis or which are the multipliers of the inconsistency certificate. The second problem that has to be solved is that, even though result checking guarantees that on normal termination the answer given by the SMT tool will be correct, runtime errors or non-termination are clearly undesirable. It remains to be seen how these situations can be avoided without much computational effort.

On the other hand, unlike with OR solvers, this approach has the advantage that theory propagation could be applied by properly extending the result checking and error recovery policies. This would take the best of the two worlds: first, the efficiency of floating-point arithmetic; and second, the possibility to convey theory information to the boolean engine.

7   Conclusions

The main contributions of this paper can be summarized as follows. First, we have explained how OR tools can be used as theory solvers for SMT by means of result checking techniques and error recovery policies. Second, we have shown that the cost of the result checking techniques is only a small fraction of the time spent in the OR solver. Third, by means of exhaustive experiments we have shown that OR tools tend to be slower than exact solvers specifically designed for the DPLL(T) framework, and thus are not adequate in the context of SMT. Finally, based on empirical results we outline a new direction of research for obtaining efficient theory solvers, which consists in combining inexact floating-point-based implementations of solvers designed for DPLL(T) with result checking and error recovery policies.

Acknowledgments. The authors would like to thank J. Cortadella, J. Carmona and J. Larrosa for technical support with CPLEX. We are also grateful to L. de Moura, P. Stuckey and the anonymous referees for insightful comments.

References

[BBC+05] Bozzano, M., Bruttomesso, R., Cimatti, A., Junttila, T., van Rossum, P., Schulz, S., Sebastiani, R.: The MathSAT 3 System. In: Nieuwenhuis, R. (ed.) CADE 2005. LNCS (LNAI), vol. 3632, pp. 315–321. Springer, Heidelberg (2005)

[BT07] Barrett, C., Tinelli, C.: CVC3. In: Damm, W., Hermanns, H. (eds.) CAV 2007. LNCS, vol. 4590, pp. 298–302. Springer, Heidelberg (2007)

[DdM06a] Dutertre, B., de Moura, L.: The YICES SMT Solver. Technical report, SRI International (2006), Available at http://yices.csl.sri.com

[DdM06b] Dutertre, B., de Moura, L.: A Fast Linear-Arithmetic Solver for DPLL(T). In: Ball, T., Jones, R.B. (eds.) CAV 2006. LNCS, vol. 4144, pp. 81–94. Springer, Heidelberg (2006)

[dMB07] de Moura, L., Bjorner, N.: Z3: An Efficient SMT Solver. Technical report, Microsoft Research, Redmon (2007), Available at http://research.microsoft.com/projects/z3

[ILO07] ILOG. ILOG CPLEX v.11 (2007), http://www.ilog.com/products/cplex

[Mak07] Makhorin, A.: GLPK 4.25 (GNU Linear Programming Kit) (2007), Available at http://www.gnu.org/software/glpk/

[Mar86] Maros, I.: A general Phase-I method in linear programming. European Journal of Operational Research 23(1), 64–77 (1986)

[NO05a] Nieuwenhuis, R., Oliveras, A.: Decision Procedures for SAT, SAT Modulo Theories and Beyond. In: Sutcliffe, G., Voronkov, A. (eds.) LPAR 2005. LNCS (LNAI), vol. 3835, pp. 23–46. Springer, Heidelberg (2005)

[NO05b] Nieuwenhuis, R., Oliveras, A.: DPLL(T) with Exhaustive Theory Propagation and its Application to Difference Logic. In: Etessami, K., Rajamani, S.K. (eds.) CAV 2005. LNCS, vol. 3576, pp. 321–334. Springer, Heidelberg (2005)

[NOT06] Nieuwenhuis, R., Oliveras, A., Tinelli, C.: Solving SAT and SAT Modulo Theories: From an abstract Davis–Putnam–Logemann–Loveland procedure to DPLL(T). Journal of the ACM, JACM 53(6), 937–977 (2006)

[Sch87] Schrijver, A.: Theory of Linear and Integer Programming. Wiley, Chichester (1987)

[TR05] Tinelli, C., Ranise, S.: SMT-LIB: The Satisfiability Modulo Theories Library (2005), http://goedel.cs.uiowa.edu/smtlib/

[YM06] Yu, Y., Malik, S.: Lemma Learning in SMT on Linear Constraints. In: Biere, A., Gomes, C.P. (eds.) SAT 2006. LNCS, vol. 4121, pp. 142–155. Springer, Heidelberg (2006)

Random Instances of W[2]-Complete Problems: Thresholds, Complexity, and Algorithms

Yong Gao*

Department of Computer Science
Irving K. Barber School of Arts and Sciences
University of British Columbia Okanagan
Kelowna, Canada V1V 1V7
yong.gao@ubc.ca

Abstract. The study of random instances of NP complete and coNP complete problems has had much impact on our understanding of the nature of hard problems as well as the strength and weakness of well-founded heuristics. This work is part of our effort to extend this line of research to intractable parameterized problems. We consider instances of the threshold dominating clique problem and the weighted satisfiability under some natural instance distribution. We study the threshold behavior of the solution probability and analyze some simple (polynomial-time) algorithms for satisfiable random instances. The behavior of these simple algorithms may help shed light on the observation that small-sized backdoor sets can be effectively exploited by some randomized DPLL-style solvers. We establish lower bounds for a parameterized version of the ordered DPLL resolution proof procedure for unsatisfiable random instances.

1 Introduction

The theory of parameterized complexity and fixed-parameter algorithms is becoming an active research area in recent years [1,2]. Parameterized complexity provides a new perspective on hard algorithmic problems, while fixed-parameter algorithms have found applications in a variety of areas such as computational biology, cognitive modelling, and graph theory. Parameterized algorithmic problems also arise in many areas of artificial intelligence and satisfiability search. See, for example, the survey of Gottlob and Szeider [3].

Recently, some problems related to detecting *backdoor sets* for instances of the propositional satisfiability problem (SAT) have been studied from the perspective of parameterized complexity [4,5,6]. In particularly, the issue of the worst-case intractability versus the practical hardness of the backdoor detection problem has been raised: while the backdoor detection problem is NP-complete and/or fixed-parameter intractable for many types of backdoors, SAT solvers such as SATZ can exploit the existence of small-sized backdoors quite effectively [4,5,7].

---

* Supported in part by NSERC Discovery Grant RGPIN 327587-06.

H. Kleine Büning and X. Zhao (Eds.): SAT 2008, LNCS 4996, pp. 91–104, 2008.
© Springer-Verlag Berlin Heidelberg 2008

The study of the parameterized proof complexity of the satisfiability problem has been initiated in [8] where lower bounds on the parameterized resolution proof are established for CNF formulas that encode some first-order combinatorial principle.

The study of random instances of NP and coNP complete problems such as SAT has had much impact on our understanding of the nature of hard problems, the strength of resolution proof systems, and the strength and weakness of algorithms and well-founded heuristics [9,10,11,12,13,14,15].

This work is part of our effort to extend this line of research to intractable parameterized problems [16]. We discuss random instances of problems whose parameterized version is W[2]-complete, including instances of the dominating clique problem from the Erdös-Renyi random graph and instances of the weighted CNF satisfiability problem from a carefully-designed random distribution.

We establish lower and upper bounds on the threshold of the phase transition of the solution probability, and show that in some region of the instance space, satisfiable instances can be solved by simple algorithms in polynomial/fixed-parameter time with high probability. Since finding a solution to a satisfiable instance of the parameterized problems under our consideration can be viewed as the task of detecting backdoor sets with respect to an (extremely) naive sub-solver that simply checks whether the all-zero assignment is a satisfying assignment, the behavior of these simple algorithms may help shed light on the observation that small-sized backdoor sets can be effectively exploited by some randomized DPLL-style solvers (See our discussion at the end of Section 3.2).

For random instances in the unsatisfiable region, we establish a lower bound on the search tree size of the parameterized version of a basic resolution proof procedure — the ordered DPLL algorithm.

In the next Section, we define necessary terminologies and notation. In Section 3, we discuss the random models and the main results. Sections 4 through 6 contain the proofs of the results.

2   Preliminaries

An instance of a *parameterized decision problem* is a pair (I, k) where I is a problem instance and k is an input parameter. A standard example is the parameterized vertex cover problem where an instance (I, k) consists of a graph I and a positive integer k, and the question is to decide whether the graph has a vertex cover of at most k vertices. A parameterized problem is fixed-parameter tractable (FPT) if any instance (I, k) can be solved in $f(k)|I|^{O(1)}$ time.

Parameterized problems are inter-related by parameterized reductions, resulting in a classification of parameterized problems into a hierarchy of complexity classes $FPT \subset W[1] \subset W[2] \cdots \subset XP$. At the lowest level is the class of FPT problems. The top level XP contains all the problems that can be solved in time $f(k)n^{g(k)}$. It is widely believed that the inclusions are strict and the notion of completeness can be naturally defined via parameterized reductions.

Domination-style problems such as the dominating set problem are representative $W[2]$-complete problems. However, the behavior of random instances of the dominating set problem is not interesting. For sparse random graphs $G(n,p)$ with $p \in o(1)$, the size of the minimum dominating set is larger than $\log n$. For dense random graphs $G(n,p)$ with $0 < p < 1$ a fixed constant, any vertex subset of size $\log_{1/(1-p)} n$ is a dominating set.

We consider the dominating clique problem in the Erdos-Renyi random graph $G(n,p)$, hoping that due to the clique constraint, random instances will have a much richer structure. The CNF formulas encoding the instances of the dominating clique problem are interesting since their structure bears similarities to that of the CNF formulas that encode instances of practical problems such as planning and model-checking, making them potential good benchmarks for the empirical study of satisfiability search algorithms [17]. Generalizing this observation, we further propose and study a random distribution defined by combining a W[1]-complete problem and a W[2]-complete problem: the weighted 2-CNF satisfiability and the general weighted CNF satisfiability.

2.1 The Threshold Dominating Clique Problem

Given a graph $G(V, E)$, we use $N(v)$ to denote the set of neighbors of a vertex $v \in V$ and use $N(U)$ to denote the open neighbor of a subset of vertices U, i.e.,

$$N(U) = \{v \in V \setminus U : N(v) \cap U \neq \phi\}.$$

The cardinality of a vertex set U is denoted by $|U|$. A clique is a subset of vertices that induces a complete subgraph. A dominating set is a subset V_D of vertices such that $N(v) \cap V_D \neq \phi$, for all $v \in V \setminus V_D$.

We use $G(n,p)$ to denote the Erdös-Renyi random graph where n is the number of vertices and p is the edge probability. In $G(n,p)$, each of the possible $\binom{n}{2}$ edges appears independently with probability p. Throughout the paper when we say "with high probability", we mean that the probability of the event under consideration is $1 - o(1)$.

Definition 1. *Let $G(V, E)$ be a graph. An α-threshold dominating clique of G is a subset of vertices $V_D \subset V$ that induces a clique such that for all $v \in V \setminus V_D$, $|N(v) \cap V_D| \geq \alpha$. A 1-threshold dominating clique is simply called a dominating clique.*

The dominating clique problem is NP-complete and W[2]-complete when parameterized by the size of the clique [1]. The α-threshold dominating clique can also be shown to be W[2]-complete by a reduction from the threshold dominating set problem (see the Appendix of [1] for the definition of the threshold dominating set problem).

2.2 The Weighted CNF Satisfiability Problem

As in the theory of NP-completeness, the propositional satisfiability problem also plays an important role in the theory of parameterized complexity. A CNF formula over a set of Boolean variables is a conjunction of disjunctions of literals. A d-clause is a disjunction of d-literals. An assignment to a set of n Boolean variables is a vector in $\{0,1\}^n$. The **weight** of an assignment is the number of the variables that are set to 1 (true) by the assignment. A representative $W[1]$-complete problem is the following weighted d-CNF satisfiability problem (weighted d-SAT):

Problem 1. **Weighted d-SAT**

Instance: A CNF formula consisting of d-clauses and a positive integer k.

Question: Is there a satisfying assignment of weight k?

Unlike the traditional satisfiability problem, the weighted 2-SAT is already $W[1]$-complete. The anti-monotone weighted d-SAT problem (the problem where each clause contains negative literatures only) is also $W[1]$-complete. The *weighted satisfiability problem* (weighted SAT) is similar to the weighted d-SAT except that there is no restriction on the length of a clause in the formula. The weighted SAT is a generic $W[2]$ complete problem.

A formal definition of a parameterized tree-like resolution proof system for weighted SAT is given in [8]. Basically, a parameterized resolution system can be regarded as a classical resolution system that has access (for free) to all clauses with more than k negated variables, where k is the parameter of the weighted SAT.

The most widely-used algorithms for the traditional satisfiability problem are variants of the Davis-Putnam-Logemann-Loveland (DPLL) procedure [18]. We consider the parameterized version of the DPLL algorithm for weighted SAT. It proceeds in the same way as the standard DPLL algorithm with the exception that a node in the search tree fails if

1. either a clause has been falsified by the partial assignment, or
2. the number of variables assigned to true in the partial assignment has exceeded k.

This way of parameterizing a proof procedure was proposed in [8].

We will provide a lower bound on a weaker version of the parameterized variants of the DPLL procedure — the parametric ordered DPLL. In the ordered DPLL [10], the variables are given a fixed order (before the algorithm starts). Except for the unit-propagation reduction steps, the variable selected to branch on is always the first one in the order that has not been assigned a value.

3   Main Results

3.1   Random Instances of Dominating Clique Problem

We use $\text{DOMC}_{\alpha,k}^{n,p}$ to denote a random instance of the α-threshold dominating clique problem parameterized by the clique size k on the random graph $G(n,p)$. The exact threshold of the phase transition of the α-threshold dominating clique problem can be established for all α by extending the proof given in [17]. The threshold for any constant α (or α up to $\epsilon \log n$ with sufficiently small $\epsilon > 0$) turns out to be the same.

Theorem 1. *Consider the random graph $G(n, p)$. For any constant $\alpha \geq 1$,*

$$\lim_n Pr\{G(n, p) \text{ has an } \alpha\text{-threshold dominating clique }\}$$

$$= \begin{cases} 0, \text{ if } p < \frac{3-\sqrt{5}}{2} \\ 1, \text{ if } p > \frac{3-\sqrt{5}}{2}. \end{cases} \tag{1}$$

The size of an α-threshold dominating clique in $G(n, p)$, if exists, turns out to be in $\Omega(\log_{1/p} n)$. As a consequence, random instances of the parameterized α-threshold dominating clique problem with a fixed parameter is with high probability unsatisfiable for any $\text{DOMC}^{n,p}_{\alpha,k}$ with $p < 1$. Due to this reason, the discussions in this subsection, especially those for the satisfiable instances, are in fact for the "LOGNP"-behavior of the problem. For future studies, one may want to consider parameterized problems that ask for a dominating clique of size $k \log_{1/\epsilon} n$ for some small constant ϵ. This difficulty largely motivates the random weighted SAT distribution to be discussed in the next subsection.

The Unsatisfiable Instances

Two exact algorithms for the dominating clique problem have been proposed [19,17]. The one proposed in [19] is shown to have a time complexity $O(1.339^n)$ while the one studied in [17] empirically works well on random graphs (In fact by adding a few simple cases, which never happen in random graphs, the algorithm studied in [17] can be shown to have a time complexity $O(1.383^n)$ by a simple analysis). In the following, we lower bound the search tree size of the ordered DPLL algorithm which is weaker than the above two branch-and-reduce algorithms, but is of interest in the study of proof complexity and logic inferences.

The parameterized dominating clique problem can be encoded as a weighted SAT problem as follows. Given a graph $G(V, E)$, we associate with each vertex with a Boolean variable. Let $\{x_1, \cdots, x_n\}$ be the set of variables corresponding to the set of vertices $V = \{v_1, \cdots, v_n\}$. The CNF formula consists of two types of clauses:

1. Anti-monotone 2-clauses. For each pair of vertices v_i and v_j such that $(v_i, v_j) \notin E$, there is a 2-clause $\overline{x}_i \vee \overline{x}_j$. This set of clauses enforces the clique constraint.
2. Monotone long clauses. For each vertex v_i, there is a clause

$$x_i \vee x_{i_1} \vee \cdots \vee x_{i_l}$$

where $\{x_{i_1}, \cdots, x_{i_l}\}$ are the neighbors of v_i. This set of clauses enforces the domination requirement.

The following theorem provides a lower bound on the size of the search tree of the parametric ordered DPLL resolution proof. Note that the result is more general than needed — we allow k to be as large as $\epsilon \log n$.

Theorem 2. *For any parameter $0 < k < \epsilon \log n$ where $\epsilon > 0$ is a small constant and any $0 < p < 1$, the size of the search tree of the parametric ordered DPLL algorithm for $DOMC_{k,\alpha}^{n,p}$ is $n^{\Omega(k)}$ with high probability.*

The Satisfiable Instances

On the positive side, we show that for any $p > \frac{1}{2}$, an α-threshold dominating clique of size $\Omega(\log n)$ in $G(n,p)$ can be found in $O(n^2)$ time with high probability.

Theorem 3. *There is an $O(n^2)$-time algorithm that with high probability, finds an α-threshold dominating clique of size $\Omega(\log n)$ in $G(n,p)$ with $p > \frac{1}{2}$.*

We consider the following greedy algorithm, G-DOMC. Except for the first α-steps, at any moment, the vertices of the graph are in one of the following groups:

1. V_C: the clique obtained so far;
2. V_W: vertices that are adjacent to every vertex in V_C;
3. $V_i, 0 \le i \le \alpha - 1$: a vertex v is in V_i if it is adjacent to exactly i vertices in V_C.
4. V_α: vertices that have been dominated by at least α vertices in V_C, but are not in V_W.

Vertices in V_α are those that have been α-threshold dominated but cannot be used to expand the current clique. Hence they play no role in the algorithm. It is easy to see that after the first α steps, vertices in V_W have been dominated by more than α vertices in V_C so that we do not need to worry about their domination. Since α is a fixed constant, it can be shown that with high probability, the algorithm will not terminate within the first α steps.

The algorithm G-DOMC repeatedly picks a random vertex in V_W to expand the current clique and updates the vertex sets V_W and V_i's accordingly, as shown in the following pseudo-code:

1. Initialization: $V_C = \phi$, $V_W = V$, and $V_i = \phi, 0 \le i \le \alpha$;
2. Repeat until either $V_i = \phi, \forall i \le \alpha$ or $V_W = \phi$
   (a) randomly pick a vertex v in V_W
   (b) $V_C = V_C \cup \{v\}$; $V_W = V_W \cap N(v)$;
   (c) For each $0 \le i \le \alpha - 1$,

$$V_i = (V_i \setminus (N(v) \cup \{v\})) \cup (V_{i-1} \cap N(v))$$

Let $X(t)$ be the size of V_W after the t-th iteration and $Y_i(t)$ be the size of V_i after the t-th iteration. Intuitively since $p > \frac{1}{2}$, each vertex is adjacent to more than half of the vertices. If we construct the clique by greedily picking one of the potential vertices, then the number of potential vertices that can be used to expand the current clique decreases at a slower rate than the number of vertices that still need to be dominated. Consequently, all the vertices will be dominated before there is no way to expand the current clique.

Formally, we will prove that

$$\mathbb{P}\left\{X(t) > n^{\delta_1} \text{ and } Y_i(t) = 0, \ \forall \ 0 \le i \le \alpha - 1\right\} > 1 - O(\frac{1}{n^\delta}) \qquad (2)$$

where $\delta > 0$ and $\delta_1 > 0$ are properly determined small constants and

$$t = -\frac{1+\delta}{\log(1-p)} \log n,$$

which guarantees that the algorithm finds an α-threshold dominating clique at step t with high probability. For the formal proof, see Section 5.

3.2   A Random Model for Weighted SAT

To have a random distribution that generates interesting instances for fixed parameters of some dominating-style problem, we propose the following model $\mathcal{M}_{n,k}^{p_1,p_2,m}$ for weighted SAT.

Definition 2. *An instances of $\mathcal{M}_{n,k}^{p_1,p_2,m}$ consists of*

1. *a collection of anti-monotone 2-clauses. Each of the potential $\binom{n}{2}$ anti-monotone clauses is included independently with probability p_1, and*
2. *m monotone clauses obtained independently in the following way: for each clause, each of the n variables appear with probability p_2.*

The number of variables is n and the input parameter is k.

It is a concern that the monotone clauses generated in the above may be trivial, either being empty or containing all the variables. This is not the case — it can be shown that for the range of m we are considering, all the clauses contain $p_2 n + o(n)$ variables with high probability.

We have the following result on the threshold of the phase transition of the solution probability.

Theorem 4. *Assume that $0 < p_1, p_2 < 1$ are fixed constants. Let $b = (1 - (1 - p_2)^k)$, and $m = c \log n$. The probability that a random instance of $\mathcal{M}_{n,k}^{p_1,p_2,m}$ has a solution is*

$$\lim_n \mathbb{P}\left\{\mathcal{M}_{n,k}^{p_1,p_2,m} \text{ has a solution }\right\} = \begin{cases} 1, \text{ if } c < -\frac{1}{\log b} \\ 0, \text{ if } c > -\frac{k}{\log b}. \end{cases}$$

For the case of $c \log b > -1$, the proof of the above theorem actually indicates that the fraction of the satisfying assignments is in a "fixed-parameter" form. As a consequence, by simply sampling the assignments of weight k, we can find a satisfying assignment of weight k. The average number of samples needed is in a "fixed-parameter" for a typical instance from $\mathcal{M}_{n,k}^{p_1,p_2,m}$. (Note however that the average is taken with respect to the sampling process only.)

Corollary 1. *Let $m = c \log n$ such that $c \log b > -1$. There is a randomized algorithm that solves the satisfiable instances of $\mathcal{M}_{n,k}^{p_1,p_2,m}$ in $2^{O(k^2)} n^{O(1)}$ time.*

Proof. Consider the algorithm that repeatedly and randomly picks an assignment of weight k until a satisfying assignment is found.

Let $a = (1 - p_1)$. From the proof of Theorem 4, we see that with high probability, an instance of $\mathcal{M}_{n,k}^{p_1,p_2,m}$ has more than $a^{\binom{k}{2}} \binom{n}{k} n^{c \log b}$ satisfying assignments.

For a typical (but fixed) instance from $\mathcal{M}_{n,k}^{p_1,p_2,m}$, the probability that a randomly-picked assignment is satisfying is

$$\frac{a^{\binom{k}{2}} \binom{n}{k} n^{c \log b}}{\binom{n}{k}} = a^{\binom{k}{2}} n^{c \log b}.$$

Thus, the average number of samples (with respect to the sampling process) required before a satisfying assignment is found is $2^{O(k^2)} n^{-c \log b}$.

To relate Corollary 1 to the backdoor set detection problem, consider the (extremely) naive sub-solver that simply checks whether the all-zero assignment is a satisfying assignment. Corollary 1 says that such a backdoor set can be found by sampling the $\binom{n}{k}$ possibilities $2^{O(k^2)} n^{-c \log b}$ times, and thus provides a theoretical support to the observation that the existence of samll-sized backdoor sets can be effectively exploited by randomized DPLL-style solvers with random restarts such as SATZ [7].

Similar to the case of the threshold dominating clique problem, a lower bound on the parametric ordered DPLL algorithm for unsatisfiable instances can be established.

Theorem 5. *Let $m = c \log n$ such that $c \log b < -k$. Then with high probability, the size of the search tree of the parametric ordered DPLL for instances of $\mathcal{M}_{n,k}^{p_1,p_2,m}$ is $n^{\Omega(k)}$.*

4   Proof of Theorem 2

Proof. We focus on the case of 1-threshold dominating clique. Let $V = \{v_1, \cdots, v_n\}$ be an ordering of the vertices and assume without loss of generality that this is also the order used by the order DPLL algorithm. Let $i = \beta n$ where $\beta > 0$ is a constant, $V_0 = \{v_1, \cdots, v_i\}$, and $U = V \setminus V_0$.

Let $\mathcal{D}$ be the collection of subsets of vertices in V_0 of size $\frac{k}{2}$ and denote by $\mathcal{N}(D)$ the set of vertices in U that are adjacent to every vertex in the vertex set $D \in \mathcal{D}$, i.e.,

$$\mathcal{N}(D) = \{u \in U \mid \forall w \in D, (u, w) \text{ is an edge}\}.$$

We say that a vertex set D in $\mathcal{D}$ is *promising* if

1. D induces a clique in $G(n, p)$, and
2. $N(v) \cap \mathcal{N}(D) \neq \phi$ for any vertex $v \in V \setminus (D \cup \mathcal{N}(D))$.

We claim that the size of the DPLL search tree is lower bounded by the number of promising vertex sets in $\mathcal{D}$. To see this, consider a subset of vertices

$$D = \{v_{i_1}, \cdots, x_{i_{\frac{k}{2}}}\} \subset V_0$$

and a path of length βn in the ordered DPLL search tree along which variables in D are assigned to true and the other variables on the path are assigned to false. Since D induces a clique, no anti-monotone clause has been falsified by the partial assignment. Since the variables in $\mathcal{N}(D)$ are those that have not been forced by the assignment to the variables in D, the fact that $N(v) \cap \mathcal{N}(D) \neq \phi$ implies that the monotone long clause enforcing the domination of vertex v is not empty yet. Therefore, this path will be explored by the ordered DPLL algorithm. This proves the claim.

To proceed, we first show that the size of $\mathcal{N}(D)$ is large with high probability. Since a vertex u is in $\mathcal{N}(D)$ if and only if it is adjacent to every vertex in D, the expected size of $\mathcal{N}(D)$ is

$$\mathbb{E}\left[|\mathcal{N}(D)|\right] = (1 - \beta)np^{\frac{k}{2}}.$$

Let $I_D(u)$ be the indicator function of the event that u is in $\mathcal{N}(D)$. Due to the independence of the edges in $G(n, p)$, the variables $\{I_D(u), u \in U\}$ are independent Bernoulli variables with mean $p^{\frac{k}{2}}$. By the Chernoff bound (see, for example, [20]), we have

$$\mathbb{P}\left\{|\mathcal{N}(D)| > \frac{1}{2}(1 - \beta)np^{\frac{k}{2}}\right\} \geq 1 - 2e^{-\frac{1}{2}(1-\beta)np^{\frac{k}{2}}}. \tag{3}$$

To complete the proof of the theorem, we show that with high probability, there are $n^{\Omega(k)}$ promising vertex sets. From Equation (3), we have for a fixed vertex set $D \in \mathcal{D}$,

$$\mathbb{P}\{D \text{ is promising} \mid D \text{ induces a clique}\}$$
$$\geq O(1)\left(1 - (1 - p)^{\frac{1}{2}(1-\beta)np^{\frac{k}{2}}}\right)^n \geq O(1) \tag{4}$$

since p is a fixed constant.

Let X be the number of vertex subsets in $\mathcal{D}$ that are promising. The expectation of X satisfies

$$O(1)\binom{\beta n}{k/2}p^{\binom{k/2}{2}} \leq \mathbb{E}[X] \leq \binom{\beta n}{k/2}.$$

Therefore $\mathbb{E}[X]$ is in $n^{\Omega(k)}$ as long as $k < \epsilon \log n$ for some $\epsilon = \epsilon(p) > 0$. To complete the proof, we apply Chebyshev's inequality

$$\mathbb{P}\left\{|X - \mathbb{E}[X]| > \frac{1}{2}\mathbb{E}[X]\right\} \leq \frac{4\mathbb{E}\left[(X - \mathbb{E}[X])^2\right]}{(\mathbb{E}[X])^2}. \tag{5}$$

and show that

$$\mathbb{E}\left[(X - \mathbb{E}[X])^2\right] = \mathbb{E}\left[X^2\right] - (\mathbb{E}[X])^2 = o(\mathbb{E}[X])^2,$$

which can be established by estimating the probability that two overlapping vertex sets D_1, D_2 in $\mathcal{D}$ are both promising. We omit the lengthy detail due to space limit.

5   Proof of Theorem 3

We prove the theorem by showing that with high probability, G-DOMC terminates with an α-threshold dominating clique of size $\Omega(\log n)$. Recall that in the algorithm G-DOMC, after the first α steps, the vertices of the graph are in one of the following groups: V_C (the current clique), V_W (vertices adjacent to every vertex in V_C), V_i (vertices dominated by exactly i vertices in V_C), and V_α (the finished vertices).

Let X_t be the size of V_W after the t-th iteration and Y_t^i be the size of V_i be the size of V_i after the t-th iteration. First, we have the following lemma

Lemma 1. *The number of vertices in V_W after the first α steps satisfies*

$$\mathbb{P}\left\{X_\alpha > \frac{1}{2}p^\alpha n\right\} \geq 1 - O(e^{-n}).$$

Due to the above lemma, we will assume that $X_\alpha > \frac{1}{2}p^\alpha n$, which further implies that $Y_t^i \leq (1 - \frac{1}{2}p^\alpha)n$ for any $0 \leq t \leq \alpha$.

Due to the assumption $p > \frac{1}{2}$, there exist small constants $\delta > 0$ and $\epsilon > 0$ such that

$$(1 - \epsilon)p > (1 - p)^{\frac{1}{1+\delta}}.$$

Let

$$t = -\frac{1 + \delta}{\log(1 - p)} \log n, \text{ and}$$

$$\delta_1 = 1 - \frac{1 + \delta}{\log(1 - p)} \log(1 - \epsilon)p > 0.$$

We will show that

$$\mathbb{P}\left\{X_t > n^{\delta_1} \text{ and } Y_t^i = 0, \forall\, 0 \leq i \leq \alpha - 1\right\} > 1 - O(\frac{1}{n^\delta}) \tag{6}$$

which indicates that with high probability, the algorithm G-DOMC terminates in t steps and finds an α-threshold dominating clique. We first consider the probability of the event $\{X_t > n^{\delta_1}\}$.

Lemma 2.

$$\mathbb{P}\left\{X_t > n^{\delta_1}\right\} \geq (1 - \frac{c}{n})^t.$$

for some fixed constant $c > 0$.

Proof. Recall that $X_t = |V_W|$ is the number of white vertices after step t. After the first α steps, the set of white vertices "evolve" on its own — no vertex in V_i's can become a white vertex and vertices in V_W have been α-dominated.

In step t, a vertex v in V_W is randomly selected, and the new V_W is formed as $V_W = V_W \cap N(v)$. Since in $G(n,p)$, the edges appear in the graph independently, by the "deferred decision" argument, we see that $\{X_t, t \geq \alpha\}$ is a Markovian chain.

Write $a_t = (1 - \epsilon)^t p^t n = n^{\delta_1}$. We have

$$\mathbb{P}\left\{X_t > n^{\delta_1}\right\} = \mathbb{P}\left\{X_t > a_t\right\}$$
$$\geq \mathbb{P}\left\{X_s > a_s \text{ for all } \alpha \leq s \leq t\right\}$$
$$= \prod_{s=\alpha}^{t} \mathbb{P}\left\{X_s > a_s \mid X_{s-1} > a_{s-1}\right\}$$

We claim that

$$\mathbb{P}\left\{X_s > a_s \mid X_{s-1} > a_{s-1}\right\} \geq \mathbb{P}\left\{Bin(p, a_{s-1}) > a_s\right\}.$$

where $Bin(p, a_{s-1})$ is a random variable that has a binomial distribution with parameters p and a_{s-1}, i.e., $Bin(p, a_{s-1})$ is the sum of a_{s-1} Bernoulli random variables with mean p. To see this, recall that X_{s-1} is the number of vertices in V_W after step $s - 1$ of G-DOMC. In step s, each of the white vertices survives with probability p, and the events that white vertices survive are mutually independent.

By the Chernoff bound on the tail probability of Bernoulli variables, we have

$$\mathbb{P}\left\{X_s > a_s \mid X_{s-1} > a_{s-1}\right\}$$
$$= \mathbb{P}\left\{X_s > (1 - \epsilon)pa_{s-1} \mid X_{s-1} > a_{s-1}\right\}$$
$$\geq 1 - e^{-\frac{\epsilon^2 p^2}{2} a(s-1)}$$

Since $a_t = (1 - \epsilon)^t p^t n$ and by the choice of t, ϵ, we see that $e^{-\frac{\epsilon^2 p^2}{2} a_{s-1}} \leq e^{-n_1^{\delta}} \in O(\frac{1}{n})$. The Lemma follows.

We now bound the conditional probability that $Y_t^i > 0$ given that $X_t > n^{\delta_1}$.

Lemma 3. *Given that $X_t > n^{\delta_1}$ (i.e., the algorithm does not terminate due to the lack of vertices to expand the clique), we have*

$$\mathbb{P}\left\{Y_t^i > 0 \text{ for some } 0 \leq i \leq \alpha - 1\right\} \leq O(\frac{\alpha}{n^\delta}).$$

Proof. Recall that after the first α steps, there will be no vertex-exchange between V_W and the V_i's. Therefore, the probabilistic behavior of the system of vertex sets $\{V_i, 0 \leq i \leq \alpha - 1\}$ is independent of the specific choice of the vertex in V_W (given that $|V_W| = X_t > 0$ so that there is always a vertex to pick).

In step t, some vertices in V_i move to V_{i+1} because they are connected to the vertex just added to the clique. Due to the same reason, there are also vertices moving from V_{i-1} to V_i. Therefore, the expectation of $Y_i(t)$ is

$$\mathbb{E}\left[Y_t^0\right] = (1-p)\mathbb{E}\left[Y_{t-1}^0\right],$$
$$\mathbb{E}\left[Y_t^i\right] = (1-p)\mathbb{E}\left[Y_{t-1}^i\right] + p\mathbb{E}\left[Y_{t-1}^{i-1}\right].$$

Write $y_t^i = \mathbb{E}\left[Y_t^i\right]$. By induction, we have

$$y_t^i \le Ct^i(1-p)^t n$$

for some constant $C > 0$. Consequently by Markov's inequality, we have

$$\mathbb{P}\left\{Y_t^i > 0\right\} \le y_t^i \le \frac{1}{n^\delta}.$$

Therefore,

$$\mathbb{P}\left\{Y_t^i > 0 \text{ for some } 0 \le i \le \alpha - 1\right\} \le \frac{\alpha}{n^\delta}.$$

The lemma follows.

Combining Lemma 2 and Lemma 3, we see that Equation (6) holds. This proves Theorem 3.

6   Proof of Theorem 4

Let S be the set of assignments of weight k. For each $s \in S$, let I_s be the indicator function of the event that s satisfies $\mathcal{M}_{n,k}^{p_1,p_2,m}$. Consider the random variable $X = \sum_{s \in S} I_s$, the number of assignments in S that satisfy $\mathcal{M}_{n,k}^{p_1,p_2,m}$. Write $a = 1 - p_1$ and $b = 1 - (1-p_2)^k$, and recall that $m = c \log n$. We have

$$\mathbb{E}\left[X\right] = \binom{n}{k} a^{\binom{k}{2}} n^{c \log b}.$$

The case of $c > -\frac{k}{\log b}$ follows from Markov's inequality. For the case of $c < -\frac{1}{\log b}$, we consider the variance $V(X)$ of $X = \sum_{s \in S} I_s$. We say that two assignments in S have i overlaps if there are exactly i variables that are set to true by both of the two assignments. Let $S(i)$ be the set of (ordered) pairs of assignments in S that have i overlaps. $V(X)$ can be written as

$$V(X) = \mathbb{E}\left[(X - \mathbb{E}\left[X\right])^2\right]$$
$$\le \mathbb{E}\left[X\right] + \sum_{i=0}^{k} \sum_{(s_1,s_2)\in S(i)} (\mathbb{E}\left[I_{s_1} I_{s_2}\right] - \mathbb{E}\left[I_{s_1}\right]\mathbb{E}\left[I_{s_2}\right]). \qquad (7)$$

For any $(s_1, s_2) \in S(0)$, it is easy to see that

$$\mathbb{E}\left[I_{s_1} I_{s_2}\right] - \mathbb{E}\left[I_{s_1}\right]\mathbb{E}\left[I_{s_2}\right] = 0.$$

Consider $(s_1, s_2) \in S(i)$ with $i > 0$. We have

$$\mathbb{E}\left[I_{s_1} I_{s_2}\right] - \mathbb{E}\left[I_{s_1}\right]\mathbb{E}\left[I_{s_2}\right] = (1 - p_1)^{\binom{2k-i}{2}}(1 - 2(1 - p_2)^k + (1 - p_2)^{2k-i})^m$$
$$- (1 - p_1)^{2\binom{k}{2}}(1 - (1 - p_2)^k)^{2m}$$

For sufficiently large n, the above can be upper bounded by

$$C(1 - p_1)^{\binom{2k-i}{2}}\left(1 - 2(1 - p_2)^k + (1 - p_2)^{2k-i}\right)^m$$
$$\leq C(1 - p_1)^{\binom{2k-i}{2}}\left(1 - (1 - p_2)^k\right)^m.$$

where $C > 0$ is a fixed constant. Therefore, we have

$$\sum_{i=1}^{k} \sum_{(s_1,s_2)\in S(i)} \left(\mathbb{E}\left[I_{s_1} I_{s_2}\right] - \mathbb{E}\left[I_{s_1}\right]\mathbb{E}\left[I_{s_2}\right]\right)$$
$$\leq Ck\binom{n}{2k-1}(1 - p_1)^{\binom{2k-1}{2}}\left(1 - (1 - p_2)^k\right)^m.$$

It follows that

$$\frac{\mathbb{V}(X)}{(\mathbb{E}\left[X\right])^2} \leq \frac{\mathbb{E}\left[X\right] + Ck\binom{n}{2k-1}\delta(1 - p_1)^{\binom{2k-1}{2}}\left(1 - (1 - p_2)^k\right)^m}{n^{2k}(1 - p_1)^{2\binom{k}{2}}\left(1 - (1 - p_2)^k\right)^{2m}}$$
$$\in O(\frac{1}{n^{1+c\log b}}).$$

For the case $c < -\frac{1}{\log b}$, write $0 < \epsilon = -c\log b < 1$. We have by Chebyshev's inequality

$$\mathbb{P}\left\{|X - \mathbb{E}\left[X\right]| > n^{\frac{1-\epsilon}{4}}\frac{1}{n^{\frac{1-\epsilon}{2}}}\mathbb{E}\left[X\right]\right\} \leq \frac{1}{n^{(1-\epsilon)/2}}.$$

Recall that $\mathbb{E}\left[X\right] = \binom{n}{k}a^{\binom{k}{2}}(1 - (1 - p_2)^k)^m$. It follows that with probability $1 - O(\frac{1}{n^{\epsilon/2}})$, we have

$$X \geq a^{\binom{k}{2}}n^{k-\epsilon} - o(n^{k-\epsilon}). \tag{8}$$

This completes the proof.

7   Conclusions

In this paper, we have studied the behavior of random instances of two W[2]-complete problems. The threshold behavior of the solution probability under the proposed random models is studied. Lower and upper bounds on the complexity of satisfiable and unsatisfiable instances are established.

It is interesting to see if the dominating clique problem on random graphs with $\frac{3-\sqrt{5}}{2} < p < \frac{1}{2}$ can be solved in polynomial time with high probability. Establishing lower bounds on the proof complexity of more general parameterized resolution proof system is a challenging future task.

There is a gap between the lower and upper bounds on the threshold of the solution probability of the random weighted SAT model. Closing the gap is interesting. Identifying more scenarios that lead to fixed-parameter tractable class of instances is perhaps even more interesting.

References

1. Downey, R., Fellows, M.: Parameterized Complexity. Springer, Heidelberg (1999)
2. Neidermeier, R.: Invitation to Fixed-Parameter Algorithms. Oxford University Press, Oxford (2006)
3. Gottlob, G., Szeider, S.: Fixed-parameter algorithms for artificial intelligence, constraint satisfaction, and database problems. The Computer Journal (to appear)
4. Dilkina, B., Gomes, C., Sabharwal, A.: Tradeoffs in the complexity of backdoor detection. In: Bessière, C. (ed.) CP 2007. LNCS, vol. 4741, pp. 256–270. Springer, Heidelberg (2007)
5. Szeider, S.: Backdoor sets for DLL subsolvers. Journal of Automated Reasoning (1-3) 73–88 (2005)
6. Nishimura, N., Ragde, P., Szeider, S.: Detecting backdoor sets with respect to horn and binary clauses. In: H. Hoos, H., Mitchell, D.G. (eds.) SAT 2004. LNCS, vol. 3542, Springer, Heidelberg (2005)
7. Williams, R., Gomes, G., Selman, B.: On the connections between heanvy-tails, backdoors, and restarts in combinatorial search. In: Giunchiglia, E., Tacchella, A. (eds.) SAT 2003. LNCS, vol. 2919, pp. 222–230. Springer, Heidelberg (2004)
8. Dantchev, S., Martin, B., Szeider, S.: Parameterized proof complexity. In: Proceedings of FOCS 2007, pp. 150–160. IEEE Press, Los Alamitos (2007)
9. Achlioptas, D., Beame, P., Molloy, M.: A sharp threshold in proof complexity. In: Proceedings of STOC 2001, pp. 337–346 (2001)
10. Beame, P., Karp, R., Pitassi, T., Saks, M.: The efficiency of resolution and Davis-Putnam procedures. SIAM Journal on Computing 31(4), 1048–1075 (2002)
11. Cheeseman, P., Kanefsky, B., Taylor, W.: Where the really hard problems are. In: Proceedings of IJCAI 1991, pp. 331–337. Morgan Kaufmann, San Francisco (1991)
12. Cook, S., Mitchell, D.: Finding hard instances of the satisfiability problem: A survey. In: Du, Gu, Pardalos (eds.) Satisfiability Problem: Theory and Applications. DIMACS Series in Discrete Mathematics and Theoretical Computer Science, vol. 35, American Mathematical Society (1997)
13. Gent, I., Walsh, T.: Analysis of heuristics for number partitioning. Computational Intelligence 14(3), 430–451 (1998)
14. Gomes, C., Fernandez, C., Selman, B., Bessiere, C.: Statistical regimes across constrainedness regions. Constraints 10(4), 313–337 (2005)
15. Gao, Y., Culberson, J.: Consistency and random constraint satisfaction models. Journal of Artificial Intelligence Research 28, 517–557 (2007)
16. Gao, Y.: Random instances of parameterized complete problems: phase transitions and complexity, Tech. rep., Computer Science, University of British Columbia Okanagan (2007)
17. Culberson, J., Gao, Y., Anton, C.: Phase transitions of dominating clique problem and their implications to heuristics in satisfiability search. In: Proceedings of IJCAI 2005, pp. 78–83 (2005)
18. Davis, M., Logemann, G., Loveland, D.: A machine program for theorem proving. Communications of the ACM 5, 394–397 (1962)
19. Kratsch, D., Liedloff, M.: An exact algorithm for the minimum dominating clique problem. Theoretical Computer Science 385(1-3), 226–240 (2007)
20. Alon, N., Spencer, J.H.: The Probabilistic Method. Wiley, Chichester (2000)

Complexity and Algorithms for Well-Structured k-SAT Instances

Konstantinos Georgiou[1] and Periklis A. Papakonstantinou[1,2]

[1] University of Toronto, Dept. of Computer Science, Toronto, ON, M5S 3G4, Canada
[2] University of Toronto, Dept. of Mathematics, Toronto, ON, M5S 2E4, Canada
{cgeorg,papakons}@cs.toronto.edu

Abstract. This paper consists of two conceptually related but independent parts. In the first part we initiate the study of k-SAT instances of bounded diameter. The diameter of an ordered CNF formula is defined as the maximum difference between the index of the first and the last occurrence of a variable. We investigate the relation between the diameter of a formula and the tree-width and the path-width of its corresponding incidence graph. We show that under highly parallel and efficient transformations, diameter and path-width are equal up to a constant factor. Our main result is that the computational complexity of SAT, MAX-SAT, #SAT grows smoothly with the diameter (as a function of the number of variables). Our focus is in providing space efficient and highly parallel algorithms, while the running time of our algorithms matches previously known results. Our results refer to any diameter, whereas for the special case where the diameter is $O(\log n)$ we show NL-completeness of SAT and NC^2 algorithms for MAX-SAT and #SAT.

In the second part we deal directly with k-CNF formulas of bounded tree-width. We describe algorithms in an intuitive but not-so-standard model of computation. Then we apply constructive theorems from computational complexity to obtain deterministic time-efficient and simultaneously space-efficient algorithms for k-SAT as asked by Alekhnovich and Razborov [1].

1   Introduction

SAT, MAX-SAT and #SAT are among the most fundamental and well-studied problems in theoretical computer science, all intractable in the most general case: SAT is NP-complete [9], MAX-SAT is NP-hard to approximate within some constant [3], while #SAT is hard for #P [32]. The intractability of SAT, MAX-SAT and #SAT soon led to the study of restricted versions based on hidden structures of formulas and in particular on the so-called width restrictions. In this work, first we introduce a natural structural width parameter directly defined on k-CNF formulas that we call diameter. We consider SAT, MAX-SAT and #SAT and parameterize them with respect to diameter, giving space-efficient and parallel algorithms. Second, given the tree decomposition of the incidence graph of a formula, we show how to decide SAT in simultaneously efficient time and space.

H. Kleine Büning and X. Zhao (Eds.): SAT 2008, LNCS 4996, pp. 105–118, 2008.

Parameterizing SAT instances using width parameters follows the more general study of NP-hard graph problems initiated by Lipton and Tarjan [19]. Along these lines, Robertson and Seymour [24,25] introduced tree-width that has been widely used to parameterize the complexity of many NP-hard problems, see e.g. surveys [5,17]. When it comes to SAT, a CNF formula can be associated with many underlying graphs and for each one of them a number of width parameters can be defined e.g. tree-width, path-width, clique-width, branch-width and cluster-width (for a comparison see [20]). There are numerous works parameterizing SAT with respect to width parameters. In what follows, due to space limitations, our exposition is far from being complete.

Khanna and Motwani [18] considered MAX-SAT for formulas of constant tree-width, while [2] exploits the same structural property for SAT. Deciding SAT has been proved fixed-parameter tractable with respect to branch-width by Alekhnovich and Razborov [1], and to tree-width by Gottlob *et al* [16] on primal graphs. Using DPLL procedures, Bacchus, Dalmao and Pitassi, [4] considered #SAT, while the same time-bound for #SAT was achieved by Samer and Szeider [27] extending [16]. Fixed-parameter tractability of SAT and #SAT has also been considered in e.g. [10,13,20,21,28]; see also [31] for a survey.

The diameter of an ordered formula formalizes the following idea: if we know that the distance between the first and last occurrence of any variable is bounded, we may be able to understand better the complexity of such restricted SAT-instances. We extend the definition to unordered formulas to be the smallest diameter over all clause-orderings. Technically, the diameter of a formula ϕ fully coincides with the bandwidth (see [7] for a survey) of the intersection graph of ϕ. In this work, we consider ordered k-CNF instances of bounded diameter, and we do not deal with the independent and well-studied problem of finding the best ordering (equivalent to bandwidth minimization) which is NP-complete.

It is worth noting that the subproblem of k-SAT instances of diameter n^ϵ, $\epsilon > 0$, where n is the number of variables, is NP-complete. In contrast we show that k-CNF formulas of $\log n$ diameter encode arbitrary NL computations. Arbitrary NL-computations are objects exhibiting highly complex interactions between their parts. Hence, it is intuitively clear that by considering instances of bounded ordered diameter we do not break the problem into independent problems (a preliminary study for a similar problem was given in [14]). Even for unordered formulas the value of the diameter is provably less informative than the width parameters in the following sense. Path-width is always upper bounded by the diameter, although the two values can be off by almost a linear factor (Lemma 2). Despite this, we prove that by a highly efficient algorithm (Theorem 2), a formula of path-width $\mathsf{d}(n)$ can be viewed as a formula of diameter $O(\mathsf{d}(n))$. Hence we (computationally) counter any undesirable properties of the diameter and we only keep its simplicity. For ordered instances of SAT, MAX-SAT and #SAT of bounded diameter we design space-efficient and time-efficient algorithms, showing that the complexity of all three problems grows smoothly with respect to the diameter. If in particular the instances are of sufficiently small diameter, we present algorithms that in addition are highly parallel. A strong point of this work is that these algorithms appear to have quite intuitive descriptions. To the best of our knowledge this is the first work that simultaneously gives efficient time and space fixed parameter tractability bounds or even deals with parallelization issues for SAT, MAX-SAT and #SAT.

Additional motivation for the study of SAT with respect to simultaneously time and space tractability is explicitly given by Alekhnovich and Razborov [1]. Given instances of bounded branch-width $w(n)$ and given a decomposition, they decide SAT in time $n^{O(1)}2^{O(w(n))}$ and in space $n^{O(1)}2^{O(w(n))}$; they further ask whether it is possible to reduce the space to polynomial preserving time efficiency. The last part of our paper goes in a fashion independent to the study of diameter. A consequence of our study is a new algorithm that matches the same time-space bounds as in [1], and more importantly an algorithm that works in time $n^{O(1)}2^{O(w(n)\log n)}$ and space $n^{O(1)}$.

2 Definitions and Preliminary Results

2.1 Notation and Terminology

All logarithms are of base 2. All propositional formulas are in CNF. A k-CNF is a CNF where each clause has at most k literals, for a constant $k \in \mathbb{N}$. We denote by ϕ_π a total ordering of the clauses of ϕ. In an input, an unordered (ordered) formula ϕ (ϕ_π) is represented in the standard way as a sequence (sequence in the given order) of clauses. We consistently use n to denote the number of variables in a formula. N is used to denote the size of given inputs. The diameter of an ordered formula is always expressed as a function of the number of variables, and it is denoted by $\mathsf{d}(n)$. All circuit families are logspace or logtime uniform. $\mathrm{DEPTH}(f(N))$ is the class of languages decidable by a family of circuits in depth $f(N)$. $\mathrm{DSPACE}(f(N))$, $\mathrm{NSPACE}(f(N))$, $\mathrm{DTIME}(f(N))$ denotes the class of problems decidable in deterministic, non-deterministic space and deterministic time $f(N)$ respectively. For the function analogs of decision complexity classes we extend the notation introducing a leading F; e.g. $\mathrm{FDSPACE}(\log^2 N)$. NC^i (AC^i) is the class of languages decidable by polynomial size circuits of depth $O(\log^i N)$ where the gates are of bounded (unbounded) fan-in. We denote by $\mathrm{NL} = \mathrm{NSPACE}(\log N)$. Our notation is standard, see e.g. [11,34]. LOGCFL is the class of languages logspace reducible to Context Free Languages (see Section 2.5). When the input is a formula of n variables we abuse notation by writing $\mathrm{COMPCLASS}(f(n))$ instead of $\mathrm{COMPCLASS}(f(N))$. Since $N > n$ our containment results are slightly better than what our notation suggests. We use the term "highly parallel algorithms" to refer to circuits that are both of polynomial size and of small depth e.g. logarithmic or a square of a logarithm.

2.2 Structural Parameters of Graphs

Definition 1. *Let $G = (V, E)$ be an undirected graph. A* tree decomposition *of G is a tuple (T, X), where $T = (W, F)$ is a tree, and $X = \{X_1, \ldots, X_{|W|}\}$ with $X_i \subseteq V$ such that: (1) $\bigcup_{s=1}^{|T|} X_s = V$; (2) For all $\{i, j\} \in E$, there exist*

$t \in W$, such that both $i, j \in X_t$; (3) For all $i \in V$, the subset $\{t : i \in X_t\}$ of W forms a subtree of T. The quantity $\max_{t \in W} |X_t| - 1$ is called the width of (T, X). The tree-width of G, denoted by $TW(G)$, is the minimum width over all tree decompositions of G. The path decomposition is defined similarly; T has to be a path and the term path-width is used instead of tree-width.

Determining the optimal tree (path) decomposition is NP-hard while the problem is approximable within factor $O(\log n)$ $(O(\log^2 n))$ [6]. Tree-width is closed under the operation of graph minors and wlog we may assume that the number of nodes of the tree decomposition (T, X) of a graph G is linear, and that up to logspace transformations the degree of T is at most 3. For a survey on tree-width we cite [5].

The diameter of a formula is related to the bandwidth of graphs.

Definition 2. For a graph $G = (V, E)$, let $f : V \to \{1, 2, \ldots, |V|\}$ be an injective map. The bandwidth of G, $\mathcal{B}(G)$ is defined as $\min_f \max_{ij \in E} |f(i) - f(j)|$. In the minimum bandwidth problem we compute f witnessing $\mathcal{B}(G)$.

The bandwidth problem is NP-complete [22] and remains intractable even if the input graph is a tree of maximum degree 3 [15]. The problem is polylogarithmic approximable due to Feige [12]. See [7] for a not-so-recent survey.

2.3   Structural Parameters of Formulas

Definition 3. Let V be the set of variables of an ordered formula ϕ_π. For $x \in V$, let $f(x), l(x)$ be the index of the clause that x appears for the first and last time respectively. The ordered diameter is $\mathcal{D}(\phi_\pi) = \max_{x \in V}(l(x) - f(x))$ and the unordered diameter is $\Delta(\psi) = \min_\pi \mathcal{D}(\psi_\pi)$.

In this work we associate a k-CNF formula ϕ with two graphs. The *incidence graph* G_ϕ of ϕ is a bipartite graph. G_ϕ has a distinct vertex for each clause and each variable. A variable-vertex u_x is connected to clause-vertex u_c whenever the variable x appears in the clause c. The *clause-graph* C_ϕ of ϕ (intersection graph) arises by associating each clause with a distinct vertex. An edge connects vertices whose clauses share a variable. In [31] it is shown that the tree-width of the incidence graph is always smaller than the corresponding width parameters on other graphs appearing in the literature.

For a formula ϕ , we further define tree-width $TW(\phi)$, path-width $PW(\phi)$ and bandwidth $\mathcal{B}(\phi)$ of ϕ to be

$$TW(\phi) = TW(G_\phi), \ PW(\phi) = PW(G_\phi), \ \mathcal{B}(\phi) = \mathcal{B}(C_\phi)$$

2.4   Relations between $TW(\phi), PW(\phi), \mathcal{B}(\phi)$ and $\Delta(\phi)$

Lemma 1. For any ordered k-CNF formula ϕ_π, the following are true:
(i) $\mathcal{B}(\phi) = \Delta(\phi)$, (ii) $PW(\phi) \leq \log n \cdot TW(\phi)$, (iii) $PW(\phi) = O(\mathcal{D}(\phi_\pi))$.

Proof. (i) Follows directly from the definitions 2 and 3.

(ii) For every graph G on n vertices, $\mathcal{PW}(G) \leq \log n \cdot \mathcal{TW}(G)$.

(iii) Consider some k-CNF ordered formula ϕ_π on n variables with $\mathcal{D}(\phi_\pi) = \mathsf{d}(n)$ and set $r = \lceil m/(\mathsf{d}(n)+1)\rceil$. We decompose G_ϕ to a path of width $(k+1) \cdot \mathsf{d}(n)$. Define the path $P = v_1, v_2, \ldots, v_r$. For every i, X_i, that v_i is associated with, consists of the following two types of vertices: clause-vertices v_{c_i} corresponding to clauses c_i, for $i = (i-1) \cdot (\mathsf{d}(n)+1)+1$ to $i \cdot (\mathsf{d}(n)+1)$; variable-vertices v_x, for all variables x that are involved in clauses with vertices already in X_i. We claim that P is valid path decomposition of G_ϕ. Indeed, properties (1),(2) of definition 1 are trivially satisfied. As for the third one, consider any variable x and the associated vertex u_x of G_ϕ. By construction we only have to consider variable-vertices.

Now suppose (for the shake of contradiction) that there exist indices $i < s < j$, such that u_x is in both X_i, X_j and $u_x \notin X_s$. Then, in ϕ_π, x does not appear in any of the $\mathsf{d}(t)+1$ clauses in X_s, and therefore $\mathcal{D}(\phi_\pi) > (j-i-1) \cdot (\mathsf{d}(n)+1)$. Finally, since ϕ is k-CNF formula, for every i, $|X_i| \leq \mathsf{d}(n) + k \cdot \mathsf{d}(n)$. $\qquad\square$

Lemma 1 does not preclude the possibility that $\Delta(\phi), \mathcal{PW}(\phi)$ are related up to (say) some constant factor. Combinatorially, things are the worst possible regarding the diameter. We show that even when each variable appears a small constant number of times the gap between tree-width (path-width) and diameter is off by almost linear factor. For this we use theorem 1, p.204 from [29].

Theorem 1 (Smithline '95). *For the complete k-ary tree of height h, $\mathcal{B}(T) = \lceil k(k^h - 1)/(k-1)(2h)\rceil$*

Lemma 2. *There exists a family formulas ϕ with n variables each one appearing only 3 times, for which $\Delta(\phi) = \Omega(n/\log n)$, $\mathcal{PW}(\phi) = O(\log n)$ and $\mathcal{TW}(\phi) = 1$.*

Proof. We determine a 3-CNF formula ϕ with positive literals, by defining its incidence graph G_ϕ. We start with the rooted complete binary tree T of height $\log n'$, where $\log n'$ is even (the root has level 0). Label all nodes of T in arbitrary breadth-first-search manner starting from the root. At an even level, associate vertex i with a new variable x_i; at an odd level, associate vertex j with a new clause c_j. Define clause c_j to be the conjunction of the parental-node $x_{\lfloor j/2\rfloor}$ and the two children-nodes x_{2j}, x_{2j+1}. Set ϕ to be the conjunction of all clauses, and n the number of variable-vertices in G_ϕ. Observe that $T = G_\phi$ and $n = \Theta(n')$.

By definition $\mathcal{TW}(T) = 1$, and by Lemma 1, $\mathcal{PW}(T) \leq \log n'$. Next we argue about the bandwidth of C_ϕ. It is easy to see that if we remove edges from C_ϕ that connect clauses that appeared in T at the same level (i.e., edges that connect clause-vertices sharing in T a common ancestor), the resulting graph consists of two disconnected complete trees. Every vertex has 4 children, and height at least $\lfloor \frac{\log n' - 1}{2}\rfloor$. Theorem 1 then implies that $\mathcal{B}(C_\phi) = \Omega(n'/\log n')$. $\qquad\square$

Despite Lemma 2, we capitalize on the fact that the notions of diameter and path-width are the same up to some constant and up to a logspace transformation. It is also essential for Corollary 1 (see below) that Theorem 2 is constructive.

Theorem 2. *For any k-CNF formula ϕ, there exists an ordered k-CNF formula ϕ_π' with $\Delta(\phi') \leq \mathcal{D}(\phi_\pi') = \Theta(\mathcal{PW}(\phi))$ such that $\phi \in$ SAT iff $\phi_\pi' \in$ SAT. Moreover, given the path decomposition of ϕ, ϕ_π' can be computed in logarithmic space with respect to the size of ϕ.*

Proof. Consider the path decomposition $X_1, \ldots, X_t$ of C_ϕ with $|X_i| = \mathsf{d}(n)$. We identify the vertices in the block X_i by the corresponding clauses and variables. We construct ϕ_π' as the output of the following iterative procedure.

For every block X_i do the following: (copy-step) output all the clauses of X_i in some order; (intercalate-step) for every variable x in X_i or in the clauses of X_i, output the renaming of x, $x \leftrightarrow x'$; finally replace all appearances of x in $X_{i+1}, \ldots X_t$ by x'. We call every clause introduced in the intercalate-step intercalary. ϕ_π' is the conjunction of the clauses ordered as the output suggests. By construction ϕ is satisfiable iff ϕ_π' is satisfiable.

It is clear that the previous procedure can be implemented in logarithmic space: instead of renaming all subsequent occurrences of x, just count its previous occurrences. In a reasonable renaming, the indices of the variables do not exceed $n + n + 2k \cdot t \cdot \mathsf{d}(n)$.

Now, we calculate the ordered diameter of ϕ_π'. We distinguish between variables introduced in the copy-step and the intercalate-step. By the renamings, it is immediate that for any variable x of a clause introduced at the copy-step, the maximum distance between occurrences of x is at most $(2k + 1) \cdot \mathsf{d}(n)$.

For variables introduced in the intercalate-step we rely on the definition of path-width. Consider such a variable x introduced between blocks X_i, X_{i+1}. Variable x is (i) either a renaming of a former variable, or (ii) it is brand new variable that replaces y. Case (i) is easy to handle. For case (ii), the clause c of X where y appeared, either appears in X_{i+1} or not. If it does not appear, then by the definition of path-width, c does not appear in any subsequent block. Finally, if c appears in X_{i+1} then it will be renamed again when we consider the next block. In every case $\mathcal{D}(\phi_\pi') \leq (2k + 1) \cdot \mathsf{d}(n)$. □

Motivated by the previous observations, and for k-CNF formulas, we define

Definition 4 (Computational Problems). SAT($\mathsf{d}(n)$), MAX-SAT($\mathsf{d}(n)$) *and* #SAT($\mathsf{d}(n)$) *are the restrictions of* SAT, MAX-SAT *and* #SAT *respectively, where the instances ϕ_π are ordered formulas and obey $\mathcal{D}(\phi_\pi) \leq \mathsf{d}(n)$.*

2.5 NAuxPDAs: A Practical Model of Computation

A non-deterministic auxiliary pushdown automaton (NAuxPDA) is a generalization of a space-bounded Turing Machine (TM) extended by an unbounded stack. Cook [8] showed that every NAuxPDA bounded to work in space $s(n)$ and arbitrary time can be simulated by a TM in time $2^{O(s(n))}$. Sudborough [30] showed that LOGCFL ($\subseteq AC^1 \subseteq NC^2$) is characterized by NAuxPDAs that run simultaneously in logarithmic space and polynomial time. Using NAuxPDAs one can simulate a special form of non-deterministic recursion and from there even a special form of divide and conquer. Non-deterministic Divide and

Conquer (ND-DnC) [23] is a paradigm which simplifies the presentation of algorithms, something that recently made possible to obtain complex polynomial time algorithms whose translations into TMs are extremely complicated and unnatural. The transformation of an NAuxPDA to a TM or to parallel algorithms (e.g. circuits or PRAMs) is possible and explicit through strongly non-trivial translation theorems, see Section 4, although the resulting TM can be conceptually complicated. Among others, application of these theorems shows that ND-DnC algorithms that have simple and elegant descriptions can find practical applications through their transformations. An example of such an application is demonstrated in Section 4.

3   Solving $\mathrm{SAT}(\mathbf{d}(n))$, $\mathrm{Max\text{-}SAT}(\mathbf{d}(n))$, $\#\mathrm{SAT}(\mathbf{d}(n))$

3.1   Algorithms for $\mathbf{d}(n) = \varOmega(\log n)$

This section is devoted to $\mathsf{d}(n) = \varOmega(\log n)$. We show that SAT can be decided within non-deterministic space $O(\mathsf{d}(n))$, whereas for MAX-SAT and #SAT it suffices to use deterministic space $O(\mathsf{d}(n)^2)$. Moreover, all three problems can be solved in (deterministic) time $2^{O(\mathsf{d}(n))}$. The time-bounded and space-bounded algorithms for MAX-SAT and #SAT are obtained independently. Under the current knowledge in computational complexity we do not know how FDSPACE($\mathsf{d}^2(n)$) compares to FDTIME($2^{O(\mathsf{d}(n))}$).

Theorem 3. SAT($\mathsf{d}(n)$) $\in$ NSPACE($\mathsf{d}(n)$), MAX-SAT($\mathsf{d}(n)$), #SAT($\mathsf{d}(n)$) $\in$ FDSPACE($\mathsf{d}(n)^2$); MAX-SAT($\mathsf{d}(n)$),#SAT($\mathsf{d}(n)$) $\in$ FDTIME($2^{O(\mathsf{d}(n))}$).

The satisfiability problem $\mathbf{SAT}(\mathbf{d}(n))$
Solve-SAT (Algorithm 1) shows that SAT($\mathsf{d}(n)$) $\in$ NSPACE($\mathsf{d}(n)$). We can standardize the way the truth assignment is stored. Reserve one bit for the

---

Algorithm 1. Solve-SAT

The input is an ordered k-CNF formula ϕ_π which $\mathcal{D}(\phi_\pi) = \mathsf{d}(n)$.

- Initially, consider a window (ordered subformula) W of length $\mathsf{d}(n)$ containing the first $\mathsf{d}(n)$ clauses of ϕ_π. Guess values for all variables in W and if the guess does not satisfy W then reject.

- Iteratively do the following.

- Slide the current position of the window W one clause to the right and free the space of the variables of the first clause of W.

- Guess (and store in the freed space) truth values for the variables of the new clause in the updated W. If the updated W is not satisfied or if the new values are inconsistent with those stored in the memory then reject. Otherwise, if there are more clauses in ϕ_π to the right of W then iterate; else accept.

---

variable of each occurrence of a literal in W repeating the value for variables which appear more than once; i.e. in total we have $k \cdot d(n)$ space.

For the correctness it is easy to see that there is a computational branch which accepts iff there exists a satisfying truth assignment for ϕ_π. Details omitted from proofs are given in the full version of the paper.

The maximization problem Max-SAT($d(n)$)

We define DAG-LONGEST-PATH to be the optimization problem where given a DAG (Directed Acyclic Graphs) $G = (V, E)$ and $w : E \to \mathbb{N}$, the goal is to output the (edge-weighted) length of a longest dipath. We reduce MAX-SAT($d(n)$) in deterministic space $O(d(n))$ to DAG-LONGEST-PATH. This is a significant improvement over the natural dynamic programming time-bounded algorithm.

Lemma 3. DAG-LONGEST-PATH $\in$ FDEPTH($\log^2 N$). *Furthermore, this family of circuits has size polynomial in N. In particular, the problem is in* P.

Here is a brief justification. Power the adjacency matrix using repeated squaring, over the semiring $\mathbb{N}$ with operations $(\max, +)$ instead of $(+, \cdot)$. This way we compute all walks of length N in depth $O(\log^2 N)$.

Solve-MaxSAT (Algorithm 2) makes use of a space-efficient routine. This is the space simulation of the above longest path algorithm. It is well-known (see e.g. [34]) that DEPTH($s(N)$) $\subseteq$ DSPACE($s(N)$), $s(N) \geq \log_2 N$. That is, DAG-LONGEST-PATH $\in$ FDSPACE($\log^2 n$), and furthermore the proof of the inclusion gives us an explicit space-efficient algorithm.

Algorithm 2. Solve-MaxSAT

The input is an ordered k-CNF formula ϕ_π with $\mathcal{D}(\phi_\pi) = d(n)$. First we show how to reduce to DAG-LONGEST-PATH working in space $d(n)$ and then we compose in the standard way two space efficient algorithms.

- The graph consists of blocks of vertices. Each block is associated with a window (ordered subformula) W of length $d(n)+1$, where W starts from a distinct position (clause) in the ordered ϕ_π. The i-th block is associated with the window which starts from the i-th clause of ϕ_π. Each of the vertices of each block is associated with a distinct, satisfying truth assignment for this window. We also introduce a fresh starting vertex s and assume it is associated with an empty subformula.

- There is an edge from a vertex v in block i to every other vertex u in block $j > i$ whenever v, u are consistent. The weight of the edge (v, u) is the number of clauses in the window associated with u, satisfied by u and not (already) by v. Let us call the constructed graph as H_{ϕ_π}.

- Solve DAG-LONGEST-PATH for H_{ϕ_π}.

The reduction works in space $O(d(n))$ since we can enumerate all pairs of vertices in H_{ϕ_π} in that space. Hence, Solve-MaxSAT requires deterministic space

$O\big(\log^2(n^{O(1)}2^{O(\mathsf{d}(n))}) + \mathsf{d}(n)\big) = O(\mathsf{d}^2(n))$. The time-bounded algorithm is obtained if instead we do matrix powering using repeated squaring (or dynamic programming) to solve DAG-LONGEST-PATH.

Correctness is transparent. Let us denote by $R(u,v)$ the relation that u is consistent with v, and u is in a smaller-indexed block than v. Then, we observe that R is transitive and moreover R is represented by the edges in H_{ϕ_π}. R can be used to prove consistency of truth assignments. Any longest path contains s. We finish by an easy induction on the index of blocks for paths starting from s.

The counting problem #SAT($\mathsf{d}(n)$)

The algorithm for #SAT($\mathsf{d}(n)$) proceeds by a logspace reduction (Reduce-#SAT) (Algorithm 3) to the problem of counting paths in a DAG.

Algorithm 3. Reduce-#SAT

The input is an ordered k-CNF formula ϕ_π with $\mathcal{D}(\phi_\pi) = \mathsf{d}(n)$.

- We construct a layered directed graph. Each layer (block) is associated with a distinct position of a window (ordered subformula) W of length $\mathsf{d}(n)$; the i-th layer is associated with the window which starts from the i-th clause of ϕ_π. Each of the vertices of each layer is associated with a distinct, satisfying truth assignment for this window. We denote by L_i the subset of vertices of the i-th layer.

- There is an edge from a vertex v in layer i to every other vertex u in layer $i+1$ whenever the partial truth assignments of the two vertices are consistent.

- Add two fresh designated vertices s, t. Add an edge from s to every vertex in L_1. Let L_h be the last layer. Add an edge from each vertex $v \in L_h$ to t. Let us denote by F_{ϕ_π} the constructed graph.

Lemma 4. *The number of s-t dipaths in F_{ϕ_π} equals the number of satisfying truth assignments of ϕ_π.*

Proof. We define a mapping from the set of truth assignments of ϕ_π to the set of s-t paths in F_{ϕ_π}. Let τ be a satisfying truth assignment for ϕ_π. By definition τ satisfies all windows. For each of the corresponding partial truth assignment there exists a vertex in the corresponding layer. Since all of them extend to the same τ they are in particular consistent and thus by construction there is a directed path in F_{ϕ_π} from a vertex in the first to a vertex in the last layer.

It is not hard to see why this mapping is a function (e.g. by considering the first time that two paths split) and why it is injective. Similarly, we define an inverse injective function. □

From this point on there are two ways to count the number of s-t paths. One is to reduce to an arithmetic circuit by mapping vertices in F_{ϕ_π} to $+$ gates and then apply the results in [33]. The other way is to deal with the problem directly. The later is even cleaner. The number of layers including s and t is $2 + h$, where $h = m - \mathsf{d}(n)$. We conclude the proof of the following by repeated squaring in the semiring $\mathbb{N}$ with operations $+, \cdot$.

Theorem 4. *Let $A \in \mathbb{N}^{|V(F_{\phi_\pi})| \times |V(F_{\phi_\pi})|}$ be the adjacency matrix of F_{ϕ_π}. The number of s-t paths in F_{ϕ_π} equals the single non-zero entry of A^{1+h}. Moreover this can be computed by a polysize circuit of depth $O(\log^2 N)$.*

3.2   Strong, Constructive Extensions of the Equivalence of Theorem 2

The equivalence of Theorem 2 extends to MAX-SAT and #SAT. The details are given in the full version of this paper. For #SAT we observe that in the reduction of Theorem 2, ϕ and ϕ' have the same number of satisfying assignments. For MAX-SAT the connection is less straightforward. We modify Theorem 2 and the graph H_{ϕ_π} in Solve-MaxSAT. In the proof of Theorem 2 we omit occurrences of a clause in multiple blocks X_i's. Furthermore, it is possible to mark on ϕ' the beginning and the end of each copy-step using "dummy" clauses. Given the transformed bounded diameter formula we construct $H_{\phi_\pi'}$ by defining windows according to the previously introduced dummy clauses. Also, we omit all windows of intercalary clauses but we use their induced relations to connect the vertices.

3.3   Diameter $O(\log n)$: Parallel Algorithms and Low Complexity Classes

When $\mathsf{d}(n) = O(\log n)$ the corresponding problems are deeply buried inside P. The proof of Lemma 5 follows the lines of the standard Cook-Levin reduction modified with systematic rewritings to avoid diameter blow-up.

Lemma 5. SAT$(\log n)$ *is* NL*-complete under many-to-one logspace reductions.*

As a corollary of Theorem 3 and its proof (in particular Lemma 3 and Theorem 4) we obtain,

Lemma 6. MAX-SAT$(\log n)$, #SAT$(\log n)$ *are in the function analog of* NC^2.

Let us consider SAT, MAX-SAT and #SAT for formulas of path-width $O(\log n)$. Results of this section and of Section 3.2 derive the following corollary.

Corollary 1 (Bounded path-width). *Consider k-CNF instances of path-width $O(\log n)$ where the path decomposition is given. For these instances* SAT *is complete for* NL, *and* MAX-SAT, #SAT *are in the function analog of* NC^2.

4   Improved Results for k-CNFs of Bounded Tree-Width

Since tree-width is at worst $\log n$ smaller than path-width, the statements of Section 3 hold for tree-width when the value of the parameter is off by $\log n$ factor. Here we improve on this corollary when it comes to SAT. To that end our treatment in this section is independent to the results obtained for the diameter. We obtain an AC^1 algorithm for $\log n$ tree-width. Furthermore, by applying strongly non-trivial results from complexity theory, we provide simultaneous space and time efficiency as asked in [1] (even for the weaker notion of the tree-width of the primal graph).

4.1   Dealing Directly with Tree-Width for SAT

Given a tree decomposition of formula of tree-width $t(n)$ we design an algorithm that in particular when $t(n) = O(\log n)$ shows SAT $\in$ LOGCFL. For notational succinctness, in this section only, n corresponds to the total number of variables and clauses in a formula.

Algorithm 4. Solve-Treewidth-SAT

The input is a k-CNF formula ϕ and a tree decomposition (T, X) of width $t(n)$ and of degree at most 3 (see Section 2.2). Initially we make a call to Recurse-Treewidth-SAT$[r]$, where r is an arbitrary root of T. If the call returns then accept.

Recurse-Treewidth-SAT[root node v]

- Guess a truth assignment τ for the clauses and the variables corresponding to v. If τ does not satisfy the clauses associated with v then reject.

- If v is a leaf then return τ. Else, let u, w be the children of v

- Set $\tau_u =$ Recurse-Treewidth-SAT$[u]$ and $\tau_w =$ Recurse-Treewidth-SAT$[w]$.

- If τ is not consistent with τ_u and τ_w then reject. Else, return τ.

Solve-Treewidth-SAT can be implemented on an NAuxPDA using space $t(n)$ and time $n^{O(1)}$ (wlog the number of nodes in the decomposition is linear to the number of nodes in the graph). When the tree-width is $t(n)$ then there are at most $t(n)$ clauses and variables whose truth values are checked at each level of the recursion. Moreover, the algorithm visits each node twice.

The proof of completeness is easy and does not even rely on tree decomposition properties. For the soundness we use the tree decomposition properties and a little preparation is necessary.

Lemma 7. *Let ϕ be a k-CNF and (T, X) a tree decomposition of G_ϕ. Construct (T, X') by extending the association of each node u to be associated with all nodes corresponding to variables that appear in the clauses associated with u. Then, X' witnesses a tree-width constant times bigger than X.*

Proof. It is obvious that each set in X' is at most k times bigger than the corresponding set in X. (T, X') is a tree decomposition: Axioms (1) and (2) are easily satisfied; hence we check whether axiom (3) is satisfied too. For clause-vertices everything is as in X. For a variable-vertex y let the subtree $T_y = \{t \in T : y \in X_t\}$ and the set $T'_y = \{t \in T : y \in X'_t\}$. Let $v \in T'_y$ such that $v \notin T_y$, where $y \in C$ for a clause C. By property (2) of the definition there exists a node $u \in T_y$ which is associated with C. Moreover, there exists a path $P_{u,v}$ connecting v and u s.t. C is associated with every vertex in $P_{u,v}$. By construction of X' the vertex associated with y is also associated with every vertex in $P_{v,u}$. That is, in X' the subtree T_y is extended to include v. □

We continue with the soundness direction. Fix an input ϕ where the algorithm accepts. Fix an arbitrary accepting computational branch. We define the binary relation Q to be the (variable, truth value) pairs that the algorithm assigned to variables in this computational branch. We need to show that Q is a function and that it is a satisfying truth assignment.

Consider any two nodes u, v of the tree decomposition where at v we have $(x, True) \in Q$ and at u we have $(x, False) \in Q$. By Proposition 7 there exists $\{i, j\} \in T$ in the u-v path, such that $x \in X_i$ and $x \in X_j$ which contradicts the consistency check of the algorithm. The proof of correctness finishes by defining and applying transitive relation R referring to consistent extensions of partial truth assignments.

When $t(n) = O(\log n)$ algorithm Solve-Treewidth-SAT runs in logspace and polytime which establishes the following strong theorem.

Theorem 5. k-SAT *with tree decompositions of width* $O(\log n)$ *is in* LOGCFL.

4.2 Alekhnovich and Razborov's Question

Given a tree decomposition of width $t(n)$, the refutation algorithm of [1] runs in time and in space $O(n^{O(1)} 2^{O(t(n))})$. By applying on Solve-Treewidth-SAT the deterministic time simulation of [8] (Theorem 1, p.7) we obtain an algorithm that runs in time $2^{O(t(n))}$ and space $2^{O(t(n))}$, $t(n) = \Omega(\log n)$, which matches the time-space bounds in [1] (note that when $t(n) = O(\log n)$ we have the very strong result of Theorem 5). In fact, when $t(n) = \omega(\log n)$ we improve on [1] as well. To that end we successively apply non-trivial results from [26] and simple well-known results from structural complexity. It is worth noting that each theorem we apply is constructive and thus we successively transform Solve-Treewidth-SAT. The following theorem is a corollary of three successive transformations in [26] Theorem 3, p.375 and Theorem 5(2),5(3) p.379.

Theorem 6 (Ruzzo '81). *NAuxPDAs working in space* $s(n)$ *and time* $z(n)$ *can be simulated by a family of circuits of size* $2^{O(s(n))}$ *and depth* $O(s(n) \log z(n))$. *Furthermore, this transformation between algorithms is given explicitly.*

Theorem 6 gives a family of circuits of size $2^{O(t(n))}$ and depth $O(t(n) \log n)$ deciding SAT instances of tree-width $t(n)$. Apart from these parallel algorithms we have the following as an immediate consequence of the depth bound.

Theorem 7. SAT *instances consisting of a* k-CNF *formulas together with tree decompositions of width* $t(n)$ *can be decided in space* $O(t(n) \log n)$ *and thus simultaneously in time* $2^{O(t(n) \log n)}$. *Furthermore, if the decomposition is not given we decide in time* $2^{O(t(n) \log n)}$ *and space* $n^{O(1)}$.

5   Open Questions

Our work raises many questions which are left open. We consider as most fundamental the following four. (1) Study interrelations of SAT, MAX-SAT and

#SAT for different bounds of the diameter; e.g. can we reduce #SAT($d(n)$) to SAT($d^2(n)$)? (2) Investigate structural complexity implications by assuming SAT instances of bounded diameter to be either in P or NP-complete. (3) Improve the result of Section 4.2 by reducing the exponent in the running time. (4) Finally, we are optimistic that our research will find empirical applications.

Acknowledgments

We would like to thank Paul Medvedev for bringing to our attention the equivalence between the CNF-diameter and the bandwidth of the intersection graph, and Mohammad Moharrami for explaining tree-width-related concepts. We also thank Steve Cook for discussions on combinatorial circuits, and Phuong Nguyen and Matei David for useful suggestions on the presentation of this work.

References

1. Alekhnovich, A., Razborov, A.: Satisfiability, branch-width and Tseitin tautologies. In: FOCS, pp. 593–603 (2002)
2. Amir, E., Mcilraith, S.: Solving satisfiability using decomposition and the most constrained subproblem. In: LICS workshop on Theory and Applications of Satisfiability Testing. Electronic Notes in Discrete Mathematics (2001)
3. Arora, S., Safra, S.: Probabilistic checking of proofs: A new characterization of NP. J. ACM 45, 70–122 (1998)
4. Bacchus, F., Dalmao, S., Pitassi, T.: Algorithms and complexity results for #SAT and bayesian inference. In: FOCS, pp. 340–351 (2003)
5. Bodlaender, H.L.: A tourist guide through treewidth. Acta Cybernet 11(1-2), 1–21 (1993)
6. Bodlaender, H.L., Gilbert, J.R., Hafsteinsson, H., Kloks, T.: Approximating treewidth, pathwidth, frontsize, and shortest elimination tree. J. Algorithms 18(2), 238–255 (1995)
7. Chinn, P.Z., Chvátalová, J., Dewdney, A.K., Gibbs, N.E.: The bandwidth problem for graphs and matrices - A survey. J. Graph Theory 6(3), 223–254 (1982)
8. Cook, S.A.: Characterizations of pushdown machines in terms of time-bounded computers. J. ACM 18(1), 4–18 (1971)
9. Cook, S.A.: The complexity of theorem-proving procedures. In: STOC, pp. 151–158 (1971)
10. Courcelle, B., Makowsky, J.A., Rotics, U.: On the fixed parameter complexity of graph enumeration problems definable in monadic second-order logic. Discrete Appl. Math. 108(1-2), 23–52 (2001)
11. Du, D.-Z., Ko, K.-I.: Theory of Computational Complexity. Wiley-Interscience, New York (2000)
12. Feige, U.: Approximating the bandwidth via volume respecting embeddings. J. Comput. Syst. Sci 60(3), 510–539 (2000)
13. Fischer, E., Makowsky, J.A., Ravve, E.R.: Counting truth assignments of formulas of bounded tree-width or clique-width. Discrete Appl. Math. 155(14), 1885–1893 (2007)

14. Flouris, M., Lau, L.C., Morioka, T., Papakonstantinou, P.A., Penn, G.: Bounded and ordered satisfiability: connecting recognition with Lambek-style calculi to classical satisfiability testing. In: Math. of language 8, pp. 45–56 (2003)

15. Garey, M.R., Graham, R.L., Johnson, D.S., Knuth, D.E.: Complexity results for bandwidth minimization. SIAM J. Appl. Math. 34(3), 477–495 (1978)

16. Gottlob, G., Scarcello, F., Sideri, M.: Fixed-parameter complexity in AI and non-monotonic reasoning. Artificial Intelligence 138(1–2), 55–86 (2002)

17. Hlineny, P., Oum, S., Seese, D., Gottlob, G.: Width parameters beyond tree-width and their applications. The Computer Journal 8, 216–235 (2007)

18. Khanna, S., Motwani, R.: Towards a syntactic characterization of PTAS. In: STOC, pp. 329–337. ACM, New York (1996)

19. Lipton, R.J., Tarjan, R.E.: Applications of a planar separator theorem. SIAM J. of Comp. 9(3), 615–627 (1980)

20. Nishimura, N., Ragde, P., Szeider, S.: Solving #SAT using vertex covers. In: Biere, A., Gomes, C.P. (eds.) SAT 2006. LNCS, vol. 4121, pp. 396–409. Springer, Heidelberg (2006)

21. Oum, S., Seymour, P.: Approximating clique-width and branch-width. J. Combin. Theory Ser. B 96(4), 514–528 (2006)

22. Papadimitriou, C.H.: The NP-completeness of the bandwidth minimization problem. Computing 16(3), 263–270 (1976)

23. Papakonstantinou, P.A., Penn, G., Vahlis, Y.: Polynomial-time and parallel algorithms for fragments of Lambek Grammars (unpublished manuscript)

24. Robertson, N., Seymour, P.D.: Graph minors I. Excluding a forest. J. of Comb. Theory (Ser. B) 35, 39–61 (1983)

25. Robertson, N., Seymour, P.D.: Graph minors. II. algorithmic aspects of tree-width. J. of Algorithms 7, 309–322 (1986)

26. Ruzzo, W.L.: On uniform circuit complexity. J. Comput. System Sci. 22(3), 365–383 (1981) Special issue dedicated to Michael Machtey

27. Samer, M., Szeider, S.: A fixed-parameter algorithm for #SAT with parameter incidence treewidth. CoRR, abs/cs/061017 (2006) informal publication

28. Samer, M., Szeider, S.: Algorithms for propositional model counting. In: Dershowitz, N., Voronkov, A. (eds.) LPAR 2007. LNCS (LNAI), vol. 4790, pp. 484–498. Springer, Heidelberg (2007)

29. Smithline, L.: Bandwidth of the complete k-ary tree. Discrete Math. 142(1-3), 203–212 (1995)

30. Sudborough, I.H.: On the tape complexity of deterministic context-free languages. J. Assoc. Comput. Mach. 25(3), 405–414 (1978)

31. Szeider, S.: On fixed-parameter tractable parameterizations of SAT. In: Giunchiglia, E., Tacchella, A. (eds.) SAT 2003. LNCS, vol. 2919, pp. 188–202. Springer, Heidelberg (2004)

32. Valiant, L.G.: The complexity of computing the permanent. Theoret. Comput. Sci. 8(2), 189–201 (1979)

33. Valiant, L.G., Skyum, S., Berkowitz, S., Rackoff, C.: Fast parallel computation of polynomials using few processors. SIAM J. Comput. 12(4), 641–644 (1983)

34. Vollmer, H.: Introduction to Circuit Complexity - A Uniform Approach. Springer, Heidelberg (1999)

A Decision-Making Procedure for Resolution-Based SAT-Solvers

Eugene Goldberg

Cadence Research Labs, USA, 2150 Shattuck Ave.,10th floor, Berkeley,
California, 94704
Tel.: 1-510-647-2825, Fax:1-510-647-2801
egold@cadence.com

Abstract. We describe a new decision-making procedure for resolution-based SAT-solvers called Decision Making with a Reference Point (DMRP). In DMRP, a complete assignment called a reference point is maintained. DMRP is aimed at finding a change of the reference point under which the number of clauses falsified by the modified point is smaller than for the original one. DMRP makes it possible for a DPLL-like algorithm to perform a "local search strategy". We describe a SAT-algorithm with conflict clause learning that uses DMRP. Experimental results show that even a straightforward and unoptimized implementation of this algorithm is competitive with SAT-solvers like BerkMin and Minisat on practical formulas. Interestingly, DMRP is beneficial not only for satisfiable but also for unsatisfiable formulas.

1 Introduction

Resolution based SAT-solvers have gained great popularity due to their ability to solve very large practical CNF formulas. An important contributor to this success is conflict driven decision making (CDDM) introduced in [17] and further developed in BerkMin [7], Minisat [3], Siege and other SAT-solvers. CDDM takes into account the history of conflicts thus forcing the SAT-solver to explore the parts of the search space that have not been visited before.

Despite the obvious success of CDDM, still there are many directions to explore. In this paper, we introduce a resolution based SAT-solver whose decision making procedure employs a complete assignment further referred to as **a reference point**. We will refer to this procedure as Decision Making with a Reference Point **(DMRP)**. (We will refer to the SAT-solver employing DMRP that we describe in this paper as **DMRP-SAT**.)

The main idea of DMRP is as follows. Let F be the CNF formula to be solved. Let $\boldsymbol{p}$ be a reference point and $M(\boldsymbol{p})$ be the set of clauses of F that are falsified by $\boldsymbol{p}$. DMRP-SAT picks a clause C of $M(\boldsymbol{p})$ and tries to find a modification $\boldsymbol{p}'$ of $\boldsymbol{p}$ that satisfies C and does not falsify any clauses of F that are not in $M(\boldsymbol{p}) \setminus \{C\}$. In other words, $M(\boldsymbol{p}') \subset M(\boldsymbol{p})$.

Importantly, the search of the point $\boldsymbol{p}'$ above is done by a regular DPLL-like procedure with conflict clause learning. After $\boldsymbol{p}'$ is found, it becomes a new

H. Kleine Büning and X. Zhao (Eds.): SAT 2008, LNCS 4996, pp. 119–132, 2008.

reference point and DMRP-SAT performs a complete restart. In our previous paper [6], we described the resolution-based SAT-solver called FI that operates on complete assignments. One can view the decision making procedure of FI as a variation of CDDM. Similarly to CDDM and decision making of FI, DMRP also gives some preference to recently derived conflict clauses. At the same time, DMRP is not just a variation of CDDM.

DMRP makes it possible for a SAT-solver to monotonically reduce the number of clauses falsified by the current reference point. So, in a sense, DMRP-SAT combines the features of algorithms based on the DPLL procedure [2] and local search SAT-algorithms pioneered in [20,21]. The strategy of reducing the number of clauses falsified by a complete assignment has been successfully applied by local search procedures to various classes of satisfiable formulas with no (or "little") structure like random CNF formulas. For structured satisfiable formulas, DPLL based SAT-solvers are usually more successful due to conflict clause learning and Boolean Constraint Propagation. Our results imply that local search strategy can be successfully applied to structured formulas as well. Interestingly, DMRP-SAT works very well not only for satisfiable formulas (which is somewhat expected) but also for unsatisfiable ones.

Currently, the main drawback of DMRP in comparison to CDDM is that DMRP is more expensive. The reason is that DMRP has to maintain a particular set of clauses that is updated after assigning/unassigning a variable. (Satisfying all the clauses of this set means that a new reference point is found that falsifies fewer clauses than the original one). However, our experiments show that due to high quality of decision making, even a straightforward unoptimized implementation of DMRP-SAT can be competitive with SAT-solvers like BerkMin and Minisat.

This paper is structured as follows. In Section 2, we introduce the idea of DMRP and give an example. Section 3 describes DMRP-SAT in more detail. In Section 4, DMRP-SAT is compared with other SAT-solvers. Experimental results are presented in Section 5. We give some conclusions in Section 6.

2   Main Idea of DMRP

In this section, we describe the basic idea of Decision Making with a Reference Point (DMRP) that is implemented in the SAT-solver DMRP-SAT.

Let F be a CNF formula and p be a complete assignment to the variables of F. (For the sake of brevity, we will also call p **a point**.) A clause C of F is said to be **falsified** (**satisfied**) by p if $C(p) = 0$ (respectively $C(p) = 1$). Denote by $Vars(C)$ the set of variables of clause C. Denote by $Vars(y)$ the set of variables assigned in a partial assignment y.

Definition 1. *Let $M(p)$ be the set of clauses of F falsified by p. We will say that p' **recursively satisfies** a clause C of $M(p)$ with respect to **the reference point** p if a) $C(p') = 1$; b) $M(p') \subset M(p)$.*

The use of term "recursively" is due to the fact that, given a reference point p, when looking for the point p' above, DMRP-SAT first satisfies clause C, then satisfies "descendants" of clause C that get falsified after satisfying C and so on.

Note that if F is satisfiable, an assignment p' meeting the two conditions of Definition 1 always exists. (An assignment p' satisfying F recursively satisfies any clause C of F with respect to any reference point p falsifying C.) On the other hand, even if F is unsatisfiable , one may find an assignment p' recursively satisfying C if $|M(p)| > 1$. Then $M(p') \subset M(p)$ and $M(p') \neq \emptyset$.

The basic idea of DMRP is to look for a complete assignment recursively satisfying a clause by regular branching as in the DPLL-procedure.

Definition 2. *Let y be a partial assignment. Denote by* **modify***(p,y) the point obtained from p by flipping the assignments that are different in p and y. (So assignments to $Vars(y)$ in the point modify(p,y) are the same as in y.)*

DMRP-SAT looks for a partial assignment y such that $p' = modify(p,y)$ recursively satisfies C. To make this search efficient, DMRP-SAT maintains a set $D(C,p,y)$ of clauses that one needs to satisfy before finding a point recursively satisfying C (see Definition 3). So if $D(C,p,y) = \emptyset$, the point $p' = modify(p,y)$ recursively satisfies C. DMRP-SAT implements DMRP using the following **greedy heuristic** aimed at making $D(C,p,y)$ empty. The next assignment is picked by DMRP-SAT so as to satisfy the largest number of clauses of $D(C,p,y)$.

Definition 3 (of the set $D(C,p,y)$). *If partial assignment y is empty, then $D(C,p,y) = \{C\}$. Otherwise, $D(C,p,y)$ is defined as follows. A clause C' of F is in $D(C,p,y)$ iff 1) there is a variable $x_i \in Vars(y) \cap Vars(C')$ that is assigned differently in p and y; 2) C' is not satisfied by y.*

Proposition 1. *Let F be a CNF formula. Let p be a complete assignment and C be a clause of $M(p)$. Let y be a partial assignment satisfying C. If $D(C,p,y) = \emptyset$, then the complete assignment $p' = modify(p,y)$ recursively satisfies the clause C with respect to the reference point p.*

Proof. Assume the contrary, i.e. there is a clause C' of $M(p')$ that is not in $M(p)$ and so p' does not satisfy C recursively. Suppose the set of variables $A = Vars(C') \cap Vars(y)$ is empty. Then, $C'(p')=0$ implies $C'(p)=0$ and so C' is in $M(p)$. We have a contradiction. Now suppose that $A \neq \emptyset$. Then all the assignments of y to the variables of A have to falsify corresponding literals of C'. (Otherwise, $C'(p') = 1$). If all the variables of A are assigned identically in y and p, then $C'(p) = 0$ and so C' is in $M(p)$. Suppose that at least one variable of A is assigned differently in y and p. Then, since $D(C,p,y)$ is empty, the clause C' has to be satisfied by some assignment in y. So we have a contradiction again.

Note that $D(C,p,y) = \emptyset$ is only a sufficient condition. For example, even if $D(C,p,y) \neq \emptyset$ but all the clauses of $D(C,p,y)$ are satisfied by the reference point p, the complete assignment $modify(p,y)$ may recursively satisfy C. One can give another definition of $D(C,p,y)$ so that $D(C,p,y) = \emptyset$ is also the necessary condition for $modify(p,y)$ to recursively satisfy C. However, in the current version of DMRP-SAT, to simplify computation of $D(C,p,y)$ we use Definition 3.

Example 1. Let F be the CNF formula specified by the following seven clauses: $C_1 = x_1 \lor x_2 \lor x_3$, $C_2 = x_3 \lor \overline{x_4} \lor x_5$, $C_3 = \overline{x_1} \lor \overline{x_6}$, $C_4 = \overline{x_1} \lor x_7$,

$C_5 = x_2 \vee \overline{x_5} \vee \overline{x_7}$, $C_6 = \overline{x_2} \vee \overline{x_4} \vee \overline{x_7}$, $C_7 = \overline{x_2} \vee x_6 \vee x_4$. Let $\boldsymbol{p}=(x_1=0$, $x_2=0$, $x_3=0$, $x_4=1$, $x_5=0$, $x_6=0$, $x_7=0)$ be a reference point. The set $M(\boldsymbol{p})$ consists of clauses C_1 and C_2. In this example, we describe a run of *DMRP-solve* (see Figures 1 and 2) called by *DMRP-SAT* when looking for a point that recursively satisfies clause C_1. In this description, we use the terminology of decision levels [22]. Decision level number k consists of the decision assignment number k and all implied assignments derived in Boolean Constraint Propagation (BCP) caused by this decision assignment.

Initially, $Vars(\boldsymbol{y})=\emptyset$. So $D(C_1,\boldsymbol{p},\boldsymbol{y}) = \{C_1\}$. Suppose that *DMRP-SAT* chose variable x_1 to satisfy C_1 (the function *pick_lit* of Figure 1). That is $x_1 = 1$ is the first decision assignment made by *DMRP-solve*. Then the clause C_1 is removed from $D(C_1,\boldsymbol{p},\boldsymbol{y})$ (because it is satisfied by an assignment in $\boldsymbol{y}$). Only clauses C_3 and C_4 of F have literal $\overline{x_1}$. They are added to $D(C_1,\boldsymbol{p},\boldsymbol{y})$ because $\boldsymbol{p}$ and $\boldsymbol{y}$ have different assignments to x_1 and neither C_3 nor C_4 are satisfied by an assignment in $\boldsymbol{y}$. So for $\boldsymbol{y}= \{x_1=1\}$ the set $D(C_1,\boldsymbol{p},\boldsymbol{y})$ is equal to $\{C_3,C_4\}$.

At this point C_3 and C_4 become unit. After BCP, *DMRP-solve* derives $x_6=0$ (from clause C_3) and $x_7=1$ (from clause C_4) and removes C_3,C_4 from $D(C_1,\boldsymbol{p},\boldsymbol{y})$. Since assignment $x_6=0$ is the same in $\boldsymbol{y}$ and $\boldsymbol{p}$, no new clauses are added to $D(C_1,\boldsymbol{p},\boldsymbol{y})$ when $\boldsymbol{y}$ becomes $(x_1=1,x_6=0)$. On the other hand, the variable x_7 is assigned differently in $\boldsymbol{y}$ and $\boldsymbol{p}$. Since x_7 is in C_5 and C_6 and they are not satisfied by $\boldsymbol{y}$, these clauses are added to $D(C_1,\boldsymbol{p},\boldsymbol{y})$. So after completing BCP of decision level 1, we have $\boldsymbol{y}=(x_1=1$, $x_6=0$, $x_7 = 1)$, $D(C_1,\boldsymbol{p},\boldsymbol{y})= \{C_5,C_6\}$.

Suppose that *DMRP-solve* picks second decision assignment $x_2=1$ to satisfy C_5. Then clauses C_6 and C_7 become unit, and *DMRP-solve* derives opposite values of x_4 from them. This leads to a conflict. *DMRP-solve* derives conflict clause $C_8 = \overline{x_1} \vee \overline{x_2}$ and backtracks to decision level 1. At this level, $\boldsymbol{y} =(x_1=1$, $x_6=0$, $x_7 = 1)$ again and $D(C_1,\boldsymbol{p},\boldsymbol{y}) = \{C_5,C_6\}$. However, now *DMRP-solve* has to update $D(C_1,\boldsymbol{p},\boldsymbol{y})$ due to appearance of conflict clause C_8 by adding it to $D(C_1,\boldsymbol{p},\boldsymbol{y})$. ($C_8$ contains variable x_1 that is assigned differently in $\boldsymbol{y}$ and $\boldsymbol{p}$ and C_8 is not satisfied by $\boldsymbol{y}$.) So, $D(C_1,\boldsymbol{p},\boldsymbol{y}) = \{C_5,C_6,C_8\}$.

At decision level 1, the conflict clause C_8 becomes unit and *DMRP-solve* derives $x_2=0$ from it. Since $x_2=0$ agrees with $\boldsymbol{p}$, no new clauses need to be added to $D(C_1,\boldsymbol{p},\boldsymbol{y})$. At the same time, C_6 and C_8 are removed from $D(C_1,\boldsymbol{p},\boldsymbol{y})$ because they are both satisfied by $x_2= 0$. So $D(C_1,\boldsymbol{p},\boldsymbol{y}) = \{C_5\}$. *DMRP-solve* derives $x_5=0$ from C_5 and the latter is removed from $D(C_1,\boldsymbol{p},\boldsymbol{y})$. Since $x_5=0$ agrees with $\boldsymbol{p}$, no new clauses are added to $D(C_1,\boldsymbol{p},\boldsymbol{y})$. So, for the partial assignment $\boldsymbol{y}=(x_1=1$, $x_6=0$, $x_7=1$, $x_2=0$, $x_5=0)$, $D(C_1,\boldsymbol{p},\boldsymbol{y})$ is empty. This means, that the clause C_1 is recursively satisfied by the assignment $\boldsymbol{p}' = modify(\boldsymbol{p},\boldsymbol{y})$ where $\boldsymbol{p}' = (x_1=1$, $x_2=0$, $x_3=0$, $x_4=1$, $x_5=0$, $x_6=0$, $x_7=1)$. It is not hard to check that indeed $C_1(\boldsymbol{p}')=1$ and $M(\boldsymbol{p}') = \{C_2\}$ and so $M(\boldsymbol{p}') \subset M(\boldsymbol{p})$. Now *DMRP-solve* performs a complete restart and picks $\boldsymbol{p}'$ as the next reference point.

3   Description of DMRP-SAT

In this section, we describe DMRP-SAT in more detail.

3.1 *DMRP-SAT* (High-Level View)

```
DMRP-SAT(F)
{p=gen_ref_point(F);
  while (true)
    {C = pick_clause(M(p));
    lit = pick_lit(C,M(p));
    (ans,y)=DMRP-solve(F,C,lit,p);
    if (ans == unsat) return(unsat);
    if (ans == sat) return(sat);
    if (ans == literal) continue;
    if (ans == rec_sat)
      {p = modify(p,y);
      if (M(p) == ∅) return(sat);}
    if (ans == new_point)
      p =modify(p,y);}}
```

Fig. 1. Pseudocode of *DMRP-SAT*

```
DMRP-solve(F,C,lit,p)
{D(C,p,y) = {C};
  while (true)
    {if (D(C,p,y) == ∅)
      {restart(F);
      return(y,rec_sat);}
    make_assgn(F,lit,D(C,p,y));
    ans = BCP(F,D(C,p,y),p);
    if (ans == sat) return(sat);
    if (ans == conflict)
      {C* = gen_cnfl_clause(F);
      if (empty(C*)) return(unsat);
      if (C* == unit) p=upd_pnt(p);
      if (C* == lit̄)
        {restart(F) ;
        return(literal); }
      add_clause(F,C*);
      backtrack(F);}
    else continue; // no conflict
    if (num_of_cnfl++ > THRESH)
      {restart(F);
      return(y,new_point);}
    if (num_of_cnfl > thresh)
      {restart(F);
      continue;} }}
```

Fig. 2. Pseudocode of *DMRP-solve*

Pseudocode of *DMRP-SAT*(F) is shown in Figure 1. First, *DMRP-SAT* generates a reference point. This is done identically to initial point generation of *FI* [6]. "Decision" assignments are made by *gen_ref_point* in the order variables of F are numbered. A decision assignment is made to variable x_i only if it has not been already assigned by BCP performed after a previous decision assignment. The polarity of assignment to x_i is chosen to satisfy the largest number of clauses of F with variable x_i. After a decision assignment is made, BCP is performed. If a clause of F is falsified during BCP, it is added to $M(p)$.

The main work is done by *DMRP-SAT* in the 'while' loop. First, *DMRP-SAT* picks a clause C of $M(p)$ to be recursively satisfied. If $M(p)$ contains conflict clauses, then the clause derived most recently is chosen as C. Otherwise, *DMRP-SAT* picks a clause of $M(p)$ that has a literal occurring most frequently among clauses of $M(p)$. Then a literal *lit* of C is chosen by the *pick_lit* procedure. Namely, it chooses the literal of C that occurs most frequently among clauses of $M(p)$. When looking for a complete assignment recursively satisfying clause C, the function *DMRP-solve* called next examines only points for which *lit* evaluates to 1.

Being a DPLL-like procedure with learning, *DMRP-solve* returns answer *unsatisfiable* if an empty clause is derived. If all clauses of F are satisfied by the current partial assignment y, then *DMRP-solve* returns satisfiable. If *DMRP-solve* derives the literal $\overline{lit}$ it returns *literal*. This means, that clause C cannot by satisfied by setting literal lit to 1. Then *DMRP-SAT* starts a new iteration.

If $D(C,\boldsymbol{p},\boldsymbol{y})=\emptyset$ (where $\boldsymbol{y}$ is the current partial assignment), *DMRP-solve* returns *rec_sat* (C can be recursively satisfied). A new reference point $modify(\boldsymbol{p},\boldsymbol{y})$ is computed. If $M(\boldsymbol{p}) = \emptyset$, then $\boldsymbol{p}$ is a satisfying assignment. Otherwise, a new iteration of the 'while' loop is started. If the number of conflicts that occurred in *DMRP-solve* exceeds *THRESH*, *DMRP-solve* returns *new_pnt*. In this case, *DMRP-SAT* generates a new reference point $modify(\boldsymbol{p},\boldsymbol{y})$ (where $\boldsymbol{y}$ is the partial assignment of *DMRP-solve* when it encountered the last conflict).

3.2   DMRP-Solve

The pseudocode of *DMRP-solve*(F,C,lit,$\boldsymbol{p}$) is shown in Figure 2. On the one hand, *DMRP-solve* is a regular DPLL-like SAT-solver with conflict clause learning. In the 'while' loop, it makes a decision assignment and then runs BCP. If after BCP, all clauses of F are satisfied, then *DMRP-solve* returns *satisfiable*. If a conflict is encountered during BCP, a conflict clause C^* is generated using the 1UIP scheme ([23]). If C^* is an empty clause, *DMRP-solve* returns *unsatisfiable*. Otherwise, C^* is added to the current CNF formula, *DMRP-solve* backtracks and a new iteration starts (unless C^* is equal to $\overline{lit}$, see below). If the number of conflicts that occurred since the last restart is larger than *thresh*, *DMRP-solve* restarts [8] (i.e. backtracks to decision level 0).

On the other hand, *DMRP-solve* has a few differentiating features. If conflict clause C^* is unit and the current reference point $\boldsymbol{p}$ falsifies C^*, then $\boldsymbol{p}$ is modified by flipping the value of the variable $Vars(C^*)$. Besides, if C^* is unit and equal to $\overline{lit}$, *DMRP-solve* informs *DMRP-SAT* that such a literal was derived. (Here, lit is the literal of clause C to be satisfied recursively that was chosen by *pick_lit* of *DMRP-SAT*). For decision-making, *DMRP-solve* maintains the set $D(C,\boldsymbol{p},\boldsymbol{y})$ (see Definition 3). Before looking for a new decision assignment, *DMRP-solve* checks if $D(C,\boldsymbol{p},\boldsymbol{y})=\emptyset$. If so, it performs a restart and informs *DMRP-SAT* that clause C is recursively satisfied.

At the first decision level, *DMRP-solve* always makes the assignment satisfying the literal lit of C (and so satisfying C). At a level greater than 1, *DMRP-solve* picks the next decision assignment as follows. If the set $D(C,\boldsymbol{p},\boldsymbol{y})$ contains a conflict clause, the clause C' of $D(C,\boldsymbol{p},\boldsymbol{y})$ that was derived most recently is chosen. Then, *DMRP-solve* finds the literal of C' that is shared by the largest number of clauses of $D(C,\boldsymbol{p},\boldsymbol{y})$ and picks the assignment that satisfies this literal. If $D(C,\boldsymbol{p},\boldsymbol{y})$ does not contain conflict clauses, *DMRP-solve* makes the assignment satisfying the largest number of clauses from $D(C,\boldsymbol{p},\boldsymbol{y})$.

If the number of conflicts that occurred since *DMRP-solve* has been called is larger than *THRESH*, *DMRP-solve* performs a restart. Then *DMRP-solve* informs *DMRP-SAT* to generate the new reference point $\boldsymbol{p}=modify(\boldsymbol{p},\boldsymbol{y})$. Here $\boldsymbol{y}$ is the partial assignment *DMRP-solve* had when the last conflict occurred. The value of *THRESH* is larger than that of *thresh* used for restarts without changing the reference point.

3.3   Computation of $D(C,p,y)$

In the current implementation of *DMRP-solve*, set $D(C,p,y)$ is computed incrementally. Initially, $D(C,p,y) = \{C\}$. When making an assignment $x_i = b$, $b \in \{0,1\}$ (either decision one or derived by BCP), *DMRP-solve* does the following. First it checks if reference point p has the same assignment $x_i = b$. If so, no new clauses are added to $D(C,p,y)$. Otherwise, *DMRP-solve* examines all the clauses of $D(C,p,y)$ in which the assignment $x_i = b$ sets a literal of x_i to 0. If such a clause is neither satisfied nor it is already in $D(C,p,y)$, it is added to $D(C,p,y)$. Then *DMRP-solve* removes from $D(C,p,y)$ all the clauses that are satisfied by $x_i = b$.

When undoing the assignment $x_i = b$ above (when backtracking), *DMRP-solve* does similar updates. First, it removes from $D(C,p,y)$ the clauses that were added to $D(C,p,y)$ due to assignment $x_i = b$. Second, it returns to $D(C,p,y)$ all the clauses that were removed because they got satisfied by $x_i = b$.

3.4   Brief Discussion of DMRP and CDDM

Similar to conflict driven decision making (CDDM) introduced by Chaff, DMRP takes into account the history of conflicts. First, when picking a clause C of $M(p)$ to be satisfied recursively, DMRP gives preference to conflict clauses derived most recently. Second, next decision assignment is made to a variable of the most recently derived conflict clause C^* of $D(C,p,y)$ (if any).

At the same time, there are obvious differences. When picking next assignment, DMRP finds the literal of $Vars(C^*)$ with the largest occurrence in clauses of $D(C,p,y)$. So no preference is given to conflict clauses. Besides, no decay scheme is used for literal activity computation. So DMRP cannot be called just a variation of CDDM.

In our current implementation, DMRP is more expensive than CDDM. As we mentioned above, every time DMRP-SAT makes/unmakes an assignment (decision or implied one) it recomputes $D(C,p,y)$. So one of the directions for future research is to cut the cost of DMRP. A potential solution to this problem is to compute $D(C,p,y)$ approximately.

4   Background

In this section, we compare DMRP-SAT with other SAT-solvers. In this comparison we take into account the following four features: BCP, learning, maintaining a complete assignment, making restarts. Each of these features is arguably beneficial. BCP allows one to find "forced" assignments. Learning (e.g. conflict clause recording [22]) helps in pruning away big chunks of the search space. Maintaining a complete assignment provides some information about how far a SAT-solver is from a satisfying assignment [20,21]. Besides, having a complete assignment can be used for (implicit) identification of small unsatisfiable sub-formulas [6]. Restarts [8] alleviate the problem of SAT-solver's getting stuck in a part of the search space that does not contain satisfying assignments. At the same time, we do not claim that the more of these four features a SAT-solver has, the more advanced it is. For example, SAT-solvers that do not employ conflict clause learning (e.g. Satz [14]) work much better for random CNF formulas.

SAT-algorithms like GSAT[20] and WalkSat[21] (and many other local search algorithms [11]) operate on complete assignments and make restarts (in the sense that they pick a new initial complete assignment every once in a while). These algorithms work very well for some classes of formulas like satisfiable random formulas. However, lack of learning and BCP makes local search algorithms less efficient when applied to "highly structured" formulas. On the other hand, DPLL-like SAT-solvers like Grasp [22], SATO [24], Zchaff [17], BerkMin [7], Minisat[3], Siege and many others use learning and BCP. Most of them also employ restarts. These SAT-algorithms have been very successful in solving both satisfiable and unsatisfiable structured formulas. This success can be attributed to a) efficient conflict driven learning (introduced by GRASP), b) fast BCP (introduced by SATO and improved by Chaff) and c) conflict driven decision making (introduced by Chaff and further developed by BerkMin, Minisat, Siege and others).

There have been significant effort to combine local search algorithms and SAT-solvers based on the DPLL procedure. In [15], in every node of the search tree, a local search procedure is invoked to identify the next variable to branch on. (An important observation made in [15] is that local search can be used for identifying unsatisfiable cores.) This approach is further improved in [9] by taking into account variable dependencies. In [19], random backtracking is used to improve the scalability of the DPLL procedure. In UnitWalk [10], BCP is used to correct values of a complete assignment. The values of this complete assignment are reassigned in a random order, every variable assignment being followed by BCP. A complete local search algorithm augmented by clause generation is introduced in [4]. Clause generation is used in [4] for escaping local minima. In [6], we described the resolution-based SAT-solver called FI that operates on complete assignments. The decision making procedure of FI can be viewed as a variation of CDDM. Namely, the choice of branching variables is reduced to variables of clauses falsified by the current complete assignment.

Although only FI and DMRP-SAT have all four features mentioned above, some SAT-algorithms can be augmented with missing features (for example, one can add clause learning to UnitWalk.) However, only $DMRP\text{-}SAT$ combines a DPLL-like procedure and the "genuine" local search strategy of minimizing the set of clauses falsified by a complete assignment. Experiments show that such a local search strategy can be very useful even for highly structured formulas (both satisfiable and unsatisfiable).

There is similarity between the notion of a recursively satisfied clause and that of an autarky [16,5,13]. When looking for a partial assignment y such that $modify(p,y)$ recursively satisfies a clause C of F, one tries to satisfy clauses of F "touched" by y (like it is done when searching for an autarky). The main difference is that a clause C' of F is considered as touched by an assignment to variable x_i only if x_i is assigned differently in y and the reference point p.

5   Experimental Results

In this section, we give results of some experiments with an implementation of DMRP-SAT. The experiments were run on Intel's Xeon CPU (3.06GHz) under Linux. The main objective of experiments was to show that although currently DMRP is more expensive than conflict driven decision making, it is competitive with the latter due to high quality of decision making. To keep our implementation easily modifiable we made it very simple. In particular, it lacked many techniques commonly employed to speed up a SAT-solver (see subsection 5.1).

We tried DMRP-SAT on a large set of structured CNF formulas. Here we give data on Bounded Model Checking (BMC) [1,18,12] and equivalence checking formulas. This data is representative of the typical behavior of DMRP-SAT. For satisfiable formulas, DMRP-SAT seems to be, in general, more robust than SAT-solvers based on conflict driven decision making. This can be attributed to that, like local search algorithms, DMRP-SAT looks for a satisfying assignment "incrementally".

It is important to note that the current version of DMRP-SAT is meant just to prove that decision-making with a reference point is viable. An optimal design of DMRP-SAT (and many details such as generation of the initial reference point, the best schedule for changing reference points and so on) will be the subject of future research.

5.1   Brief Description of Implementation

DMRP-SAT is written in C++ and compiled by gcc (version 3.2.2). We used the STL library for data structures like dynamic arrays. As mentioned above, our implementation of DMRP-SAT is very simple. It does not have advanced features like fast BCP, efficient formula representation, special treatment of binary clauses and so on. For example, to check if a clause is unit in BCP, DMRP-SAT just counts the number of unassigned literals (as it was done before SATO and Chaff). The only kind of optimization we used in DMRP-SAT was lazy computation of $D(C,\boldsymbol{p},\boldsymbol{y})$. Namely, during BCP, DMRP-SAT accumulated all the new assignments of $\boldsymbol{y}$ and only when BCP was over it updated $D(C,\boldsymbol{p},\boldsymbol{y})$ *if no conflict occurred*. The reason is that in case of a conflict, recomputing $D(C,\boldsymbol{p},\boldsymbol{y})$ is a waste of time because DMRP-SAT immediately backtracks eliminating all the assignments made at the conflict decision level.

For each literal $lit(x_i)$, DMRP-SAT maintains an array with indexes of clauses of the current formula containing $lit(x_i)$. So when x_i is, say, set to 0, DMRP-SAT examines the clauses of the corresponding array to see if new unit clauses appeared. To avoid examining satisfied clauses, when $lit(x_i)$ is set to 1, all the clauses with $lit(x_i)$ unsatisfied so far are marked as satisfied. The clauses satisfied at a particular decision level are recorded together so that they can be efficiently unmarked when backtracking.

To facilitate decision making and computation of the set $D(C,\boldsymbol{p},\boldsymbol{y})$, DMRP-SAT maintains an array that marks clauses that are currently in $D(C,\boldsymbol{p},\boldsymbol{y})$. For every $lit(x_i)$ it also maintains a counter containing the number of clauses of

$D(C,\boldsymbol{p},\boldsymbol{y})$ that have $lit(x_i)$. Besides, it maintains the set of variables of clauses that are currently in $D(C,\boldsymbol{p},\boldsymbol{y})$. If this set is empty, then $D(C,\boldsymbol{p},\boldsymbol{y}) = \emptyset$ and C is recursively satisfied by $modify(\boldsymbol{p},\boldsymbol{y})$. Finally, DMRP-SAT records the indexes of clauses that are added to $D(C,\boldsymbol{p},\boldsymbol{y})$ at a particular decision level. When undoing assignments of this level, DMRP-SAT removes the recorded clauses from $D(C,\boldsymbol{p},\boldsymbol{y})$.

In all the experiments, the values of *thresh* and *THRESH* were 150 and 3000 respectively (see Figure 1 and Figure 2). That is every 150 conflicts DMRP-SAT made a restart without changing the reference point and every 3000 conflicts such a restart was accompanied by changing the reference point.

5.2   BMC and Equivalence Checking Formulas

In this subsection, we compare our implementation of DMRP-SAT with two SAT-solvers. The first SAT-solver is a version of BerkMin [7] that is very close to Forklift, the winner of the SAT-2003 contest in the industrial category (but in contrast to Forklift, it only learns conflict clauses). This version is much faster than the publicly available one (BerkMin561) on large CNF formulas. The second SAT-solver is Minisat [3], version 1.13 (a similar version of Minisat was the runner-up of the SAT-2005 contest in the industrial category). Table 1 summarizes results of BerkMin, Minisat and DMRP-SAT on a set of large BMC formulas (up to a few millions of variables). These formulas describe various properties of more than a dozen of customer designs. This set consists of 79 formulas (28 satisfiable and 51 unsatisfiable). For all three SAT-solvers, Table 1 gives the total number of conflicts (in thousands), total and median runtime for satisfiable, unsatisfiable and both types of formulas. A sample of formulas from this set are shown in Table 2 (satisfiable formulas are marked with '*').

These two tables show that, for satisfiable formulas, DMRP-SAT makes significantly fewer backtracks (conflicts). Even though BerkMin and Minisat have much faster code and DMRP is more expensive, DMRP-SAT converts the advantage in the number of conflicts into smaller run-times. For unsatisfiable BMC formulas, DMRP-SAT also has fewer conflicts, but this difference is not large enough to convert it into better performance. (However, this should change with a faster implementation.)

Table 3 gives direct evidence that DMRP-SAT indeed benefits from its decision making strategy. For a sample of satisfiable BMC formulas (from the set of 28 formulas mentioned above), this table describes the process of finding a satisfying assignment in more detail. DMRP-SAT can find a satisfying assignment in two ways (see Figures 1,2). First, it can extend the current partial assignment $\boldsymbol{y}$ so that all clauses of the CNF formula become satisfied. Second, when DMRP-SAT is successful in recursively satisfying a clause C, it may find a reference point $\boldsymbol{p}' = modify(\boldsymbol{p},\boldsymbol{y})$ such that $M(\boldsymbol{p}') = \emptyset$. (When this happens, current partial assignment $\boldsymbol{y}$ may satisfy only a fraction of clauses of F.) Interestingly, for each of 28 satisfiable BMC formulas we used, a satisfying assignment was found after recursively satisfying a clause.

Table 1. BMC formulas

category (#formulas)	BerkMin		Minisat		DMRP-SAT	
	#cnfl. $\times 10^3$	total (median) time, sec.	#cnfl. $\times 10^3$	total (median) time, sec.	#cnfl. $\times 10^3$	total (median) time, sec.
sat (28)	2,546	44,814 (246)	3,457	58,319 (619)	**333**	**9,565 (57)**
unsat (51)	2,156	28,594 **(64)**	1,355	**14,507** (80)	791	15,160 (151)
total(79)	4,702	73,408 (96)	4,812	72,826 (178)	**1,124**	**24,725 (69)**

Table 2. Sample of BMC formulas (satisfiable* and unsatisfiable)

name	#vars $\times 10^6$	#clau- ses $\times 10^6$	BerkMin		Minisat		DMRP-SAT	
			#cnfl. $\times 10^3$	time (sec.)	#cnfl. $\times 10^3$	time (sec.)	#cnfl. $\times 10^3$	time (sec.)
sched*	1.0	2.7	24	386	23	1,038	**0.07**	**2.6**
byteen*	0.2	0.6	21	**138**	60	1,074	**8.8**	245
stimulus*	0.1	0.4	7.9	**39**	49	370	**7.5**	82
ipt*	1.2	3.5	61	2,896	108	3,029	**4.8**	**205**
iqm*	2.3	7.0	308	11,704	732	16,568	**0.5**	**70**
prop3*	1.4	4.3	822	5,230	495	9,084	**77**	**2,479**
gmtx*	2.7	7.9	12	281	47	2,462	**0.05**	**7.5**
sdl*	0.4	1.2	183	551	149	**472**	75	1,659
write*	0.6	1.8	8.4	168	48	552	**1.2**	**51**
prop9*	1.0	3.0	74	898	40	429	**2.9**	**58**
unsatisfiable formulas								
always	0.2	0.8	19	45	21	213	**5.0**	**38**
page	0.2	0.8	19	**35**	19	151	14	425
mcbdm	0.3	0.8	17	144	6.2	84	**1.5**	**31**
lddata	0.2	0.5	20	**31**	55	666	18	255
cmcnt	1.2	3.6	8.5	491	**2.5**	**68**	3.0	134
iwrk	1.3	4.1	202	3,934	31	447	**6.5**	**108**
cho	0.1	0.3	**14**	**23**	15	42	31	1,308
CCC	0.3	1.1	38	199	**22**	**165**	23	1,941

The number of backtracks made before finding a satisfying assignment is reported in the second column of Table 3. The third column shows the number of clauses $|M(\boldsymbol{p})|$ falsified by the original reference point $\boldsymbol{p}$. The number of cases when a clause of $M(\boldsymbol{p})$ was recursively satisfied is given in the fourth column. The size of the longest chain of events when a clause was recursively satisfied with fewer than $THRESH=3000$ backtracks is shown in the fifth column. (Recall that when the number of backtracks exceeds 3000, DMRP-SAT makes a restart and picks a new reference point $\boldsymbol{p'}$. Usually $|M(\boldsymbol{p})| < |M(\boldsymbol{p'})|$.) The last column

Table 3. Statistics on recursively satisfied clauses

name	#confl.	size of initial $M(p)$	#cases of rec. sat. a clause	#longest chain	$|y|/|\mathit{Vars}(F)|$ when $M(p)=\emptyset$ %
sched	67	1	1	1	18
byteen	8,824	543	255	255	3.5
stimulus	7,518	276	29	29	1.8
data	15,521	1,034	212	114	77
ifreeq	3,426	615	438	438	1.9
ipt	4,750	775	601	601	0.8
prop3	77,127	44	29	6	76
muls	556	104	69	69	1.4
T1	64	2	1	1	67
TX	77,934	8	7	3	96
HP-4850	17,932	62	8	7	1.0
HP-974	2,092	1	1	1	44
write	1,175	149	87	87	0.9
prop9	2,892	1	1	1	31
SUN-442	17	1	1	1	95
SUN-443	2,010	3,999	2,000	2,000	1.6

gives the size of y (in percent of $|\mathit{Vars}(F)|$) when a satisfying assignment $p' = modify(p,y)$ was found.

Table 3 shows that DMRP-SAT indeed successfully used the "local search strategy" of minimizing the set of falsified clauses to find satisfying assignments. For example, for the formula *byteen*, the original reference point falsified 543 clauses. Then after 255 cases of recursively satisfied clauses a satisfying assignment was found. At this point, only 3.5% of the variables were assigned in the partial assignment y. So, in the case of formula *byteen*, DMRP-SAT kept monotonically reducing the size of $M(p)$ until a satisfying assignment was found. For some formulas (like *data*), the value of *THRESH* was exceeded and a new reference point was generated (possibly more than once). In such cases the size of the longest chain is smaller than the number of cases when a clause was recursively satisfied. It is worth mentioning that DMRP-SAT had a lot cases of recursively satisfying clauses of $M(p)$ for unsatisfiable formulas too.

Finally, Table 4 summarizes results of experiments with satisfiable equivalence checking CNF formulas. Each formula F of Table 4 describes equivalence checking of a combinational circuit N_1 with a circuit N_2 obtained from N_1 by optimization. If N_1 and N_2 are functionally equivalent (inequivalent), then F is unsatisfiable (respectively satisfiable). We manually introduced detectable bugs to the circuit N_2. So all equivalence checking formulas of Table 4 were satisfiable. The first column of Table 4 identifies circuit N_1 (*des* stands for *design*) and gives the number of bugs introduced in circuit N_2 (each bug corresponds to a separate satisfiable formula). Second and third columns give the size of the formula F describing equivalence checking of N_1 and N_2 without any bugs. (Introducing a bug does not affect the formula size much.)

Table 4. Equivalence checking (satisfiable formulas)

name (#bugs)	#vars ×10³	#clau- ses ×10³	BerkMin		Minisat		DMRP- SAT	
			#cnfl. ×10³	total time (sec.)	#cnfl. ×10³	total time (sec.)	#cnfl. ×10³	total time (sec.)
des_1 (7)	4.7	53	271	120	110	143	**11**	**56**
des_2(8)	2.4	24	229	40	**95**	**10**	162	197
des_3(7)	2.7	29	169	39	**46**	**3**	83	114
des_4(5)	1.0	9.8	54	5	14	**0.4**	**13**	6
des_5 (5)	1.9	20	80	8	55	4	**37**	25
des_6(7)	9.5	106	2,484	2,624	1,327	1,783	**88**	**389**
des_7 (4)	1.6	16	75	11	39	2	15	8
des_8(4)	3.5	39	69	**17**	111	53	**52**	116
Total			3,431	2,864	1,797	1,998	**461**	**911**

Results of Table 4 show again that DMRP-SAT generated fewer conflicts than BerkMin and Minisat and this advantage was converted into better summary performance. Although DMRP-SAT did not have smaller number of backtracks for all designs, it showed more robust behavior. In particular, it relatively easily solved the equivalence checking formulas generated off the design des_7 that contained a multiplier. We also applied DMRP-SAT to unsatisfiable equivalence checking formulas (no bugs in N_2). DMRP-SAT again had very good performance in terms of the number of backtracks and run-times. For the lack of space we omit these results.

6   Conclusions

We introduce a new decision making strategy DMRP (decision making with a reference point) for resolution-based SAT-solvers. DMRP allows a DPLL-like procedure to pursue a local search strategy. Experiments show that our SAT-solver DMRP-SAT implementing DMRP works well for both satisfiable and unsatisfiable structured formulas. In the current implementation, DMRP is more expensive than conflict driven decision making introduced by Chaff. In our future research we will work on reducing the cost of DMRP. At the same time, even a straightforward and unoptimized implementation of DMRP-SAT shows very good performance due to high quality of decision-making.

References

1. Biere, A., Cimatti, A., Clarke, E., Strichman, O., Zhu, Y.: Bounded Model Checking (a book chapter). In: Zelkovitz, M. (ed.) Advances in computers, vol. 58, Elsevier, Amsterdam (2003)
2. Davis, M., Longemann, G., Loveland, D.: A Machine program for theorem proving. Communications of the ACM 5, 394–397 (1962)

3. Een, N., Sorensson, N.: An extensible SAT-solver. In: Giunchiglia, E., Tacchella, A. (eds.) SAT 2003. LNCS, vol. 2919, pp. 503–518. Springer, Heidelberg (2004)
4. Fang, H., Ruml, W.: Complete Local Search for Propositional Satisfiability. In: Proc. of 19th National Conference on Artificial Intelligence, pp. 161–166 (2004)
5. Gelder, A.V.: Autarky pruning in propositional model elimination reduces failure redundancy. J. of Autom. Reasoning 23(2), 137–193 (1999)
6. Goldberg, E.: Determinization of resolution by an algorithm operating on complete assignments. In: Biere, A., Gomes, C.P. (eds.) SAT 2006. LNCS, vol. 4121, pp. 90–95. Springer, Heidelberg (2006)
7. Goldberg, E., Novikov, Y.: BerkMin: a Fast and Robust SAT-Solver. In: DATE 2002, Paris, pp. 142–149 (2002)
8. Gomes, C.P., Selman, B., Kautz, H.: Boosting Combinatorial Search Through Randomization. In: Proc. AAAI 1998 (1998)
9. Habet, D., Li, C.M., Devendeville, L., Vasquez, M.: A hybrid approach for SAT. In: Int. Conf. on Principles and Practice of Constraint Programming, pp. 172–184 (2002)
10. Hirsch, E.A., Kojevnikov, A.: UnitWalk: A new SAT solver that uses local search guided by unit clause elimination. Annals of Math. and Artif. Intell. 43(1-4), 91–111 (2005)
11. Hoos, H., Stutzle, T.: Stochastic Local Search: Foundations and Applications. Morgan Kaufmann, San Francisco (CA) (2004)
12. Katz, J., Hanna, Z., Dershowitz, N.: Space-efficient Bounded Model Checking. In: DATE 2005, pp. 686–687 (2005)
13. Kullmann, O.: Investigations on autark assignments. Discrete Applied Mathematics 107, 99–137 (2000)
14. Li, C.M.: A constrained-based approach to narrow search trees for satisfiability. Information processing letters 71, 75–80 (1999)
15. Mazure, B., Sais, L., Gregoire, R.: Boosting complete techniques thanks to local search methods. Annals of Math. and Artif. Intell. 22, 319–331 (1998)
16. Monien, B., Speckenmeyer, E.: Solving satisfiability in less than 2^n steps. Discrete Applied Mathematics 10, 287–295 (1985)
17. Moskewicz, M., Madigan, C., Zhao, Y., Zhang, L., Malik, S.: Chaff: Engineering an Efficient SAT Solver. In: DAC 2001 (2001)
18. Prasad, M., Biere, A., Gupta, A.: A survey of recent advances in SAT-based formal verification. STTT 7(2), 16–173 (2005)
19. Prestwich, S.: Local search and backtracking vs. non-systematic backtracking. In: AAAI Fall Symposium on Using Uncertainty Within Computation, North Falmouth, Cape Cod, MA, November 2-4, 2001, pp. 109–115 (2001)
20. Selman, B., Levesque, H., Mitchell, D.: A New Method for Solving Hard Satisfiability Problems. In: AAAI 1992, pp. 440–446 (1992)
21. Selman, B., Kautz, H.A., Cohen, B.: Noise strategies for improving local search. In: AAAI 1994, Seattle, pp. 337–343 (1994)
22. Silva, J.P.M., Sakallah, K.A.: GRASP: A Search Algorithm for Propositional Satisfiability. IEEE Transactions of Computers 48, 506–521 (1999)
23. Zhang, L., Madigan, C., Moskewicz, M., Malik, S.: Efficient Conflict Driven Learning in a Boolean Satisfiability Solver. In: ICCAD 2001 (2001)
24. Zhang, H.: SATO: An efficient propositional prover. In: International Conference on Automated Deduction, July 1997, pp. 272–275 (1997)

Online Estimation of SAT Solving Runtime[*]

Shai Haim and Toby Walsh

NICTA and UNSW
{shai.haim,toby.walsh}@nicta.com.au

Abstract. We present an online method for estimating the cost of solving SAT problems. Modern SAT solvers present several challenges to estimate search cost including non-chronological backtracking, learning and restarts. Our method uses a linear model trained on data gathered at the start of search. We show the effectiveness of this method using random and structured problems. We demonstrate that predictions made in early restarts can be used to improve later predictions. We also show that we can use such cost estimations to select a solver from a portfolio.

1   Introduction

Modern SAT solvers present several challenges for estimating their runtime. For instance, clause learning repeatedly changes the problem the solver faces. Estimation of the size of the search tree at any point should take into consideration the changes that future learning clauses will cause. As a second example, restarting generates a new search tree which again makes prediction hard. Our approach to these problems is to use a machine learning based on-line method to predict the cost of the search by observing the solver's *behaviour* at the start of search.

Previous methods include the Weighted Backtrack Estimator, the Recursive Estimator ([5]) and the SAT Progress Bar ([6]) that do not support backjumping or restarts, and the BDD-based Satometer ([1]) which doesn't provide an estimate for the size of the decision tree. Machine learning has also been used to estimate search cost. Horovitz et al used a Bayesian approach to classify CSP and SAT problems according to their runtime [4]. Whilst this work is close to ours, there are some significant differences. For example, they used SATz-Rand which does not use clause learning. Xu et. al [9] used machine learning to tune empirical hardness models [7]. The only non-static features used were generated by probes of DPLL and stochastic search. Their method gives an estimate for the distribution of runtimes and not, as here, an estimate for a specific run. Finally, an online machine learning method has been used for QBF solvers [8].

2   Linear Model Prediction (LMP)

We predict the size of subtrees to follow from the subtrees explored in the past. Given a problem $\mathcal{P} \in E$, when E is an ensemble of problems, we first train the

---

[*] The second author is funded by DCITA and the ARC through Backing Australia's Ability and the ICT Centre of Excellence program.

H. Kleine Büning and X. Zhao (Eds.): SAT 2008, LNCS 4996, pp. 133–138, 2008.

model using a subset of problems $\mathcal{T} \subset E$. For every training example $t \in \mathcal{T}$, we create a feature vector $x_t = \{x_{t,1}, x_{t,2}, \ldots, x_{t,k}\}$. We select features by removing those with the smallest standardised coefficient until no improvement is observed based on the standard AIC (Akaike Information Criterion). We then search for and eliminate co-linear features in the set.

Using ridge linear regression, we fit our coefficient vector w to create a linear predictor $f_w(x_i) = w^T x_i$. We chose ridge regression since it is quick and simple, and generally yields good results. We predict the log of the number of conflicts as runtimes vary significantly. Since the feature vector is computed online, we do not want it to add significant cost to search. It therefore only contains features that can be calculated in (amortized) constant time. We define the *observation window* to be that part of search where data is collected. At the end of the observation window, the feature vector is computed and the model queried for an estimation.

The feature vector measures both problem structure and search behaviour. Since data gathered at the beginning of a restart tends to be noisy, we do not open the observation window immediately. To keep the feature vector of reasonable size, we use statistical measures of features (that is, the minimum over the observation window, the maximum, the mean, the standard deviation and the last value recorded). The list of features is shown in Table 1. The only feature that takes more than constant time to calculate is the *log(WBE)* feature. This is based on the Weighted Backtrack Estimator [5]. This estimates search tree size

Table 1. The feature vector used by linear regression to construct prediction models

Feature	init	Observation Window min	max	mean	SD	last
Number of *variables* (*var*)	✓					
Number of *clauses* (*cls*)	✓					
cls/var	✓	✓	✓	✓	✓	✓
var/cls	✓	✓	✓	✓	✓	✓
Fraction of Binary Clauses	✓			✓	✓	✓
Fraction of Ternary Clauses	✓			✓	✓	✓
Avg. Clause Size	✓			✓	✓	✓
Search Depth (from assignment stack)			✓	✓	✓	
Search Depth (in corresponding binary tree)[a]			✓	✓	✓	
Backjump Size			✓	✓	✓	
Learnt Clause Size		✓	✓	✓	✓	
Conflict Clause Size		✓	✓	✓	✓	
Fraction of assigned vars before backtracking (*abb*)		✓	✓	✓	✓	
Fraction of assigned vars after backtracking (*aab*)		✓	✓	✓	✓	
aab.mean/abb.mean		✓	✓	✓	✓	
abb.mean/aab.mean		✓	✓	✓	✓	
log(WBE)		✓	✓	✓	✓	✓

[a] See [3] for further details.

using the weighted sum: $\frac{\sum_{d \in D} prob(d)(2^{d+1}-1)}{\sum_{d \in D} prob(d)}$ where $prob(d) = 2^{-d}$ and D is the multiset of branches lengths visited. In [3], we extended WBE to support conflict driven backjumping. As the new method requires $O(d)$ time and space, we only compute it every d conflicts. To deal with quick restarts, we wait until the observation window fits within a single restart. In addition, we exploit estimates from earlier restarts by augmenting the feature vector with all the search cost predictions from previous restarts.

3   Experiments

We ran experiments using MiniSat [2], a state-of-the-art solver with clause learning, an improved version of VSIDS and a geometrical restart scheme. We used a geometrical factor of 1.5, which is the default for MiniSat. A geometrical factor of 1.2 gave similar results. We used three different ensembles of problems.

- *rand:* 500 satisfiable and 500 unsatisfiable random 3-SAT problems with 200 to 550 variables and a clause-to-var ratio of 4.1 to 5.0.
- *bmc:* 250 satisfiable and 250 unsatisfiable software verification problems generated by CBMC[1] for on a binary search algorithm, using different array sizes and number of loop unwindings. To generate satisfiable problems, faulty code that causes memory overflow was added. These problems create a very homogeneous ensemble.
- *fv:* 56 satisfiable and 68 unsatisfiable hardware verification problems distributed by Miroslav Velev[2]. This is less homogeneous than the other ensembles.

Since training examples can be scarce, we restricted our training set to no more than 500 problems, though we had far fewer for the hard verification problems. In the first part of our experiments, when restarts were turned off, many of the hardware verification problems were not solved. Our results in this part will only compare the other datasets. When restarts were enabled, all three data sets were used. In all experiments we used 10-fold cross validation, never using the same instance for both training and testing purposes. We measured prediction quality by observing the percentage of predictions within a certain factor of the correct cost (the *error factor*). For example, 80% for error factor 2, denotes that for 80% of the instances, the predicted search cost was within a factor of 2 of the actual cost.

3.1   Search without Restarts

We queried our predictor at different points of the search, ranging from 2000 to 50000 backtracks. Comparisons of the performance of LMP for the *rand* and *bmc* data set are presented in Figure 1. Satisfiable problems are harder to predict for both *rand* and *bmc* datasets, due to the abrupt way in which search terminates with open nodes.

---

[1] http://www.cs.cmu.edu/ modelcheck/cbmc/
[2] http://www.miroslav-velev.com/sat_benchmarks.html

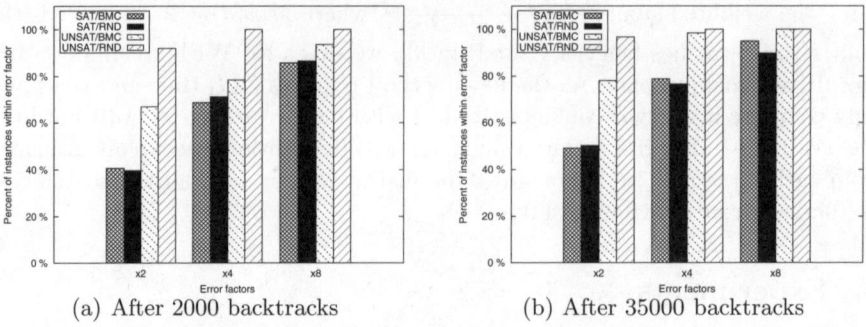

(a) After 2000 backtracks   (b) After 35000 backtracks

Fig. 1. Quality of prediction, without restarts, for the *rand* and *bmc* datasets

3.2 Search with Restarts

With restarts, we have to use smaller observation windows to give a prediction early in search as many early restarts are too small. Figure 2 compares the quality of prediction of LMP for the 3 different datasets. The quality of estimates improves with the *bmc* data set when restarts are enabled. We conjecture this is a result of restarts before the observation window reducing noise.

In order to see if predictions from previous restarts improve the quality of prediction, we opened an observation window at every restart. The window size is $max(1000, 0.01 \cdot s)$ and starts after a waiting period of $max(500, 0.02 \cdot s)$, when s is the size of the current restart. At the end of each observation window, two feature vectors were created. The first (x_r) holds all features from Table 1, while the second $(\hat{x}_r)$ is defined as $\hat{x}_r = \{x_r\} \cup \{f_{w_1}(x_1), f_{\hat{w}_2}(\hat{x}_2), \ldots, f_{\hat{w}_{r-1}}(\hat{x}_{r-1})\}$. Figure 3 compares the two methods. We see that predictions from earlier restarts improve the quality of later predictions but not greatly.

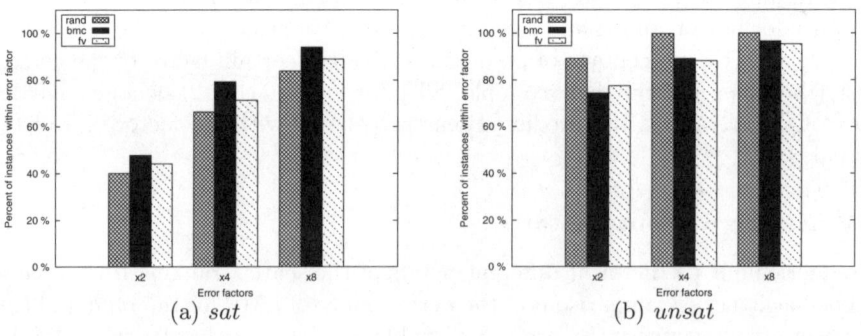

(a) *sat*   (b) *unsat*

Fig. 2. Quality of prediction for the 3 different datasets when using restarts (after 2000 backtracks in the *query restart*)

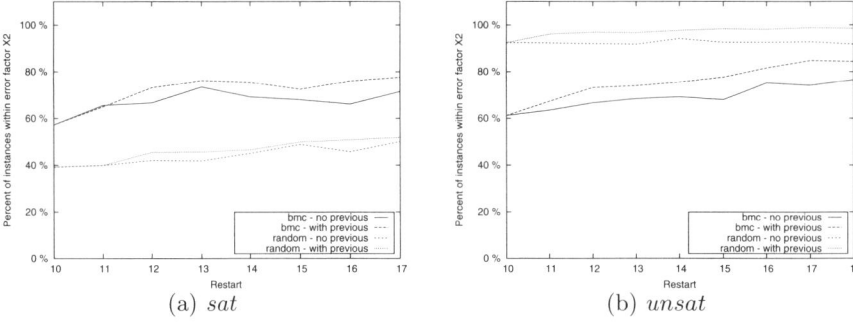

Fig. 3. The effect of using predictions from previous restarts. We compare the quality of prediction, through restarts, using two datasets (*bmc*,*rand*). The plots represent the percentage of instances within a factor of 2 from the correct size.

3.3 Solver Selection Using LMP

In our final experiment, we used these estimations of search cost to improve solver performance. We used two different versions of MiniSat. Solver A used the default MiniSat setting (geometrical factor of 1.5), while solver B used a geometrical factor of 1.2. The challenge is to select which is faster at solving a problem instance.

Table 2 describes the percentage improvement achieved by each of the following strategies. All values are fractions of the cost of solving the entire dataset, picking a solver randomly for each problem, with equal probability. Hence, for each dataset, $average(A, B) = 1$:

- *best:* Use an oracle to indicate which solver will solve the problem faster $(min(A, B))$.
- *LMP (oracle):* Use both solvers until each reaches the end of its observation window and generate a prediction, using two different models for *sat* and *unsat*. Use a satisfiability oracle to indicate which model should be queried. Terminate the solver that is predicted to be worse.
- *LMP (two models):* Use both solvers until each reaches the end of its observation window and generate a prediction, using two different models for *sat* and *unsat*. Query both models and use the geometric mean as the prediction[3]. Terminate the solver that is predicted to be worse.

These results show that for satisfiable problems, where solver performance varies most, our method reduces the total cost. For unsatisfiable problems, where solver performance does not vary as much, our method does not improve search cost. However, as performance does not change significantly on unsatisfiable instances, the overall impact of our method on satisfiable and unsatisfiable problems is positive.

---

[3] We found this method to yield more accurate runtime estimations than using one model for both *sat* and *unsat* instances. For further details see [3].

Table 2. Improvement in total search cost using different schemes

Dataset		Best	LMP (oracle)	LMP(two models)
rand	sat	0.591	0.930	0.895
	unsat	0.925	1.009	1.014
fv	sat	0.333	0.828	0.832
	unsat	0.852	1.006	1.033
bmc	sat	0.404	0.867	0.864
	unsat	0.828	0.997	1.004

References

1. Aloul, F., Sierawski, B., Sakallah, K.: Satometer: How much have we searched? In: Design Automation Conf., pp. 737–742. IEEE, Los Alamitos (2002)
2. Een, N., Sorensson, N.: An extensible SAT-solver. Theory and Applications of Satisfiability Testing, 502–518 (2003)
3. Haim, S., Walsh, T.: SAT Solving Cost Estimation using Online Techniques, Technical Report 0805, UNSW, Australia (February 2008)
4. Horvitz, E., Ruan, Y., Gomes, C., Kautz, H., Selman, B., Chickering, M.: A Bayesian approach to tackling hard computational problems. In: Proc. the 17th Conf. on Uncertainty in Artificial Intelligence (UAI 2001) (2001)
5. Kilby, P., Slaney, J., Thiébaux, S., Walsh, T.: Estimating Search Tree Size. In: Proc. of the 21st National Conf. of Artificial Intelligence, AAAI, Menlo Park (2006)
6. Kokotov, D., Shlyakhter, I.: Progress bar for sat solvers (unpublished manuscript) (2000), http://sdg.lcs.mit.edu/satsolvers/progressbar.html
7. Leyton-Brown, K., Nudelman, E., Shoham, Y.: Learning the Empirical Hardness of Optimization Problems: The Case of Combinatorial Auctions. In: Proc. of the 8th Int. Conf. on Principles and Practice of Constraint Programming, pp. 556–572. Springer, Heidelberg (2002)
8. Samulowitz, H., Memisevic, R.: Learning to Solve QBF. In: Proc. of 22nd Conf. on Artificial Intelligence (AAAI 2007) (2007)
9. Xu, L., Hoos, H.H., Leyton-Brown, K.: Hierarchical Hardness Models for SAT Principles and Practice of Constraint Programming, 696–711 (2007)

A Max-SAT Inference-Based Pre-processing for Max-Clique

Federico Heras and Javier Larrosa⋆

Universitat Politècnica de Catalunya,
Gran Capità 1-3,
08034 Barcelona, Spain

Abstract. In this paper we propose the use of two resolution-based rules for the Max-SAT encoding of the Maximum Clique Problem. These rules simplify the problem instance in such a way that a lower bound of the optimum becomes explicit. Then, we present a pre-processing procedure that applies such rules. Empirical results show evidence that the lower bound obtained with the pre-processing outperforms previous approaches. Finally, we show that a branch-and-bound Max-SAT solver fed with the simplified problem can be boosted several orders of magnitude.

Keywords: Max-SAT, Max-clique, Inference.

1 Introduction

Given an undirected graph, the *maximum clique Problem* (*Max-Clique*) calls for finding a maximum-sized complete subgraph, that is, a subgraph whose vertices are pairwise adjacent. The *Max-Clique* is a prominent combinatorial optimization problem with many applications such as bioinformatics [10, 23, 14] and computer vision [3] to name a few. From the recent literature, there are two types of algorithms to handle the Max-Clique problem. The first one is formed by *branch and bound* algorithms that solve the problem to optimality [11, 25, 22]. The second one is formed by *stochastic local search* solvers that cannot prove optimality, but empirical results show that they return quite accurate upper bounds [24, 5]. Both types of algorithms have a graph as input and they apply techniques that exploit the structure of such graph.

In this paper, we focus on the Max-SAT encoding of the Max-Clique problem and we exploit its properties. We introduce two simplification rules for the Max-Clique problem based on the resolution rule for Max-SAT [16] and we apply them in a preprocessing procedure. The result of the pre-process is an equivalent Max-SAT formula with an explicit lower bound of the optimum. Afterwards, we give the pre-processed instance to the state-of-the-art Max-SAT solver MINIMAXSAT [13]. Empirical results indicate that our pre-processing generates very powerful initial lower bounds. Besides, fetching the Max-SAT solver with the simplified formula can boost the search process in several problem instances.

⋆ Research funded by project TIN2006-15387-C03-0.

H. Kleine Büning and X. Zhao (Eds.): SAT 2008, LNCS 4996, pp. 139–152, 2008.

The structure of this paper is the following. Section 2 introduces all the preliminary notation and concepts about Max-SAT and how to encode the Max-Clique problem as Max-SAT. Then, Section 3 presents the two simplification rules that are used in the pre-processing introduced in Section 4. Section 5 includes the experimental investigation and the related work can be found in Section 6. Finally, Section 7 presents some concluding remarks and points out our future work.

2 Preliminaries

2.1 The Max-SAT Framework

The following notation and terminology has been borrowed from [16]. In the sequel X is a set of boolean variables taking values over the set $\{t, f\}$, which stands for *true* and *false*, respectively. A *literal* is either a variable (e.g. x) or its negation (e.g. $\bar{x}$). We will use $l_1, l_2, l_3, \ldots$ to denote literals and $var(l)$ to denote the variable related to l (namely, $var(x) = var(\bar{x}) = x$). A *clause* $C = l_1 \vee l_2 \vee \ldots \vee l_k$ is a disjunction of literals such that $\forall_{1 \le i,j \le k,\ i \ne j}\ var(l_i) \ne var(l_j)$. The size of a clause, noted $|C|$, is the number of literals that it has. $var(C)$ is the set of variables that appear in C (namely, $var(C) = \{var(l)|l \in C\}$). We refer to a clause as *positive (negative)* if all its literals appear in the positive (negative) polarity. An assignment satisfies a clause iff it satisfies one or more of its literals. If variable x is instantiated to t, literal x is satisfied and literal $\bar{x}$ is falsified. Similarly, if variable x is instantiated to f, literal $\bar{x}$ is satisfied and literal x is falsified. The *empty clause*, noted $\square$, cannot be satisfied. Sometimes it is convenient to think of clause C as its equivalent $C \vee \square$. An *assignment* is an instantiation of a subset of X. The assignment is *complete* if it instantiates all the variables (otherwise it is partial).

A *weighted clause* is a pair (C, w) such that C is a classical clause and w is the cost of its falsification. In this paper we assume costs being natural numbers. A *weighted formula* in conjunctive normal form (CNF) is a set of weighted clauses. The cost of an assignment is the sum of weights of all the clauses that it falsifies.

As shown in [16], the De Morgan rule cannot be used in Max-SAT. Instead, the following rule should be repeatedly used until CNF is achieved:

$$(A \vee \overline{l \vee C}, w) \equiv \{(A \vee \bar{C}, w), (A \vee \bar{l} \vee C, w)\}$$

Following [16], we assume without loss of generality the existence of a known upper bound $\top$ of the optimal solution ($\top$ is a strictly positive natural number). A *model* is a complete assignment with cost less than $\top$. A Max-SAT instance is a pair $(\mathcal{F}, \top)$ and the task of interest is to find a model of minimum cost, if there is any. We say that two weighted formulas are equivalent, $\mathcal{F} \equiv \mathcal{F}'$, if the cost of their optimal assignment is the same or if neither of them has a model.

Observe that any weight $w \ge \top$ indicates that the associated clause *must be necessarily satisfied*. Thus, we can replace w by $\top$ without changing the problem. Consequently, we can assume all costs in the interval $[0..\top]$. A clause with weight $\top$ is called *mandatory* (or *hard*), otherwise it is called *non-mandatory* (or *soft*).

Let u and w be two costs. Their sum is defined as,

$$u \oplus w = min\{u + w, \top\}$$

in order to keep the result within the interval $[0..\top]$. If $u \geq w$, their subtraction is defined as,

$$u \ominus w = \begin{cases} u - w & : & u \neq \top \\ \top & : & u = \top \end{cases}$$

Essentially, $\ominus$ behaves like the usual subtraction except in that $\top$ is an absorbing element.

The identification of mandatory clauses with the $\top$ symbol allows to extend some well-known simplification rules from SAT to Max-SAT such as *addition* $\{(A, u), (A, w)\} \equiv \{(A, u \oplus w)\}$ or *subsumption* $\{(A, \top), (A \vee B, w)\} \equiv \{(A, \top)\}$.

A weighted CNF formula may contain $(\Box, w)$. Since $\Box$ cannot be satisfied, w is added to the cost of any assignment. Therefore, w is an explicit *lower bound* of the optimal model. When the lower bound and the upper bound have the same value (i.e., $(\Box, \top) \in \mathcal{F}$) the formula does not have any model and we call this situation an *explicit contradiction*.

The notion of resolution can be extended to weighted formulas as follows,

$$\{(x \vee A, u), (\bar{x} \vee B, w)\} \equiv \begin{cases} (A \vee B, m), \\ (x \vee A, u \ominus m), \\ (\bar{x} \vee B, w \ominus m), \\ (x \vee A \vee \bar{B}, m), \\ (\bar{x} \vee \bar{A} \vee B, m) \end{cases}$$

where A and B are arbitrary disjunctions of literals and $m = min\{u, w\}$.

$(x \vee A, u)$ and $(\bar{x} \vee B, w)$ are called the *prior clashing clauses*. $(A \vee B, m)$ is called the *resolvent*. $(x \vee A, u \ominus m)$ and $(\bar{x} \vee B, w \ominus m)$ are called the *posterior clashing clauses*. $(x \vee A \vee \bar{B}, m)$ and $(\bar{x} \vee \bar{A} \vee B, m)$ are called the *compensation clauses*.

Example 1. *If we apply resolution to the following clauses* $\{(x_1 \vee x_2, 3), (\bar{x}_1 \vee x_2 \vee x_3, 4)\}$ *(with* $\top = 5$*) we obtain* $\{(x_2 \vee x_2 \vee x_3, 3), (x_1 \vee x_2, 3 \ominus 3), (\bar{x}_1 \vee x_2 \vee x_3, 4 \ominus 3), (x_1 \vee x_2 \vee \overline{(x_2 \vee x_3)}, 3), (\bar{x}_1 \vee \bar{x}_2 \vee x_2 \vee x_3, 3)\}$. *The first and fourth clauses can be simplified. The second clause can be omitted because its weight is zero. The fifth clause can be omitted because it is a tautology. Therefore, we obtain the equivalent formula* $\{(x_2 \vee x_3, 3), (\bar{x}_1 \vee x_2 \vee x_3, 1), (x_1 \vee x_2 \vee \bar{x}_3, 3)\}$.

2.2   Inference-Based Simplification Rules

A Max-SAT problem can be solved to optimality with a pure inference algorithm, namely, an algorithm that only applies the resolution rule [4, 16]. However, such an algorithm has exponential space requirements and it is not used in practice. A natural alternative is to use only restricted forms of resolution that simplify the formula and use search afterwards. The application of a *simplification rule* is simply the application of a limited number of resolution steps. Current Max-SAT solvers apply simplification rules at each node of a search tree. Their main objective is to simplify the problem instance and to make explicit a lower bound (i.e. create new empty clauses). The following example shows the application of two steps of resolution that lead to increase the lower bound.

Example 2. *Consider a weighted formula* $\{(x_1 \vee x_2, 3), (\bar{x}_1 \vee x_2, 2), (\bar{x}_2, 1)\}$ *(with* $\top = 5$*). Suppose we apply the resolution rule between the first and the second clause. We obtain* $\{(x_1 \vee x_2, 1), (x_2, 2), (\bar{x}_2, 1)\}$*. Now, we apply the resolution rule between the second and the third clause so that the lower bound is increased* $\{(x_1 \vee x_2, 1), (x_2, 1), (\Box, 1)\}$*. Observe that the three formulas are equivalent, but the last one is more explicit and presumably simpler.*

2.3 Encoding the Min-Vertex-Covering and Max-Clique as Max-SAT

Definition 1. *Given a graph* $G = (V, E)$*, a* vertex covering *is a set* $U \subseteq V$ *such that for every edge* (v_i, v_j) *either* $v_i \in U$ *or* $v_j \in U$*. The size of a vertex covering is* $|U|$*. The* minimum vertex covering *(Min-Vertex-Covering) problem consists in finding a covering of minimal size.*

The *minimum vertex covering* problem is a well-known NP-Hard problem and it is well-known that it can be naturally formulated as (weighted) Max-SAT. We associate one variable x_i to each graph vertex. Value *true* (respectively, *false*) indicates that vertex x_i belongs to U (respectively, to $V - U$). There is a binary weighted clause $(x_i \vee x_j, \top)$ for each edge $(v_i, v_j) \in E$. It specifies that at least one of these vertices must be in the covering because there is an edge connecting them. There is a unary clause $(\bar{x}_i, 1)$ for each variable x_i, in order to specify that it is preferred not to add vertices to U. $\top$ must be set to a sufficiently large number. Note that different weights in unary and binary clauses are required to express the relative importance of each type of clauses.

Definition 2. *Given a graph* $G = (V, E)$*, a* clique *is a set* $U \subseteq V$ *such that for every vertex* $v \in U$*,* v *is connected to all the vertices in* U*. The size of a clique is* $|U|$*. The* maximum clique *problem (Max-Clique) consists in finding a clique of maximal size.*

The *maximum clique* problem is a well-known NP-Hard problem. As noted in [11], finding the maximum clique of a graph $G = (V, E)$ is equivalent to finding a minimum vertex covering of the complementary graph $\bar{G}$. Given a graph $G = (V, E)$, its complementary graph is denoted by $\bar{G} = (V, \bar{E})$. It is constructed with the same set of vertices V and $(v_i, v_j) \in \bar{E}$ iff $(v_i, v_j) \notin E$. Hence, we can model Max-Clique problems as Minimum Vertex Covering problems over the complementary graph. Observe that the maximum size of the maximum clique is equivalent to $|V| - s$, where s is the size of the minimum vertex covering.

Note that the Max-SAT encoding of the Max-Clique problem only contains negative unit soft clauses and positive binary hard clauses.

3 Two Simplification Rules

In this section we present two simplification rules that can be executed frequently in the Max-SAT encoding of a Max-Clique problem. The correctness of both rules can be easily established with a sequence of resolution steps.

3.1   Star Rule

The *Star Rule* [21] can be used to create new empty clauses from a long clause and a set of appropriate unit clauses.

$$\{(\overline{x_{i1}} \vee \overline{x_{i2}} \vee \ldots \vee \overline{x_{ik}}, w_0), (x_{i1}, w_1), (x_{i2}, w_2), \ldots, (x_{ik}, w_k)\} \equiv$$

$$\begin{cases} (\Box, m), (\overline{x_{i1}} \vee \overline{x_{i2}} \vee \ldots \vee \overline{x_{ik}}, w_0 \ominus m), \\ (x_{i1}, w_1 \ominus m), (x_{i2}, w_2 \ominus m), \ldots, (x_{ik}, w_k \ominus m), \\ (x_{i1} \vee \overline{x_{i2}} \vee \overline{x_{i3}} \vee \ldots \vee \overline{x_{ik}}, m), \\ (x_{i2} \vee \overline{x_{i3}} \vee \overline{x_{i4}} \vee \ldots \vee \overline{x_{ik}}, m), \\ (x_{i3} \vee \overline{x_{i4}} \vee \overline{x_{i5}} \vee \ldots \vee \overline{x_{ik}}, m), \\ \ldots, \\ (x_{ik-1} \vee x_{ik}, m) \end{cases}$$

where $m = min\{w_0, w_1, \ldots, w_k\}$.

The new empty clause is added to the possibly existing one, which produces a lower bound increment. It is clear that the Star Rule may be effective when a large number of unit clauses are available.

Example 3. *Consider the initial formula* $\{(\bar{x}_1 \vee \bar{x}_2, 1), (x_1, 1), (x_2, 1)\}$. *In this example, we show each step of resolution needed to obtain the same result provided by the Star Rule. First, we apply the resolution rule between the first and the third clauses and we obtain* $\{(x_1 \vee x_2, 1), (x_1, 1), (\bar{x}_1, 1)\}$. *Then, we apply the resolution rule between the second and the third clauses to obtain* $\{(x_1 \vee x_2, 1), (\Box, 1)\}$.

3.2   Unit Rule

The original *Unit Rule* can be used to create new unit clauses from a long clause and a set of appropriate binary hard clauses. Given a subset of variables $\{x_{i1}, x_{i2}, \ldots, x_{ik}, x_j\} \subseteq X$, consider the following subset of binary hard clauses:

$$Bin(x_{i1}, x_{i2}, \ldots, x_{ik}, x_j) = \left\{ (x_{i1} \vee x_j, \top), (x_{i2} \vee x_j, \top), \ldots, (x_{ik} \vee x_j, \top) \right\}$$

The *Unit Rule* has the form,

$$\{(\overline{x_{i1}} \vee \overline{x_{i2}} \vee \ldots \vee \overline{x_{ik}}, w), Bin(x_{i1}, x_{i2}, \ldots, x_{ik}, x_j)\} \equiv$$

$$\left\{ (\overline{x_{i1}} \vee \overline{x_{i2}} \vee \ldots \vee \overline{x_{ik}} \vee \overline{x_j}, w), Bin(x_{i1}, x_{i2}, \ldots, x_{ik}, x_j), (x_j, w) \right\}$$

Example 4. *Consider the initial formula* $\{(\bar{x}_1 \vee \bar{x}_2, 1), (x_1 \vee x_3, \top), (x_2 \vee x_3, \top)\}$. *In this example, we show each step of resolution needed to obtain the same result provided by the Unit Rule. First, we apply the resolution rule between the first and the second clauses and we obtain* $\{(\bar{x}_2 \vee x_3, 1), (\bar{x}_1 \vee \bar{x}_2 \vee \bar{x}_3, 1), (x_1 \vee x_3, \top), (x_2 \vee x_3, \top)\}$. *Then, we apply the resolution rule between the first and the last clauses to obtain* $\{(x_3, 1), (\bar{x}_1 \vee \bar{x}_2 \vee \bar{x}_3, 1), (x_1 \vee x_3, \top), (x_2 \vee x_3, \top)\}$.

4   Pre-processing

In this section we show a pre-process that exploits the synergy between the Unit and the Star rules. The unit rule generates unit positive clauses from negative clauses and binary positive hard clauses. This unit clauses are used by the star rule which transforms them into empty clauses, which means an increment of the lower bound. The pre-process works in a on-demand manner: it triggers the unit rule only if it can guarantee that it will allow the subsequent execution of the star rule.

Before introducing the details of the pre-processing, we present a useful definition.

Definition 3. *A negative clause* $(C, w) = (\overline{x_{i1}} \vee \overline{x_{i2}} \vee \ldots \overline{x_{ik}}, w)$ *is* unit-related *with respect to* x', *and it is noted* $(C, w)_{x'}$, *if and only if* $Bin(x_{i1}, x_{i2}, \ldots, x_{ik}, x') \in \mathcal{F}$.

Observe that we can always apply the Unit Rule to a clause C unit-related with respect to literal x in order to generate a new positive unit soft clause (x, w).

The basic idea of the pre-processing is to generate the appropriate unit clauses with the *Unit Rule* so that we can apply the *Star Rule* later in order to increase the lower bound. The final objective is to increase as much as possible the lower bound.

The pre-processing is shown in Algorithm 1. It iterates over all the negative clauses (line 1). For each negative clause (C, w_0) the algorithm wants to obtain one unit clause for each literal in C. To do so, for each literal l_i in C the algorithm seeks a clause (C', w_i) unit-related with respect to l_i (i.e. $(C', w_i)_{l_i}$) and store it in the structure S. Note that all the negative clauses inserted in S must be different, and they must be also different from the initial (C, w_0) (lines 2-5). If it succeeds in finding unit-related clauses for each literal in C (line 6), then the algorithm applies the two simplification rules. First, for each pair in structure S, it applies the Unit Rule in order to create the corresponding unit clause (lines 7,8). Once all unit clauses have been generated, the algorithm proceeds to apply the Star Rule (line 9).

Recall that this process is applied to Max-clique problems (the original formula contains negative soft units and positive hard binary clauses). Therefore, one can easily see that, at any point of the execution of algorithm 1, each negative clause (C, w) is in $\mathcal{F}$ because either (i) (C, w) is an initial unit soft clause in $\mathcal{F}$ or (ii) (C, w) was generated by some application of the Unit Rule. This observation leads to a nice property:

Lemma 1. *Within the pre-processing algorithm, all the compensation clauses in* $\mathcal{F}$ *generated by the Star Rule are subsumed by binary hard clauses in* $\mathcal{F}$.

Proof-Sketch 1. *Consider the case in which all clauses have weight 1 (as it happens in the Max-Clique Problem). Suppose that the Star Rule is applied to an arbitrary subset of clauses in* $\mathcal{F}$:

$$\{(\overline{x_{i1}} \vee \overline{x_{i2}} \vee \ldots \vee \overline{x_{ik}}, 1), (x_{i1}, 1), (x_{i2}, 1), \ldots, (x_{ik}, 1)\}$$

Observe that $\{(\overline{x_{i1}} \vee \overline{x_{i2}} \vee \ldots \vee \overline{x_{ik}}, 1)$ *is in* $\mathcal{F}$ *because the following set of Unit Rules were applied (in reverse order):*

$$\{(\overline{x_{i1}} \vee \overline{x_{i2}} \vee \ldots \vee \overline{x_{ik-1}}, 1), Bin(x_{i1}, x_{i2}, \ldots, x_{ik-1}, x_{ik})\}$$

Algorithm 1. Algorithm to transform the Max-Clique problem into an equivalent but simpler one. Note that each application of the Unit Rule (line 5) generate a new clause to be considered in the main Loop 1

Procedure MC-Preprocessing($\mathcal{F}$)

1   **foreach** $(C, w_0) = (\overline{l_1} \vee \overline{l_2} \vee \ldots \vee \overline{l_k}, w_0) \in \mathcal{F}$ **do**

2      $S := \emptyset$;

3   **foreach** $l_i \in C$ **do**

4   **if** $\exists (C', w_i)$ s.t. $(C', w_i) \neq (C, w_0) \wedge (C', w_i) \notin S \wedge (C', w_i)_{l_i}$ **then**

5        $\quad \lfloor S := S \cup ((C', w_i), l_i)$

6   **if** $|S| = k$ **then**

7   **foreach** $((C', w_i), l_i) \in S$ s.t. $(C', w_i) = (\overline{l'_1} \vee \overline{l'_2} \vee \ldots \vee \overline{l'_p}, w_i)$ **do**

8        $\quad \lfloor$ Apply Unit Rule to $\{(C', w_i), \text{Bin}(l'_1, l'_2, \ldots, l'_p, l_i) \}$;

9        $\lfloor$ Apply Star Rule to $\{(l_1, w_1), (l_2, w_2), \ldots, (l_k, w_k), (C, w_0)\}$;

$$\{(\overline{x_{i1}} \vee \overline{x_{i2}} \vee \ldots \vee \overline{x_{ik-2}}, 1), Bin(x_{i1}, x_{i2}, \ldots, x_{ik-2}, x_{ik-1})\}$$

$$\cdots$$

$$\{(\overline{x_{i1}}, 1), Bin(x_{i1}, x_{i2})\}$$

Precisely, the sets of binary clauses used at each application of the Unit Rule are enough to subsume all the compensation clauses produced by the initial Star Rule.

5 Empirical Results

In this section we present the benchmarks and the algorithms we tested in our empirical evaluation.

5.1 Benchmarks

In our experiments we consider instances that have been used in several works related with the Max-Clique problem.

- Random graph instances for which we solved the Max-Clique problem [16]. They were submitted to Max-SAT Evaluations 2006 and 2007 [2]. A *random graph* is defined by two parameters $\langle n, e \rangle$ where n is the number of nodes and e is the number of edges. Edges are randomly decided using a uniform probability distribution. Those random instances have a fixed number of nodes (150) and the graph density is varied.
- The 66 Max-Clique instances from the DIMACS challenge [15] [1] [11,25,24,5]. They were also submitted to Max-SAT Evaluations 2006 and 2007 [2].

---

[1] ftp://dimacs.rutgers.edu/pub/challenge/graph/benchmarks/clique

- Ke Xu's Max-Clique instances with hidden optimum solutions [30, 29] which are advocated to be very difficult to solve. We considered the following publicly available sets of instances: frb10, frb15, frb20, frb25 and frb30.
- 11 Max-Clique real instances, provided by J.S. Sokol, corresponding to the protein structure alignment problem transformed into the maximum clique problem as described in [10, 16]. In this problem, the goal is to compute a score of similarity between two proteins based on a particular knowledge of their respective tri-dimensional structure.

5.2 Experiments Considered

First, we compare the lower bound obtained with our new pre-processing with respect to previous lower bounds. Second, we study the effect of feeding a Max-SAT solver with the pre-processed instance. The results are presented in plots and tables. In most of the tables, the common columns are: Problem, Nodes and Density that refer to the name of the instance, the number of nodes and the density of the graph, respectively. In tables and plots, OPT refers to value of the optimal solution. For each experiment, additional information is presented. Execution times are presented in seconds. All the experiments were performed on a 3.2 Ghz Intel Pentium with 1 GB and Linux.

5.3 Comparison of the New Lower Bound

Let LB-NEW be the lower bound obtained with the new pre-processing. Our first aim is to compare LB-NEW with respect to previous lower bounds for *Max-SAT* and *WCSP* which are deeply related to our approach. The best current lower bound in Max-SAT solvers is LB-UB [18]. The best current lower bounds for WCSP solvers are EDAC* [8] and OSAC [7]. We did preliminary experiments and observed that the lower bounds computed with LB-UB and EDAC* were similar. Hence, we simply present the results for LB-UB. We also tested WCSP solvers with OSAC [7] but we discarded it for the final experiments because most of its executions were aborted due to a time limit of 600 seconds. See Section 6 for a more detailed explanation of these lower bounds and their relationships. In this experiment we will also report the time needed by the novel pre-processing and we will refer to it as Pre-Time.

Figure 1 reports the results of applying our preprocessing to the max-clique problem of random graphs with 150 nodes and varying number of edges. Note that instances with low graph density have an associated Max-SAT encoding containing a large number of binary hard clauses. The figure shows three values which are average results of 30 instances per point. Two main observations can be extracted. First, the lower bound of the novel preprocessing LB-NEW is quite powerful when the graph density is low, and it is quite near to the optimal solution value. However, it loses accuracy as the graph density increases. The second observation is that LB-NEW is much more accurate than LB-UB.

Figure 2 shows results for the 66 DIMACS instances. The plot reports lower bound gains (as (LB-NEW - LB-UB)/ LB-UB) versus problem density. It can be seen that LB-NEW is typically $60 - 80\%$ higher than LB-UB. The effect of the problem density in this small sample of instances is not very clear, but again it seems that the benefits of LB-NEW are more important in low density graphs.

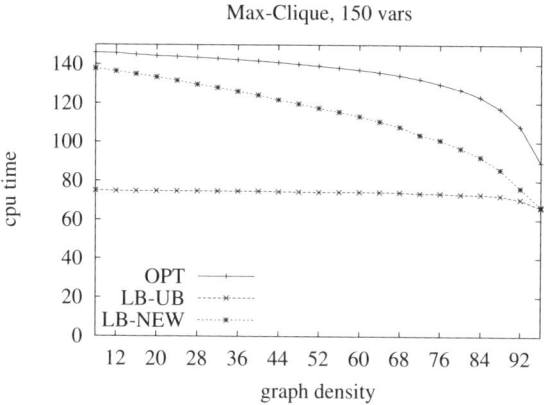

Fig. 1. Lower bounds computed as a pre-processing compared with optimal solutions for random graphs

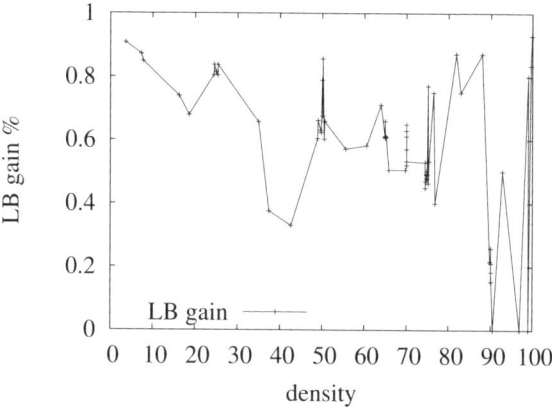

Fig. 2. Lower bound increment as (LB-NEW - LB-UB)/ LB-UB for the 66 Dimacs instances

Observe the time of the pre-processing Time-Pre, the new lower bound LB-NEW, the lower bound LB-UB and the value of the optimal solution OPT in Figures 4 and 5. For the instances with hidden optimum solutions (Figure 4), the time required by the pre-processing is negligible and LB-NEW clearly improves LB-UB. For the protein alignment instances (Figure 5) our pre-processing can be computed in less than 5 seconds for all the instances and LB-NEW is very close to the optimal solution OPT. It is worth to observe that LB-NEW almost doubles LB-UB.

5.4   Feeding a Max-SAT Solver with the Pre-processed Instance

In our second experiment we analyze how the execution time of a Max-SAT solver can be reduced by feeding it with the pre-processed instance. Then we also compare it to specific Max-Clique solvers. However most of the Max-Clique solvers in the literature are not publicly available. The only Max-Clique solver available is DF-MAX [6, 15] that has been used in most of the comparisons of previous works. While DF-MAX is an old algorithm, it is still very efficient on sparse graphs [27]. In that context, we used MINIMAXSAT [13], the overall best branch and bound Max-SAT solver in the 2007 Max-SAT Evaluation [2] and we will refer to it simply as MS. When we feed MS with the pre-processed instance we will refer to it as MS+Pre.

First, we compared the execution times of MS and MS+Pre in the random graphs of Figure 1 but no significant differences were found in execution time.

In Figure 3 we compare the number of solved instances by the Max-SAT solver MS with respect to the rest of specific Max-Clique solvers in the 66 DIMACS instances. For this experiment, we considered a time limit of 2.5 hours so that we can compare with other solvers via a normalization process. The results indicate that the Max-SAT solver MS lays in the middle, being the constraint programming approach [25] and the coloring approach [11] the best current solvers for the DIMACS instances. But, we observed that giving a larger time limit of 8 hours 4 more instances can be solved with MS and MS+Pre, precisely the group of brock400*. We observed that MS+Pre was able to solve the instances san200_0.7_1 and san200_0.7_2 in 0.14 seconds and 116.31 seconds, while MS required 363.47 seconds and 6046.06 seconds, respectively. However, MS was able to solve the instance san400_0.9_1 within 5 minutes, while MS+Pre was not able to solve such an instance. In the other 63 instances, both approaches performed quite similar.

Solver	Solved Instances
Regin [25]	52
Fahle [11]	45
MS	39
MS + Pre	38
Wood [28]	38
Ostergard [22]	36
DF-MAX	31

Fig. 3. Solved instances for the 66 graphs from the Second DIMACS challenge [15] within a time limit of 2.5 hours

From the previous results, it seems that our new pre-processing generates a powerful initial lower bound but it does not allow important speed-ups for a Max-SAT solver. In what follows, we show that such speed-ups occur in some specific instances.

Observe the results in Figures 4 and 5. They include the execution time of MS, MS+Pre and DF-MAX. The time limit was set to 4 hours.

Regarding the instances with hidden optimum solutions (Figure 4), MS and DF-MAX performed quite similar and solved exactly the same number of instances. Differently, MS+Pre is about two orders of magnitude faster than the other approaches. Furthermore, it is able to solve all the instances of the frb25 set, and one instance of the frb30 set.

Problem	nodes	density	Time-Pre	MS	DF-MAX	MS+NEW	LB-UB	LB-NEW	OPT
frb10-6-1	60	65.71	0.00	0.00	0.00	0.00	30	45	50
frb10-6-2	60	64.63	0.00	0.00	0.00	0.00	30	43	50
frb10-6-3	60	66.05	0.00	0.00	0.00	0.00	30	43	50
frb10-6-4	60	63.95	0.00	0.00	0.00	0.00	30	44	50
frb10-6-5	60	64.24	0.00	0.00	0.00	0.00	30	43	50
frb15-9-1	135	72.22	0.00	1.65	1.00	0.10	67	106	120
frb15-9-2	135	72.21	0.00	1.11	1.00	0.22	67	102	120
frb15-9-3	135	72.40	0.00	1.31	1.00	0.14	67	105	120
frb15-9-4	135	72.32	0.00	1.05	1.00	0.25	67	105	120
frb15-9-5	135	71.69	0.00	1.58	1.00	0.37	67	105	120
frb20-11-1	220	76.61	0.00	295.94	418.00	2.40	110	173	200
frb20-11-2	220	76.86	0.01	267.54	412.00	29.39	110	172	200
frb20-11-3	220	76.70	0.00	411.83	483.00	4.84	110	175	200
frb20-11-4	220	76.87	0.00	490.59	450.00	5.58	109	174	200
frb20-11-5	220	76.75	0.00	556.99	698.00	16.72	110	173	200
frb25-13-1	325	79.87	0.02	-	-	2212.94	162	260	300
frb25-13-2	325	79.83	0.01	-	-	583.69	162	261	300
frb25-13-3	325	79.72	0.01	-	-	247.34	162	256	300
frb25-13-4	325	80.18	0.01	-	-	432.19	162	260	300
frb25-13-5	325	80.04	0.02	-	-	930.17	161	259	300
frb30-15-1	450	82.28	0.02	-	-	-	225	360	-
frb30-15-2	450	82.24	0.02	-	-	-	224	364	-
frb30-15-3	450	82.28	0.02	-	-	-	225	363	-
frb30-15-4	450	82.28	0.02	-	-	-	225	366	-
frb30-15-5	450	82.31	0.03	-	-	5772.32	225	366	420

Fig. 4. Instances with hidden optimum solutions. 4 hours of time limit.

Problem	nodes	density	Time-Pre	MS	DF-MAX	MS+NEW	LB-UB	LB-NEW	OPT
1bpi–2knt	2436	15.06	3.80	-	528.00	564.99	1218	2372	2407
1bpi–5pti	3016	15.36	5.62	-	2525.00	456.18	1508	2945	2974
1knt–1bpi	2494	14.86	3.80	-	430.00	375.45	1247	2429	2464
1knt–2knt	1806	14.76	2.02	358.93	19.00	87.22	903	1751	1767
1knt–5pti	2236	15.15	3.19	-	226.00	13441.60	1118	2179	2208
1vii–1cph	171	10.88	0.01	0.03	0.00	0.19	85	158	165
2knt–5pti	2184	15.28	2.96	-	238.00	327.04	1092	2124	2156
3ebx–1era	2548	14.72	4.04	-	886.00	865.05	1274	2483	2517
3ebx–6ebx	1768	14.45	1.99	1532.70	88.00	94.56	884	1717	1740
6ebx–1era	1666	14.35	1.89	2705.69	45.00	104.35	833	1616	1646
sandiaprot	2279	14.83	3.30	-	189.00	7710.05	1139	2220	2248

Fig. 5. Protein structure alignment problem transformed into Max-Clique. 4 hours of time limit.

Regarding the protein alignment instances (Figure 5) the novel lower bound LB-NEW is very close to the optimal solution OPT. MS can solve only four instances within the time limit, while MS+Pre is able to solve all the instances and 9 of them in less than fifteen minutes. The performances of DF-MAX and MS+Pre are quite similar in most of the instances. Recall that DF-MAX is still the best specific solver for low density instances like those. Observe that in [16] a Max-SAT solver called MAX-DPLL was able to solve 10 of the 11 instances. MAX-DPLL applies at each node of the search tree the *cycle rule* which can be seen as a very limited version of our more general approach. However, the current implementation of the cycle rule has severe memory limitations [16].

6   Related Work

There exists a handful of specific Max-Clique *branch and bound* solvers [6, 15, 22, 11, 25, 27]. They mainly differ in their bounds. The best current upper bounds are based on constraint programming techniques [25] and on approximate graph coloring techniques [11, 27].

Besides specific algorithms, Max-Clique can also be solved with generic solvers. In the following, we review Max-SAT and *WCSP* approaches.

Current complete algorithms for Max-SAT are also branch and bound algorithms. In that context, the upper bound ub is the cost of the best complete assignment found so far and the lower bound (lb) is the sum of the weights of the clauses in the original formula violated by the current partial assignment plus an *underestimation* of the cost of extending the current partial assignment. lb and ub are used to avoid visiting useless parts of the search tree when $lb \geq ub$. Most of them compute underestimations based on detecting *inconsistent subsets*: Given a WCNF formula $\mathcal{F}$, an *inconsistent subset* is a subset of clauses of $\mathcal{F}$ such that at least one of the clauses is always unsatisfied by any assignment of the variables. When an inconsistent subset is detected, two approaches are possible:

- Relaxation: Remove the clauses involved in the inconsistent subset from the formula and increase the underestimation [18].
- Inference: Apply the resolution rule to create an equivalent formula with new empty clauses [16].

Best current Max-SAT solvers use unit propagation (UP) to detect inconsistent subsets and then they apply a mixture of the previous two approaches [13, 20].

The *star rule* [21] captures the following inconsistent subset that can be also detected via UP:

$$\{(\overline{x_1} \vee \overline{x_2} \vee \ldots \vee \overline{x_k}, w_0), (x_1, w_1), (x_2, w_2), \ldots, (x_k, w_k)\}$$

If we relax the formula, the underestimation can be increased by $min\{w_0, w_1, \ldots, w_k\}$. This was applied during search in [26, 1] restricted to $k = 2$ and in general in [18, 20].

Following the inference-based approach, we have precisely the same transformation presented in Section 3. It was applied in [12, 20] restricted to $k = 2$ and in general in [13].

Let S be the largest subset of hard binary clauses of the Max-Clique problem with no literals in common among them. For each clause $(x_i \lor x_j, \top)$ in S and their respective unit clauses $(x_i, 1)$ and $(x_j, 1)$, we can relax the formula by removing them and increasing the underestimation by 1. Hence, we can obtain a resulting lower bound of $LB_\mathcal{P} = |S|$. It is precisely the best lower bound that can be computed via UP and then relaxing the formula like LB-UB used in Section 5. A lower bound based on detecting inconsistent subsets via UP and then transforming the formula may be greater than $LB_\mathcal{P}$. However, in practice it is always quite near to $LB_\mathcal{P}$. The main observation of our work is that the binary hard clauses of the Maximum Clique problem lead to generate lots of unit clauses that can be used later by the star rule once and again.

The Max-SAT problem was reformulated as a *Weighted Constraint Satisfaction Problem (WCSP)* [17] in [9, 8] with boolean variables and weighted constraints. Current WCSP solvers apply an inference-based method called *Existential Directional Arc Consistency (EDAC*)* [8]. The application of EDAC* in a WCNF formula only affects to unit and binary clauses that are replaced by other unary and binary clauses and a weighted empty clause. A lower bound based on EDAC* is always equal or lower than $LB_\mathcal{P}$. The most relevant difference of our approach is that we allow the creation of larger arity clauses.

7   Conclusions and Future Work

In this paper we present a pre-processing based on the Max-SAT resolution rule for the Maximum Clique problem (and other related problems). We show empirically how it can be used to obtain a powerful initial lower bound and, in some cases, how the search process of a complete systematic search algorithm can be boosted several orders of magnitude. The following step of our research plan will consist in powering a branch and bound algorithm with our new algorithm at each node of the search tree. This may lead to solve the Maximum Clique problem in a very efficient way. To do so, we are currently working on the necessary data-structures so that the algorithm can be performed in a very fast way.

References

1. Alsinet, T., Manyà, F., Planes, J.: Improved exact solvers for weighted Max-SAT. In: Bacchus, F., Walsh, T. (eds.) SAT 2005. LNCS, vol. 3569, pp. 371–377. Springer, Heidelberg (2005)
2. Argelich, J., Li, C.M., Manyá, F., Planes, J.: The first and second max-sat evaluations. Journal on Satisfiability, Boolean Modeling and Computation (to appear, 2008)
3. Balas, E., Yu, C.S.: Finding a maximum clique in an arbitrary graph. SIAM Journal of Computation 15(4), 1054–1068 (1986)
4. Bonet, M.L., Levy, J., Manyà, F.: Resolution for max-sat. Artificial Intelligence 171(8-9), 606–618 (2007)
5. S.B.: A new trust region technique for the maximum weight clique problem. Discrete Applied Mathematics 154(15), 2080–2096 (2006)
6. Carraghan, R., Pardalos, P.: An exact algorithm for the maximum clique problem. Operations Research Letters 9, 375–382 (1990)

7. Cooper, M.C., de Givry, S., Schiex, T.: Optimal soft arc consistency. In: Proceedings of IJCAI, pp. 68–73 (2007)
8. de Givry, S., Heras, F., Larrosa, J., Zytnicki, M.: Existential arc consistency: getting closer to full arc consistency in weighted csps. In: Proceedings of IJCAI (2005)
9. de Givry, S., Larrosa, J., Meseguer, P., Schiex, T.: Solving max-sat as weighted csp. In: Rossi, F. (ed.) CP 2003. LNCS, vol. 2833, pp. 363–376. Springer, Heidelberg (2003)
10. Barnes, E., Strickland, D.M., Sokol, J.S.: Optimal protein structure alignment using maximum cliques. Operations Research 53, 389–402 (2005)
11. Fahle, T.: Simple and fast: Improving a branch-and-bound algorithm for maximum clique. In: Möhring, R.H., Raman, R. (eds.) ESA 2002. LNCS, vol. 2461, pp. 485–498. Springer, Heidelberg (2002)
12. Heras, F., Larrosa, J.: New inference rules for efficient max-sat solving. In: AAAI (2006)
13. Heras, F., Larrosa, J., Oliveras, A.: Minimaxsat: A new weighted max-sat solver. In: Marques-Silva, J., Sakallah, K.A. (eds.) SAT 2007. LNCS, vol. 4501, pp. 41–55. Springer, Heidelberg (2007)
14. Ji, Y., Xu, X., Stormo, G.D.: A graph theoretical approach for predicting common RNA secondary structure motifs including pseudoknots in unaligned sequences. Bioinformatics 20(10), 1603–1611 (2004)
15. Johnson, D.S., Trick, M.: Second DIMACS implementation challenge: cliques, coloring and satisfiability. DIMACS Series in Discrete Mathematics and Theoretical Computer Science, vol. 26, AMS (1996)
16. Larrosa, J., Heras, F., de Givry, S.: A logical approach to efficient Max-SAT solving. Artificial Intelligence, an international journal 172, 204–233
17. Larrosa, J., Schiex, T.: Solving weighted csp by maintaining arc-consistency. Artificial Intelligence 159(1-2), 1–26 (2004)
18. Li, C.M., Manyà, F., Planes, J.: Exploiting unit propagation to compute lower bounds in branch and bound Max-SAT solvers. In: van Beek, P. (ed.) CP 2005. LNCS, vol. 3709, pp. 403–414. Springer, Heidelberg (2005)
19. Li, C.M., Manya, F., Planes, J.: Detecting disjoint inconsistent subformulas for computing lower bounds for Max-SAT. In: Proceedings of AAAI (2006)
20. Li, C.M., Manyà, F., Planes, J.: New inference rules for max-sat. Journal of Artificial Intelligence Research 30, 321–359 (2007)
21. Niedermeier, R., Rossmanith, P.: New upper bounds for maximum satisfiability. Journal of Algorithms 36(1), 63–88 (2000)
22. Ostergard, P.R.J.: A fast algorithm for the maximum clique problem. Discrete Applied Mathematics 120, 197–207 (2002)
23. Pevzner, P.A., Sze, S.: Combinatorial approaches to finding subtle signals in DNA sequences. In: ISMB, pp. 269–278 (2000)
24. Pullan, W.J., Hoos, H.H.: Dynamic local search for the maximum clique problem. J. Artif. Intell. Res (JAIR) 25, 159–185 (2006)
25. Régin, J.-C.: Using constraint programming to solve the maximum clique problem. In: Rossi, F. (ed.) CP 2003. LNCS, vol. 2833, pp. 634–648. Springer, Heidelberg (2003)
26. Shen, H., Zhang, H.: Study of lower bounds for Max-2-SAT. In: AAAI (2004)
27. Tomita, E., Seki, T.: An efficient branch-and-bound algorithm for finding a maximum clique. In: Calude, C.S., Dinneen, M.J., Vajnovszki, V. (eds.) DMTCS 2003. LNCS, vol. 2731, pp. 278–289. Springer, Heidelberg (2003)
28. Wood, D.: An algorithm for finding maximum cliques in a graph. Operations Research Letters 21, 211–217 (1997)
29. Xu, K., Boussemart, F., Hemery, F., Lecoutre, C.: A simple model to generate hard satisfiable instances. In: Proceedings of IJCAI, pp. 337–342 (2005)
30. Xu, K., Li, W.: Many hard examples in exact phase transitions with application to generating hard satisfiable instances. CoRR, cs.CC/0302001 (2003)

SAT, UNSAT and Coloring

Kazuo Iwama[*]

School of Informatics, Kyoto University, Kyoto 606-8501, Japan
iwama@kuis.kyoto-u.ac.jp

In this survey, we study recent developments on the CNF satisfiability problem. The first one is about deterministic (and of course exponential-time) algorithms for k-SAT. The most recent improvement for $k = 3$ and 4 is based on the nontrivial combination of the Schöning's local search algorithm and the backtrack-type algorithm by Paturi, Pudlák, Saks, and Zane. This approach is due to Iwama and Tamaki and the current fastest algorithm, based on the same method, is due to Rolf, which runs in time $O(1.32216^n)$.

The second topic is on the inapproximability of MAX-3SAT and related problems, based on the famous PCP Theory. Recently, there was an important progress in this field, the Unique Games Conjecture (UGC), by Khot. UGC implies several optimal inapproximability results, such as Vertex Cover.

The third one is the proof complexity of unsatisfiable formulas. Whether or not extended Frege systems, the most powerful proof systems ever known, are polynomially bounded is a most important open question in this field. Pitassi and Urquhart proved that the above open question is equivalent to whether the Hajós calculus, which is a simple, nondeterministic procedure for generating non-3-colorable graphs, is polynomially bounded. Thus, the famous open question in proof complexity is beautifully linked to the open question in graph theory; in order to prove superpolynomial lower bounds for the extended Frege systems, it now suffices to find a "hard example" from the set of non-3-colorable graphs. Thanks to the long and extensive research history of graph theory and graph algorithms, this is hopefully easier than finding a hard example from the set of formulas. Recently Iwama and Tamaki made another step toward this direction by showing that it still suffices if Hajós calculus is restricted to within the class of planar graphs, not only for the final graph but also intermediate ones.

---

[*] Supported in part by Scientific Research Grant, Ministry of Japan, 16300002 and 19200001.

H. Kleine Büning and X. Zhao (Eds.): SAT 2008, LNCS 4996, p. 153, 2008.
© Springer-Verlag Berlin Heidelberg 2008

Computation of Renameable Horn Backdoors*

Stephan Kottler, Michael Kaufmann, and Carsten Sinz

Eberhard Karls Universität Tübingen, Wilhelm–Schickard–Institute, Tübingen,
Germany

Abstract. Satisfiability of real-world SAT instances can be often decided
by focusing on a particular subset of variables - a so-called Backdoor Set.
In this paper we suggest two algorithms to compute Renameable Horn
deletion backdoors. Both methods are based on the idea to transform
the computation into a graph problem. This approach could be used as
a preprocessing to solve hard real-world SAT instances. We also give some
experimental results of the computations of Renameable Horn backdoors
for several real-world instances.

1 Introduction

It is a well known phenomenon that SAT instances evolving from industrial
applications can be solved much faster than this could be expected from the
theoretical point of view. This allows current state-of-the-art solvers for dealing
with instances that consist of up to hundreds of thousands variables. To decide
satisfiability for industrial instances it is often sufficient to focus on a particular and primarily small subset of variables - a so-called *backdoor set*. In the
groundbreaking work [20] Williams, Gomes and Selman already gave examples
of instances with approximately 6,700 variables and nearly 440,000 clauses that
exhibit backdoor sets with only 12 variables. Ruan, Kautz and Horvitz showed
empirically that an extension of the concept of backdoor sets is a good predictor
for the hardness of SAT problems [16]. Moreover, Interian showed that random
3-SAT instances exhibit backdoor sets with 30% to 65% of all variables [10].

Knowing a small backdoor set for an instance in advance could speed up
the solving process extraordinarily. However, according to the work of Szeider
[18], it is in general not possible to decide in reasonable time whether a given
SAT instance exhibits a backdoor with limited size with respect to a DPLL
based subsolver (see [7,6]). Throughout this paper we consider a variant of strong
backdoors (see [20]), so-called *deletion backdoors* [13,19]:

A backdoor is defined with respect to a base class C of formulas that can be recognized and solved in polynomial time. $B \subset V$ is a *deletion backdoor* if the formula
$F - B$ belongs to C, where $F - B$ denotes the result of removing all occurrences
(both positive and negative) of the variables in B from the clauses of formula F.

---

* This work was partly supported by DFG-SPP 1307, project "Structure-based Algorithm Engineering for SAT-Solving".

H. Kleine Büning and X. Zhao (Eds.): SAT 2008, LNCS 4996, pp. 154–160, 2008.

Nishimura, Ragde and Szeider proved that every deletion backdoor is a strong backdoor, if the base class $\mathcal{C}$ is *clause-induced* ($F \in \mathcal{C} \Rightarrow F' \in \mathcal{C}$ for all $F' \subseteq F$) [13]. For the computation of backdoors with base class Horn and 2-SAT the same authors proved fixed-parameter tractability. Hence, the question whether a formula exhibits a Horn backdoor (respectively a Binary backdoor) with at most k variables can be answered in time that is only exponential in k but not in the number of variables [12]. Moreover, Interian approximated backdoors with respect to the base classes Horn and 2-SAT for random 3-SAT instances [10].

In this article we study the computation of Renameable Horn backdoors. Thus, the base class $\mathcal{C}$ is Renameable Horn. A formula is Horn, if every clause contains at most one positive literal and it is Renameable Horn (RHorn) if it can be renamed to a Horn formula by flipping the literals of some variables.

Paris et al. used a two phase approach to compute RHorn backdoors as a preprocessor in a modification of the zChaff SAT solver [15]. In a first step the algorithm tries to increase the number of Horn clauses by flipping the literals of some variables in a local search manner. Secondly, variables are chosen for the backdoor in a greedy fashion to make all non-Horn clauses become Horn.

In a recent work Dilkina, Gomes and Sabharwal formulated linear programs to compute optimal RHorn backdoors [9]. An important result is that smallest RHorn backdoors can be exponentially larger than general strong backdoors.

2 Two Approaches to Compute RHorn Backdoors

The computation of RHorn backdoors of both approaches presented in this article is based on an equivalent graph problem. The second approach approximates the minimum RHorn backdoor with an approximation ratio that is equal to the size of a so-called *conflict loop* in the graph. Furthermore, when using the second algorithm, the exact approximation ratio is known as soon as the RHorn backdoor is computed. In the following subsection we briefly describe how to transform the problem of finding RHorn backdoors to a problem on directed graphs.

Renameable Horn Backdoors as Graph Problem. For a given formula F we create a so-called *dependency graph* $G = (V_G, E_G)$ with $2 * |\mathcal{V}|$ vertices. Each variable v_i entails two vertices k_i^0 and k_i^1 that represent the facts that variable v_i has to be renamed (k_i^0) respectively must not be renamed (k_i^1) in order to make F a Horn formula. The directed edges of G represent the implications of renaming or not renaming variables, according to the clauses of F. A RHorn dependency graph can be created in time $O(m * \text{size\_of\_max\_clause}^2)$ by traversing all possible pairs of literals for each of overall m clauses.

Lewis introduced a method to decide whether a given formula F belongs to class Renameable Horn [11]. The conditions that have to be satisfied to rename F to a Horn formula are formulated as a 2-SAT instance S. It was proved that F is Renameable Horn iff S is satisfiable [11]. The described dependency graph corresponds to the implication graph in [3], that could be used to solve the 2-SAT instance S. In difference to the algorithm in [3] our computations do not deal with strongly connected components. However, the following properties of implication graphs that are needed for our algorithms can be found in [3,4,14] or derived straightforwardly from these results.

Definition 1. *We call a vertex k_i^q ($q \in \{0,1\}$) a conflict vertex if there is a path from k_i^q to $k_i^{(q \oplus 1)}$. A variable $x_i \in \mathcal{V}$ has a conflict loop if k_i^0 and k_i^1 are both conflict vertices.*

Corollary 1. *If there is no path from k_i^q to $k_i^{(q \oplus 1)}$ then none of the vertices that can be reached from k_i^q is a conflict vertex.*

Lemma 1. *A formula F is Renameable Horn iff there exists no variable that has a conflict loop in the dependency graph.*

Corollary 2. *If variable $x_i \in \mathcal{V}$ does not have a conflict loop than neither vertex k_i^0 nor vertex k_i^1 can be involved in a conflict loop of any other variable.*

According to Lemma 1 the task to compute a RHorn backdoor can be accomplished by destroying all conflict loops in the appropriate dependency graph. In particular, we aim to delete a minimal amount of variables from the Boolean formula such that the deletion of the according vertices and their incident edges results in a dependency graph without any conflict loops. We call the set of variables involved in a conflict loop a *conflict set*. It is important to notice that a conflict loop and its conflict set do not necessarily have to have the same size.

A heuristic to destroy all conflict loops. The first approach mainly considers small conflict sets and variables that occur in many of these conflict sets. The implementation is based on the function COMPUTECONFLICTSETS($G, \mathcal{U}$) that computes one conflict set for each variable in $\mathcal{U} \subseteq \mathcal{V}$ with respect of the dependency graph G. For each variable x_i in $\mathcal{U}$ a conflict loop is computed by checking whether there is a path from vertex k_i^0 to k_i^1 and vice versa. All variables that occur in one of the two computed conflict paths constitute the conflict set for variable x_i. If, on the other hand there is no path from k_i^q to $k_i^{(q \oplus 1)}$ ($q \in \{0,1\}$) then we know by Corollary 1 that none of the vertices $R_i \subseteq V_G$ that can be reached from vertex k_i^q can be a conflict vertex. By Corollary 2 we can disregard the according variables ($\mathcal{R}$) of the vertices in R_i for the remaining computation. Thus, for each variable in $\mathcal{R}$ both representing vertices and their incident edges can be deleted from the dependency graph.

The entire computation of a RHorn backdoor starts with creating the dependency graph and computes a small conflict set for each variable of the formula. It starts with an empty backdoor set $\mathcal{B}$ and chooses greedily one variable for $\mathcal{B}$ that occurs most frequently in all known conflict sets S. Ties are broken in favor of variables that occur in small conflict sets. We have applied different strategies to choose a variable for the backdoor, but none of them clearly outperformed the described one. The according vertices of the chosen variable and their incident edges are removed from the dependency graph. New conflict sets are then computed for those variables whose conflict loops were destroyed. At this point the graph may shrink rapidly for some instances, due to the simplification rules in procedure COMPUTECONFLICTSETS. The algorithm terminates as soon as all conflict loops are destroyed.

Approximating minimal Renameable Horn Backdoors. The second approach to compute Reanameable Horn backdoors basically adapts the idea of [8] to approximate a weighted FEEDBACK VERTEX SET in a directed graph. The Algorithm is divided into two phases. In the first phase conflict loops in the graph are destroyed by always taking all related variables of a chosen conflict loop for the backdoor. In the second phase the algorithm tries to shrink the backdoor by reinserting the related vertices and edges of some backdoor variables, ensuring that no conflict loops are created. Since an optimal backdoor has to contain at least one variable of any conflict set, it is evident that always taking all variables of any remaining conflict set into the backdoor (phase 1) already approximates the optimal RHorn backdoor by a factor that is smaller or equal to the size of the biggest chosen conflict set. Due to the fact that in the second phase of the algorithm the found backdoor can be only improved the approximation ratio applies for the entire Algorithm. To keep the approximation ratio small it is reasonable to choose as small as possible conflict loops in the first phase.

Using the reachability data structure introduced in [8] the time to destroy all conflict loops in a given graph can be bounded by $O(|\mathcal{V}|^3)$ for both algorithms. However, for industrial SAT instances the computation benefits from the fact that on the one hand, at the beginning, small conflict paths are computed which requires clearly less than the worst-case bound. On the other hand, at the end, the number of vertices and edges has substantially decreased (see section 3).

3   Some Experimental Results

In order to get an idea of the sizes of backdoors of real-world SAT instances we computed RHorn backdoors for several instances in [1,2,17]. For the computation of Horn and Binary backdoors it turned out that a simple greedy strategy yields the best results for most instances. For Binary backdoors we always choose that variable for the backdoor that reduces the size of the most clauses with more than two literals, terminating as soon as all clauses are binary. The computation of Horn backdoors can be done analogously. A few results are listed in Fig. 1.

Especially the last two rows, the results for the two instances *eq.atree.braun** have to be emphasized. Though relatively small, both instances could not be solved by any solver in the sat competition 2007 within the allowed time (10,000 seconds). Our heuristic found a RHorn backdoor with 761 variables for the instance *eq.atree.braun.13** in less than four minutes. Although a solving process cannot examine all 2^{761} Renameable Horn instances, this still reduces the amount of 'relevant' variables by more than 62%.

It is also worth to mention the good results for instances from Car Configuration [17]. E.g., the optimal RHorn backdoors for the two instances *C208_FA** contain 4.51% resp. 7.46% of all variables [9]. For these instances the heuristic found backdoors with 4.73% resp. 8.21% of all variables (lines 2,3).

An alternative approach to compute RHorn backdoors was used in [15] as a preprocessor in a modified zChaff SAT solver. For the most instances that are given in [15] our algorithm could discover slightly smaller RHorn backdoors.

Instance	# Vars	# Cls	Binary	Horn	RHorn
C169_FW	1402	1982	56	59	**2**
C208_FA_SZ_120	1608	5278	161	168	**76**
C208_FA_UT_3254	1876	7334	419	434	**154**
apex7_gr_rcs_w5.shuffled.cnf	1500	11695	900	832	**635**
dp10s10.shuffled.cnf	7759	23004	2005	3256	**1543**
vda_gr_rcs_w9.shuffled.cnf	6498	130997	5054	4695	**4262**
cnf-r4-b4-k1	9528	59248	8363	4569	**576**
comb3	4774	16331	1641	2095	**1119**
dp06u05	2055	6053	560	889	**457**
ezfact256_1	49153	324873	**24092**	32936	35998
f2clk_30	20458	59559	7109	7813	**3338**
par32-4	3176	10313	**463**	1658	1290
cnf-r4-b1-k1.1-03-416	2424	14812	2113	1141	**174**
f2clk_40-03-424	27568	80439	9597	10562	**4514**
eq.atree.braun.12.unsat	1694	5726	686	1003	**647**
eq.atree.braun.13.unsat	2010	6802	822	1194	**761**

Fig. 1. Computation of different Backdoors. The three rightmost columns indicate the sizes of the found backdoors with base classes 2-SAT, Horn and RHorn. The smallest backdoor of an instance is highlighted with bold font. Due to space limitations information about runtime etc. are omitted.

However, for a few instances like e.g. *dp10s10** with 8,372 variables our heuristic did considerably better: The local search strategy found a backdoor with 2,635 variables [15], whereas the described heuristic discovered a backdoor with 1,543 variables. The reason for this might be that unlike the local search approach, a computation based on the dependency graph is mainly independent of the number of renamings that have to be made to make the remaining instance $F - \mathcal{B}$ Horn.

A further interesting aspect when analyzing the computation of RHorn backdoors for industrial SAT instances is the simplification of the dependency graph. In Fig. 2 it can be observed that the simplification of the dependency graphs for instances of the same family behaves similar. For easy instances like those of

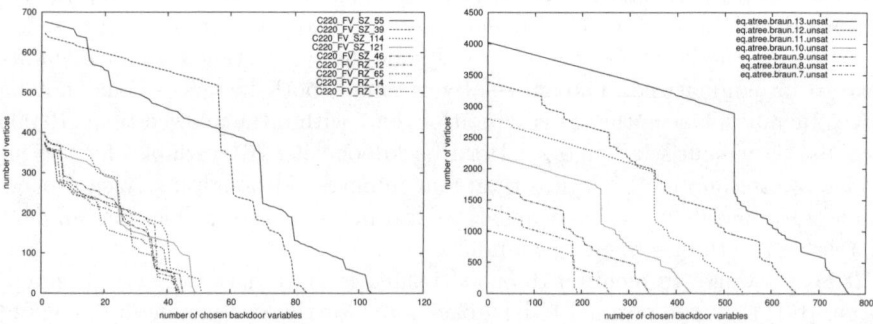

Fig. 2. Simplification of dependency graphs for different families of SAT instances from [2,17]. The y-axis indicate the number of vertices in the graph and the x-axis indicate the number of variables that are chosen for the present backdoor.

the family $C220\_FV^*$ (left plot) there are several break downs where numbers of vertices can be disregarded and hence deleted according to Corollary 2. On the other hand the computation of backdoors for the very hard real-world instances of the family $eq.atree.braun^*$ (right plot) nearly behaves like the computations for generated instances. Applying Corollary 2 is practically impossible in the first two-thirds of the computation.

4 Conclusions and Further Work

In this paper we have presented two approaches to compute RHorn backdoors for CNF formulas in polynomial time. Both approaches are based on the idea to destroy conflict loops in a RHorn dependency graph. We think that this idea could be used as a preprocessing step for solving small but hard real-world SAT instances in order to drastically reduce the amount of variables to consider for the solving process. For the more general case where a minimal amount of variables has to be deleted in order to make a 2-SAT instance satisfiable, the idea of destroying conflict loops in the implication graph of [3] can be adapted.

Furthermore, it is still an open problem if the computation of a minimum RHorn backdoor is fixed parameter tractable. In 2007 the FEEDBACK VERTEX SET problem in directed graphs was proved to be in FPT [5]. It might be possible to adapt their approach for RHorn dependency graphs to remove vertices in order to destroy all conflict loops in the graph.

References

1. Dimacs,
   ftp://dimacs.rutgers.edu/pub/challenge/satisfiability/benchmarks/
2. The international SAT competition (2002-2007),
   http://www.satcompetition.org
3. Aspvall, B., Plass, M.F., Tarjan, R.E.: A linear-time algorithm for testing the truth of certain quantified boolean formulas. Inf. Proc. Lett. 8, 121–123 (1979)
4. Buresh-Oppenheim, J., Mitchell, D.G.: Minimum witnesses for unsatisfiable 2CNFs. In: Biere, A., Gomes, C.P. (eds.) SAT 2006. LNCS, vol. 4121, Springer, Heidelberg (2006)
5. Chen, J., Liu, Y., Lu, S.: Directed feedback vertex set problem is fpt. In Structure Theory and FPT Algorithmics for Graphs, Digraphs and Hypergraphs (2007)
6. Davis, M., Logemann, G., Loveland, D.: A machine program for theorem-proving. Commun. ACM 5(7), 394–397 (1962)
7. Davis, M., Putnam, H.: A computing procedure for quantification theory. J. ACM 7(3), 201–215 (1960)
8. Demetrescu, C., Finocchi, I.: Combinatorial algorithms for feedback problems in directed graphs. Inf. Process. Lett. 86(3), 129–136 (2003)
9. Dilkina, B., Gomes, C.P., Sabharwal, A.: Tradeoffs in the complexity of backdoor detection. In: Bessière, C. (ed.) CP 2007. LNCS, vol. 4741, Springer, Heidelberg (2007)
10. Interian, Y.: Backdoor sets for random 3-sat. In: Giunchiglia, E., Tacchella, A. (eds.) SAT 2003. LNCS, vol. 2919, Springer, Heidelberg (2004)

11. Lewis, H.R.: Renaming a set of clauses as a horn set. J. ACM 25, 134–135 (1978)
12. Nishimura, N., Ragde, P., Szeider, S.: Detecting backdoor sets with respect to Horn and Binary clauses. In: H. Hoos, H., Mitchell, D.G. (eds.) SAT 2004. LNCS, vol. 3542, Springer, Heidelberg (2005)
13. Nishimura, N., Ragde, P., Szeider, S.: Solving #SAT using vertex covers. Acta Informatica 44(7-8), 509–523 (2007)
14. Papadimitriou, C.H.: Computational Complexity. Addison-Wesley, Reading (1994)
15. Paris, L., Ostrowski, R., Siegel, P., Sais, L.: Computing horn strong backdoor sets thanks to local search. In: ICTAI 2006, IEEE Computer Society, Los Alamitos (2006)
16. Ruan, Y., Kautz, H.A., Horvitz, E.: The backdoor key: A path to understanding problem hardness. In: AAAI, pp. 124–130 (2004)
17. Sinz, C.: SAT benchmarks (2003), http://www-sr.informatik.uni-tuebingen.de/~sinz/DC
18. Szeider, S.: Backdoor sets for dll subsolvers. J. Autom. Reasoning 35, 73–88 (2005)
19. Szeider, S.: Matched formulas and backdoor sets. In: Marques-Silva, J., Sakallah, K.A. (eds.) SAT 2007. LNCS, vol. 4501, Springer, Heidelberg (2007)
20. Williams, R., Gomes, C., Selman, B.: Backdoors to typical case complexity. In: IJCAI (2003)

A New Bound for an NP-Hard Subclass of 3-SAT Using Backdoors[*]

Stephan Kottler, Michael Kaufmann, and Carsten Sinz

Eberhard Karls Universität Tübingen, Wilhelm–Schickard–Institute,
Tübingen, Germany

Abstract. Knowing a *Backdoor Set B* for a given SAT instance, satisfiability can be decided by only examining each of the $2^{|B|}$ truth assignments of the variables in B. However, one problem is to efficiently find a small backdoor up to a particular size and, furthermore, if no backdoor of the desired size could be found, there is in general no chance to conclude anything about satisfiability.

In this paper we introduce a complete deterministic algorithm for an NP-hard subclass of 3-SAT, that is also a subclass of Mixed Horn Formulas (MHF). For an instance of the described class the absence of two particular kinds of backdoor sets can be used to prove unsatisfiability. The upper bound of this algorithm is $O(p(n) * 1.427^n)$ which is less than the currently best upper bound for deterministic algorithms for 3-SAT and MHF.

1 Introduction and Definitions

The boolean satisfiability problem (SAT) is one of the well known hard problems in theoretical computer science. Even when restricting the number of literals in each clause to a maximum of three (3-SAT), deciding satisfiability of a given instance is known to still be NP-complete. From the theoretical point of view the upper bound to solve 3-SAT could be improved steadily (see [13]). From the practical point of view we know by experience that many SAT instances evolving from real-world applications can be solved within nearly linear time. This is often due to some hidden structure that facilitates the solving process enormously. One possibility to measure this structure, namely *Backdoor Sets*, was introduced in 2003 by Williams, Gomes and Selman [16]. On the one hand it was shown that small backdoor sets are often related to real-world instances [16,12], on the other hand minimal backdoors of randomized, hence unstructured 3-SAT instances contain from 30% to 65% of all variables [6].

We use backdoors not as a measure of structure but rather to guide an algorithm for an NP-hard subclass of 3-SAT and Mixed Horn Formulas (MHF). MHF denotes the set of all SAT instances in conjunctive normal form where each clause is either Horn or binary [10,11].

---

[*] This work was partly supported by DFG-SPP 1307, project "Structure-based Algorithm Engineering for SAT-Solving".

H. Kleine Büning and X. Zhao (Eds.): SAT 2008, LNCS 4996, pp. 161–167, 2008.
© Springer-Verlag Berlin Heidelberg 2008

Strong Backdoor Sets. We use the definition of strong backdoor sets that is given in [8]. Note that there are also *weak backdoor sets* [16,8], however, they are not relevant for this paper. A backdoor is defined with respect to a class $\mathcal{C}$ of formulas that can be recognized and solved in polynomial time. A set $\mathcal{B}$ of variables $\mathcal{V}$ of a boolean formula F is a *strong backdoor set* of F with respect to $\mathcal{C}$ (strong $\mathcal{C}$-backdoor) if $F[\tau] \in \mathcal{C}$ for every truth assignment $\tau : \mathcal{B} \mapsto \{0, 1\}$. $F[\tau]$ denotes the result of removing all clauses that contain a literal x with $\tau(x) = true$ and removing all literals y with $\tau(y) = false$ from F.

We particularly use a variant of strong backdoors, so-called *deletion backdoors* [9,14]: $\mathcal{B}$ is a deletion backdoor if the formula $F - \mathcal{B}$ belongs to $\mathcal{C}$, where $F - \mathcal{B}$ denotes the result of removing all occurrences (both positive and negative) of the variables in B from the clauses of formula F. Every deletion backdoor is a strong backdoor, if class $\mathcal{C}$ is *clause-induced* ($F \in \mathcal{C} \Rightarrow F' \in \mathcal{C}$ for all $F' \subseteq F$) [9]. In this paper we solely deal with the two clause-induced classes Horn and 2-SAT as base classes of backdoors.

Parameterized Algorithms. Constitute one possible approach to cope with computational intractability [7]. One basic idea of parameterized algorithms is to ask whether a given NP-hard problem has a solution that can be bounded by some non-negative integer parameter k. If a problem is *fixed-parameter tractable* this question can be solved in time that is only exponential in k but not in the size of the original problem. Since our approach rather applies than creates parameterized algorithms we refer the reader to [7] for a formal definition and more detailed information on parameterized complexity.

2   A NP-Hard Subclass of 3-SAT

Definition 1. *Let* 2*-CNF *be the subclass of* 3-SAT *with the restriction that any clause* C *with* $|C| = 3$ *must only contain negative literals.*

Theorem 1. 2*-CNF *is NP-complete.*

Proof. The definition of 2*-CNF as a subclass of 3-SAT $\in$ NP directly implies 2*-CNF to be in NP. The NP-completeness of 2*-CNF can be shown by the polynomial time reduction 3-SAT $\leq_p$ 2*-CNF.

Let F be a boolean formula represented in 3-SAT. We need to specify a formula $F' \in 2^\star$-CNF such that F' is satisfiable if and only if F is satisfiable. Let therefore $\mathcal{C}^p$ denote the set of all clauses C of F with $|C| = 3$ and C containing at least one positive literal. Let $\mathcal{C}^n := \mathcal{C} \setminus \mathcal{C}^p$ denote the remaining clauses. Moreover, let $\mathcal{V}^p \subseteq \mathcal{V}$ denote those variables of F which occur positively in at least one clause of $\mathcal{C}^p$. In order to transform a formula $F \in$ 3-SAT into $F' \in 2^\star$-CNF, all clauses $\mathcal{C}^n$ can be adopted unchanged ($\mathcal{C}'^n := \mathcal{C}^n$) for F'. For every variable $x_i \in \mathcal{V}^p$ we introduce one variable $x_i^\star$ and one clause $(x_i^\star \vee x_i)$ in F'. We refer to the set of these added clauses as $\mathcal{C}'^a$. Furthermore, all clauses in $\mathcal{C}^p$ are modified to clauses $\mathcal{C}'^p$ by replacing each occurrence of a positive literal x_i by the (negative) literal $\overline{x_i^\star}$. Hence, F' belongs to class 2*-CNF and it is:

$$F \text{ is satisfiable} \Leftrightarrow F' \text{ is satisfiable}$$

'$\Rightarrow$': If F is satisfiable there exists a model M_F (set of satisfying literals). We create an according model for F' by initializing $M_{F'} := M_F$. Moreover, for all variables $x_i \in \mathcal{V}^p$ we apply the following rule:

$$M_{F'} = \begin{cases} M_{F'} \cup \{\overline{x_i^*}\} & \text{if (positive) literal } x_i \in M_F \\ M_{F'} \cup \{x_i^*\} & \text{otherwise} \end{cases}$$

With this, all clauses $\mathcal{C}'^n \cup \mathcal{C}'^a$ are satisfied. Let now C' be any arbitrary clause in the remaining set of clauses $\mathcal{C}'^p$. There exists at least one literal $l_j \in M_F$ (of variable x_j) which satisfies the according clause $C \in \mathcal{C}^p$. Due to the initialization it is $l_j \in M_{F'}$. In case l_j is a negative literal, C' also contains l_j and hence is satisfied. In case l_j is a positive literal, C' contains the literal $\overline{x_j^*}$ that was chosen for $M_{F'}$ and hence satisfies C'. Consequently all clauses in F' are satisfied.

'$\Leftarrow$': F' is satisfiable by the assignment of the literals in $M_{F'}$. Initializing $M_F := \{l \in M_{F'} : l \text{ belongs to } F\}$ satisfies at least all clauses in $\mathcal{C}^n$. Now consider any clause $C \in \mathcal{C}^p$: Since the according clause $C' \in \mathcal{C}'^p$ is satisfied there exists at least one literal $l \in M_{F'}$ satisfying C'. In case l belongs to F then $l \in M_F$ and thus, $C \in F$ is satisfied. If, on the other hand, l does not belong to F then l must be an added and negated variable of the form $\overline{x_i^*}$, whereas clause C contains literal $x_i \in F$. Since $M_{F'}$ is a model for F', in particular the added clause $(x_i^* \vee x_i) \in F'$ is also satisfied by $M_{F'}$. Hence, literal x_i has to be contained in $M_{F'}$ and so it is also contained in M_F. With this M_F is a model for F.

Since for any positive literal in F we added at most one new variable and one new clause to F', the reduction is polynomial. Thus, it is NP-complete to decide whether a given formula $F \in 2^*\text{-CNF}$ is satisfiable.   □

Note that an alternative proof could adapt the idea to prove NP-hardness for MHF [10]. It turns out that $2^*\text{-CNF} \subset \text{MHF}$ encodes the problem to decide whether the vertices of a graph can be colored with at most three different colors such that no vertices with the same color are connected by an edge.

3   A Backdoor–Driven Approach

Based on the concept of backdoor sets we can specify a simple deterministic algorithm to decide satisfiability for arbitrary formulas of the class 2^*-CNF. The main algorithm is listed in Alg. 1 and is explained in detail in this section.

In the second line we define a constant c whose value solely depends on the runtime of two parameterized algorithms we use as subroutines further below. The particular value will become more clear when analyzing the complexity of the algorithm. In line 3 we first consider all clauses C^+ of F that consist of exactly two positive literals. Note that with F being an instance of class 2^*-CNF, any clause within the set $\{F \setminus C^+\}$ contains at most one positive literal and thus these clauses are all Horn clauses. In the next line we aim to find the smallest possible set of variables $\mathcal{B}^+$ such that every clause in C^+ contains at

---

Algorithm 1. A Backdoor-driven 2*-CNF Solver

---

1 **Function** bd_solve(F)
2     $c \leftarrow \log_{4.151}(2.0755) \approx 0.513$
3     $C^+ \leftarrow \{(x_i \vee x_j) \in F : x_i, x_j \text{ positive}\}$
4   Choose minimum $\mathcal{B}^+ \subseteq \mathcal{V}$, such that $\forall\, C \in C^+ \;\exists\, x_i \in \mathcal{B}^+ : x_i \in C$
5   **if** $|\mathcal{B}^+| \leq c * |\mathcal{V}|$ **then**
6   ⌊ **return** Solve F by using the Horn-Backdoor $\mathcal{B}^+$
7     $C^- \leftarrow \{(\overline{x_h} \vee \overline{x_i} \vee \overline{x_j}) \in F : \overline{x_h}, \overline{x_i}, \overline{x_j} \text{ negative}\}$
8   Choose minimum $\mathcal{B}^- \subseteq \mathcal{V}$, such that $\forall\, C \in C^- \;\exists\, x_i \in \mathcal{B}^- : \overline{x_i} \in C$
9   **if** $|\mathcal{B}^-| \leq (1 - c) * |\mathcal{V}|$ **then**
10   ⌊ **return** Solve F by using the Binary-Backdoor $\mathcal{B}^-$
11   ⌊ **return** F Unsatisfiable

---

least one variable of the set $\mathcal{B}^+$. Since by definition all clauses within C^+ are binary clauses the problem to find the smallest possible set $\mathcal{B}^+$ can be seen as a VERTEX-COVER-problem:

Understanding binary clauses as edges and the variables of the two literals of each clause as the endpoints of an edge, our task responds to find the smallest set of endpoints to cover each edge. It is easy to verify that the set of variables $\mathcal{B}^+$ constitutes a deletion backdoor with the base class $\mathcal{C}$ = Horn: Each clause of the instance $F - \mathcal{B}^+$ contains at most one positive literal. For complexity reasons we target to determine a set $\mathcal{B}^+$ of the size not greater than $c * |\mathcal{V}|$.

If the instance F does not contain a Horn-backdoor of the desired size, we then consider the set of all clauses (C^-) consisting of three literals. Recall that these literals are all negative. In line 8 we aim to find a smallest possible set of variables $\mathcal{B}^-$ such that each clause within the set C^- contains at least one (negative) literal of the variables within the set $\mathcal{B}^-$. This task corresponds to a 3-HITTING-SET problem (see [6]): Clauses of the set C^- can be seen as subsets of three items (variables) each. For $\mathcal{B}^-$ we search for the smallest set of items to hit each subset in C^-. Note that any clause in $F \setminus C^-$ consists of at most two literals. Hence it is clear that the set of variables $\mathcal{B}^-$ constitutes a deletion backdoor with base class $\mathcal{C}$ = 2-SAT: The instance $F - \mathcal{B}^-$ belongs to class 2-SAT, since from each clause with three literals (C^-) at least one is removed. Again for complexity reasons we focus on finding a Binary-backdoor with size not greater than $(1 - c) * |\mathcal{V}|$.

When reaching line 11 we know that there exists neither a set of variables $\mathcal{B}^+$ nor a set $\mathcal{B}^-$ with the desired size. In this case we can conclude unsatisfiability of F. Since the considered clauses within C^+ solely consist of positive literals we need to set the values of at least $|\mathcal{B}^+|$ variables to true in order to satisfy all the clauses in C^+. Analogously the size $\mathcal{B}^-$ indicates the number of variables whose values have to be set to *false* in order to satisfy all clauses within the set C^-. This is impossible with $|\mathcal{B}^+|$ being greater than $c * |\mathcal{V}|$ and $|\mathcal{B}^-|$ being greater than $(1 - c) * |\mathcal{V}|$ at the same time.

A similar argument to prove unsatisfiability of big random 3-SAT instances has been used by Franco and Swaminathan in [5]. The authors show that an approximation algorithm for 3-HITTING-SET can determine bounds on how many variables must be set to *true* and how many must be set to *false*.

Complexity of the Algorithm

It is easy to verify that satisfiability of a boolean formula of class 2^*-CNF can be decided by the algorithm described above. In this subsection we analyze the complexity of Algorithm 1. In particular we have to focus on the following four computationally intensive tasks of the algorithm:

1. In order to compute the set of variables $\mathcal{B}^+$ a VERTEX COVER problem has to be solved (line 4). There are several good approximation algorithms to deal with VERTEX COVER problems. However, in our case we need to know exactly the minimum set of variables to cover all clauses in C^+ which cannot be achieved by using approximation methods. Considering the fact that we are only interested in a variable set $\mathcal{B}^+$ up to a particular size, we can make use of a parameterized algorithm.

Given a graph $G = (V_G, E_G)$ the parameterized VERTEX COVER problem asks if there is a subset of vertices $C \subseteq V_G$ with k or fewer vertices such that each edge in E_G has at least one of its endpoints in C. According to [7] there are algorithms solving the parameterized VERTEX COVER in time $O(k * |V_G| + 1.29^k)$. Since in our case the parameter k is given by $c * |\mathcal{V}| = 0.513 * |\mathcal{V}|$ and $|V_G| = |\mathcal{V}| = n$ the complexity of this task can be limited by $O(n^2 + 1.14^n)$.

2. Solving the instance F by using a Horn-backdoor $\mathcal{B}^+$ with at most $c * n = 0.513 * n$ variables (line 6) may in the worst case imply to examine all possible truth assignments of the variables in $\mathcal{B}^+$. More precisely this might mean that for each of the $2^{0.513n} = 1.427^n$ truth assignments a Horn instance has to be solved. The satisfiability of a Horn instance can be decided in linear time by applying for example the algorithm described in [3]. Concluding, the complexity of this part is limited by $O(1.427^n * |F|)$.

3. Analogously, to determine the set $\mathcal{B}^+$, we can use a parameterized algorithm in order to solve the 3-HITTING-SET problem to detect whether there is a set $\mathcal{B}^-$ with at most $(1 - c) * |\mathcal{V}| = (1 - 0.513) * |\mathcal{V}|$ variables (line 8).

Given a collection Q of subsets of size at most three of a finite set S and a non-negative integer k, the parameterized 3-HITTING-SET problem asks if there is a subset $S' \subseteq S$ with $|S'| \leq k$ which allows S' to contain at least one element from each subset in Q [7]. Algorithms to solve this problem have been steadily improved in the last years. In 2004 Fernau published an algorithm for the parameterized 3-HITTING SET problem bounded by $O(2.179^k + |S|)$ [4]. Wahlström recently improved this result and gave an algorithm with an upper bound $O(p(n) * 2.0755^k)$ with a polynomial $p(n)$ [15]. With $k := (1 - c) * |\mathcal{V}| = 0.487 * n$ in our case the complexity can be bounded by $O(1.427^n * p(n))$.

4. To determine satisfiability of F by using the Binary-backdoor $\mathcal{B}^-$ with at most $(1 - c) * n = 0.487 * n$ variables (line 10) may in the worst case imply to solve a 2-SAT instance for each possible truth assignment of the variables in $\mathcal{B}^-$. Since 2-SAT can be solved in linear time [1,3] the complexity of this part can be limited by $O(1.402^n * |F|)$.

With this, the complexity of Algorithm 1 is bounded by $O(1.427^n * p(n))$. Hence, the idea of considering two different types of backdoors yields a good upper bound for the special class 2^*-CNF $\subset$ 3-SAT. This bound is slightly better than the bound $O(2^{0.5284n}) = O(1.4423^n)$ to solve the more general class MHF [10]. Just for comparison, the currently best deterministic algorithm for 3-SAT has an upper bound of $O(1.473^n)$ [2].

4   Conclusion

Based on the concept of backdoor sets we have bounded the complexity to decide satisfiability for 2^*-CNF $\subset$ 3-SAT. The complexity for our algorithm mainly depends on the runtime to solve parameterized 3-HITTING SET problems.

It would be interesting to study whether the idea to compute different minimum backdoors of a SAT instance can be used to generate algorithms for further NP-hard subclasses of SAT or MHF.

References

1. Aspvall, B., Plass, M.F., Tarjan, R.E.: A linear-time algorithm for testing the truth of certain quantified boolean formulas. Inf. Proc. Lett. 8, 121–123 (1979)
2. Brüggemann, T., Kern, W.: An improved deterministic local search algorithm for 3-sat. Theor. Comput. Sci. 329(1-3), 303–313 (2004)
3. del Val, A.: On 2-SAT and renamable horn. In: AAAI: 17th National Conference on Artificial Intelligence, AAAI / MIT Press (2000)
4. Fernau, H.: A top-down approach to search-trees: Improved algorithmics for 3-hitting set. Electronic Colloquium on Computational Complexity, TR04-073 (2004)
5. Franco, J., Swaminathan, R.: On good algorithms for determining unsatisfiability of propositional formulas. Discrete Appl. Math. 130(2), 129–138 (2003)
6. Interian, Y.: Backdoor sets for random 3-sat. In: SAT (2003)
7. Niedermeier, R.: Invitation to Fixed-Parameter Algorithms. Universität Tübingen (October 2002)
8. Nishimura, N., Ragde, P., Szeider, S.: Detecting backdoor sets with respect to horn and binary clauses. In: SAT (2004)
9. Nishimura, N., Ragde, P., Szeider, S.: Solving #SAT using vertex covers. In: Biere, A., Gomes, C.P. (eds.) SAT 2006. LNCS, vol. 4121, pp. 396–409. Springer, Heidelberg (2006)
10. Porschen, S., Speckenmeyer, E.: Worst Case Bounds for Some NP-Complete Modified Horn-SAT Problems. In: H. Hoos, H., Mitchell, D.G. (eds.) SAT 2004. LNCS, vol. 3542, pp. 251–262. Springer, Heidelberg (2005)
11. Porschen, S., Speckenmeyer, E.: Satisfiability of mixed Horn formulas. Discrete Applied Mathematics 155(11), 1408–1419 (2007)

12. Ruan, Y., Kautz, H.A., Horvitz, E.: The backdoor key: A path to understanding problem hardness. In: AAAI, pp. 124–130 (2004)
13. Schöning, U.: A probabilistic algorithm for k-sat and constraint satisfaction problems. In: Symposium on Foundations of Computer Science (1999)
14. Szeider, S.: Matched Formulas and Backdoor Sets. In: Marques-Silva, J., Sakallah, K.A. (eds.) SAT 2007. LNCS, vol. 4501, pp. 94–99. Springer, Heidelberg (2007)
15. Wahlström, M.: Algorithms, measures, and upper bounds for satisfiability and related problems. PhD thesis, Linköping University, Dissertation no 1079 (2007)
16. Williams, R., Gomes, C., Selman, B.: Backdoors to typical case complexity. In: IJCAI (2003)

Improvements to Hybrid Incremental
SAT Algorithms

Florian Letombe and Joao Marques-Silva

School of Electronics and Computer Science, University of Southampton, UK
{fl,jpms}@ecs.soton.ac.uk

Abstract. Boolean Satisfiability (SAT) solvers have been successfully applied to a wide range of practical applications, including hardware model checking, software model finding, equivalence checking, and planning, among many others. SAT solvers are also the building block of more sophisticated decision procedures, including Satisfiability Modulo Theory (SMT) solvers. The large number of applications of SAT yields ever more challenging problem instances, and motivate the development of more efficient algorithms. Recent work studied hybrid approaches for SAT, which involves integrating incomplete and complete SAT solvers. This paper proposes a number of improvements to hybrid SAT solvers. Experimental results demonstrate that the proposed optimizations are effective. The resulting algorithms in general perform better and, more importantly, are significantly more robust.

1 Introduction

Motivated by significant improvements to Boolean Satisfiability (SAT) solvers over the last decade, SAT has been applied to a large number of areas, including model checking [2], model finding, planning, bioinformatics, and security, among many others. The widespread use of SAT in so many areas generates a large number of challenging problems instances, many of which modern SAT solvers are not capable of solving. This in turn, motivates the development of ever more effective SAT solvers. Nevertheless, recent years have seen a slowdown in improvements made to SAT solvers. As a result, a number of alternatives has been considered, one of which is the use of hybrid incremental SAT solvers [5], that build on existing SAT algorithms that are effective in solving different types of problems. The hybrid incremental SAT solver combines the power of local search (LS) SAT solvers and of conflict-driven clause learning (CDCL) SAT solvers. These more complex algorithms are expected to provide additional performance improvements, in general not as the first choice SAT solver, but as a reliable alternative SAT solver for more complex SAT instances, with the goal of increased robustness in industrial settings.

This paper develops a number of optimizations to the original hybrid incremental SAT algorithm. The proposed optimizations provide better performance and, more importantly, significantly improve the robustness of SAT solvers. A comprehensive experimental evaluation on industrial SAT problem instances provides evidence that the proposed optimized hybrid incremental SAT solver is more robust than other existing SAT solvers.

H. Kleine Büning and X. Zhao (Eds.): SAT 2008, LNCS 4996, pp. 168–181, 2008.

The paper is organised as follows. The next section provides a necessarily brief perspective on SAT solvers (CDCL and LS SAT solvers), as well as the original hybrid incremental SAT solver [5]. Afterwards, section 3 presents several optimizations to the basic hybrid incremental SAT algorithm. A comprehensive experimental evaluation is summarized in section 4. Additional related work is briefly surveyed in section 5. Finally, the paper concludes in section 6.

2   Boolean Satisfiability Solvers

This section provides a quick overview of Boolean Satisfiability solvers, including Conflict-Driven Clause Learning (CDCL) SAT solvers, Local Search (LS) SAT solvers, and the recent generation of hybrid incremental SAT solvers. CDCL SAT solvers are widely used for solving large industrial problem instances. LS SAT solvers are used in less applied contexts, but have also been used for developing branching heuristics for complete solvers.

Most propositional decision procedures assume the input problem to be in conjunctive normal form (CNF). Moreover, the SAT problem is defined as follows. A formula Σ in CNF is represented as a set of clauses, each clause is a set of literals, and each literal is either a positive or negative propositional variable in V. Moreover, a formula is interpreted as a conjunction of clauses, and a clause is interpreted as a disjunction of literals. For example, $(x_1 \vee \neg x_2 \vee x_3) \wedge (x_4 \vee \neg x_5)$ is represented as $\{\{x_1, \neg x_2, x_3\}, \{x_4, \neg x_5\}\}$. The SAT problem consists in finding an assignment (also called a model) for a subset of V satisfying each clause in a CNF formula or proving that no such assignment exists.

2.1   CDCL and LS SAT Solvers

CDCL SAT solvers follow the organization of the DPLL algorithm [3], but integrate a number of effective techniques, including clause learning [17], lazy data structures [21] and search restarts [10]. CDCL SAT solvers have evolved from the original solvers [17], which essentially proposed clause learning for SAT, to the more recent CDCL SAT solvers, that also integrate lazy data structures and search restarts [21,9,4]. A number of these concepts, used in the following sections, are briefly reviewed below (see [17,21,9,4] for additional detail). A CDCL SAT solver is usually organized into three main engines [17,21,4]: the decision engine, used for branching; the deduction engine, used for unit propagation and identification of unsatisfied clauses (or *conflicts*); and the diagnosis engine, used for clause learning. A *decision level* is associated with each assigned variable. Decision levels measure the depth of the search tree in terms of the number of variables the SAT algorithm has branched on. Variables can be assigned a Boolean value, either resulting from a decision (or branching step), or as the result of unit propagation [3]. Variables assigned as the result of unit propagation are said to be *implied*. With each implied variable the SAT algorithm also associates a *reason* or *antecedent*, representing the clause that explains why the variable is implied. The set of assigned variables and associated reasons implicitly represent the *implication graph* [17]. Finally, the process of *clause learning* consists of traversing the implication graph from a given unsatisfied clause using the reasons of implied variable assignments, and recording unsatisfied literals assigned at decision levels less than the current one. The resulting set of recorded literals is then used to create a new clause, which serves for backtracking non-chronologically, and for preventing the same conflict from occurring again during the search process.

Local search is a meta-heuristic for solving computationally hard optimization problems. It can be used on problems that can be formulated as finding a solution maximizing a criterion among a number of candidate solutions. Local search algorithms move from solution to solution in the space of candidate solutions (the search space) until a solution deemed optimal is found or a time bound is reached. In SAT a candidate solution is a truth assignment, and the target is to maximize the number of satisfied clauses by the assignment. In this case, the final solution is of use only if it satisfies all clauses. Local search for solving SAT became notorious with GSAT [25] and is often very effective at finding models of satisfiable formulas [13]. LS starts by assigning random values to all the variables. If the assignment satisfies all clauses, the algorithm stops and returns it. Otherwise, the value of a variable is changed, and the process is repeated. The variable to change is the one that minimizes the number of unsatisfied clauses in the new assignment. If no assignment satisfying all clauses has been found after a fixed number of iterations (called cutoff or number of flips), the algorithm starts again with a new random assignment. The algorithm terminates either when a model of the formula has been found or when the number of restarts exceeds a fixed number.

Selman *et al.* have also proposed improvements to GSAT, including WalkSAT, whose main differences to GSAT are the addition of random noise, and the step of selecting variables to be flipped from unsatisfied clauses [24]. Many other new heuristics have been proposed, including among others HSAT [8], Novelty and R-Novelty [19], Novelty+ and R-Novelty+, Adaptive Novelty+ [12], and g2wsat (including adaptg2wsat+) [15]. Local search SAT solvers are incomplete, and so cannot prove unsatisfiability. It should be noted that recent work has shown how to use local search for proving unsatisfiability [22], but then the resulting algorithm can no longer prove satisfiability.

2.2 Hybrid Incremental SAT Solvers: The hbisat Algorithm

Recent work has addressed hybrid solutions for SAT, where both LS and CDCL SAT solvers cooperate to solve a target problem instance. This section overviews hbisat (for HyBrid Incremental SAT solver) [5], given its promising experimental results. The motivation for the hbisat algorithm is to combine the power of LS SAT solvers for finding solutions of satisfiable formulas and the power of CDCL SAT solvers for proving formulas to be unsatisfiable. Albeit past work focused on using assignments suggested by LS solvers to help CDCL solvers selecting decision variables and assignments, hbisat proposed the opposite: partial models computed by the CDCL solver serve to initialize the LS solver truth assignment. Clauses not satisfied by the LS solver are sent to the CDCL solver, trying to either identify an unsatisfiable sub-formula in the clauses sent to the CDCL SAT solver, or satisfying the clauses in the LS solver. Preliminary experimental results suggest this approach can be effective [5]. The `hbisat` procedure is summarized in the not underlined part of algorithm 1.1. Note that Σ_Γ is initially empty, and α is first randomly initialized.

A proof that `hbisat` algorithm is sound and complete can be found in [5]. Clearly, if the LS procedure LSSOLVE can find a truth assignment, then the initial formula is satisfiable. Otherwise, some clauses, chosen from those unsatisfied (or *broken*) during local search, are sent to the CDCL solver. Besides these, a few additional clauses can also be sent to the CDCL solver, subject to a number of *criteria* outlined below. In the algorithm description, the clauses sent to the CDCL solver are denoted $\Sigma_\Lambda^{Criteria}$, and computed with GETCLAUSESTOSEND procedure. If the CDCLSOLVE procedure concludes that the sub-formula is unsatisfiable, then the original problem instance is also unsatisfiable, and the algorithm terminates. Alternatively, if the CDCLSOLVE procedure concludes that the sub-formula is satisfiable, the computed assignment, obtained with procedure GETMODEL, serves to initialize the next iteration of the LS solver. In the `hbisat` algorithm the following *criteria* are used to decide which clauses are sent to the CDCL solver [5].

1. Unsatisfied clauses, i.e. clauses that the LS solver was not able to satisfy;
2. Clauses containing the most flipped variable during local search;
3. Clauses with two or more of its literals having opposite polarities to literals in broken clauses.

Note that the last two criteria are empirically only applied after the first four calls to function ISSATISFIABLE. Moreover, all remaining clauses are sent to the CDCL solver if one of the two following additional criteria is satisfied [6]:

1. Less than 1% or less than 50 clauses remain to be sent to the CDCL solver;
2. The number of learnt clauses in the CDCL solver is greater than 20% of the total number of clauses.

For each execution of `hbisat` algorithm, the operation $\Sigma_\Gamma \leftarrow \Sigma_\Gamma \cup \Sigma_\Lambda^{Criteria}$ denotes that the clauses identified by the above criteria and added to the clauses in the CDCL SAT solver. Finally, the algorithm allows for three hundred iterations, i.e. recursive calls to procedure ISSATISFIABLE. Afterwards, all remaining clauses are sent to the CDCL SAT solver to solve the problem.

The `hbisat` algorithm is illustrated in figure 1 (top). A set of clauses is associated with each solver, and the solid black squared portions represent clauses identified by the above criteria. These are the clauses sent to the CDCL solver.

3   New Hybrid Incremental SAT Algorithms

This section proposes several optimizations to the `hbisat` algorithm. All optimizations are included in a new hybrid incremental SAT solver, `hinotos`[1], that can be configured to also implement the original `hbisat` algorithm.

---

[1]   `hinotos` denotes Hybrid Incremental SAT for NOTOS, a LTL model checker.

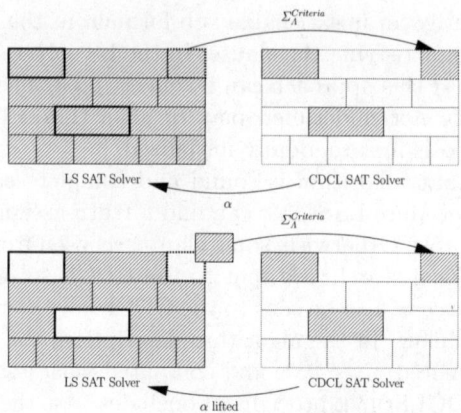

Fig. 1. Solver interaction in `hbisat` (top) and in `hinotos` (bottom)

3.1 Variable Lifting and Blocking Clauses

Modern SAT solvers identify complete assignments. The main reason for this is motivated by the use of lazy data structures, which prevent having knowledge of clause state [21]. Most often, a reasonable number of branching decision made by SAT solvers are irrelevant for satisfying all clauses of a problem instance. *Variable lifting* is the process of removing literals or, equivalently, variables from a satisfying assignment such that for each valuation of the lifted variables, the formula is still satisfied [23].

Example 1 (Variable lifting). Define $V = \{x_1, x_2, x_3, x_4, x_5, x_6\}$ and the following propositional formula Σ over V:

$$(x_1 \vee \neg x_2 \vee x_6) \wedge (x_1 \vee \neg x_2 \vee \neg x_4) \wedge (x_1 \vee \neg x_2 \vee x_4) \wedge (x_3 \vee x_4 \vee x_5) \wedge (x_1 \vee x_2 \vee x_6).$$

Whereas the lifted assignment $\{x_1, x_3\}$ is enough to satisfy the formula, a CDCL solver will identify a complete assignment, e.g. $\{x_1, \neg x_2, x_3, \neg x_4, \neg x_5, x_6\}$. Σ is satisfied, independently of assignments to the other variables $\{x_2, x_4, x_5, x_6\}$.

In `hbisat` complete assignments are sent from the CDCL SAT solver to the LS SAT solver at each step. This can be ineffective, since the LS solver will be unable to identify the assignments that are *relevant* from the ones that are *not relevant*. As a result, variable lifting can be used effectively in hybrid incremental algorithms as shown in the next section. For the experimental results presented in this paper, the variable lifting procedure consists of simply scanning the watched literals in every clause and selecting one of the watched literals as the one that satisfies the clause. As a result, variables not used for satisfying clauses can be lifted.

Another effective technique used in model checking is the use of *blocking clauses* [20] for quantification (or equivalently for model enumeration). Blocking clauses prevent previously computed satisfying partial assignments from being recomputed again during the search, and are created after variable lifting is applied. In hybrid incremental SAT algorithms, the use of blocking clauses guarantees that no part of the search tree is visited more than once.

3.2   Optimized Interaction between the LS and CDCL Solvers

A detailed analysis of the original `hbisat` algorithm reveals that in several situations the same search space can be re-visited, and that the number of times this can happen can be arbitrarily large, up to the limit imposed by the number of times the LS solver is called. The LS solver can repeatedly re-visit the same complete assignments, independently of the information provided by the CDCL solver. In contrast, the CDCL solver may have to redo parts of the search space, motivated by the fact that the LS solver may force the CDCL solver to reconsider branching decisions already made. These are the main sources of inefficiency in the `hbisat` algorithm.

One optimization that addresses these two problems is to guarantee that the clauses in the two solvers are distinct. Each clause sent by the LS solver to the CDCL solver is *removed* from the LS solver. Hence, at any stage, the original clauses are divided into two sets, one associated with the LS solver and the other with the CDCL solver, and these two sets form a partition of the original set of clauses. As a result, the overhead of the LS solver is *effectively reduced* at each iteration of the algorithm due to the reduced number of clauses. Moreover, assignments communicated by the CDCL solver must be *respected* by the LS solver. As a result, variables assigned by the CDCL solver are said to be *tabu* to the LS solver, and will be untouched by the LS solver. Clearly, this idea can be effective only if variable lifting is applied; otherwise the CDCL solver would assign all variables, and the LS solver would be unable to flip any variable.

Figure 1 (bottom) illustrates how `hinotos` implements the interaction between the LS and the CDCL solvers. The clauses sent to the CDCL solver, and removed from the LS solver, are shown in solid black squares.

3.3   Additional Criteria for Moving Clauses

Besides the optimizations outlined in the previous sections, two additional criteria are used for moving clauses to the CDCL SAT solver. The first criterion ensures that a minimal amount of clauses is sent into the CDCL solver. In the original algorithm [5], at least one clause, picked randomly, was ensured to be sent at each step of the algorithm. Since 1% of clauses (or 50 clauses) are considered to be "simple" enough for the CDCL solver, this same amount of clauses is also considered each time clauses are moved to the CDCL solver. As a result, instead of the 300 iterations proposed in the original `hbisat` algorithm, the new criterion implies that at most 100 iterations are executed. The second criterion for moving clauses (from the LS solver to the CDCL solver) is a natural consequence of the organization of the `hinotos` algorithms. If all variables become tabu in the LS solver, then the LS solver becomes irrelevant, and all clauses are sent to the CDCL solver.

Algorithm 1.1. hbisat and <u>hinotos</u>

function IsSATISFIABLE
Input: Σ_Λ and Σ_Γ two CNF formulas, with $\Sigma = \Sigma_\Lambda \cup \Sigma_\Gamma$ and $\Sigma_\Lambda \cap \Sigma_\Gamma = \emptyset$;
    α an assignment.
Output: **True** if Σ is satisfiable, **False** otherwise.
begin

> TABU(Σ_Λ, α); /* New tabu variables to LS solver */
> **if** LSSOLVE(Σ_Λ, α) = **SAT then return True**; /* Solution found with LS */
> $\Sigma_\Lambda^{Criteria} \leftarrow$ GETCLAUSESTOSEND(Σ_Λ);
> $\Sigma_\Gamma \leftarrow \Sigma_\Gamma \cup \Sigma_\Lambda^{Criteria}$;
> $\Sigma_\Lambda \leftarrow \Sigma_\Lambda \setminus \underline{\Sigma_\Lambda^{Criteria}}$; /* Sent clauses removed from LS solver database */
> **if** CDCLSOLVE$\underline{(\Sigma_\Gamma)}$ = **SAT then**
>
>> **if** $\Sigma_\Gamma = \Sigma$ **then return True**;
>> /* CDCL solver has found a model to be used by LS solver */
>> $\alpha \leftarrow$ GET<u>LIFTED</u>MODEL(Σ_Γ);
>> **return** IsSATISFIABLE(Σ_Λ, Σ_Γ, α);
>
> **else return False**; /* CDCL solver proved sub-formula to be unsatisfiable */

end

3.4 The hinotos Algorithm

The hinotos algorithm implements the original hbisat algorithm (shown in Algorithm 1.1 without the underlined parts) as well as the optimizations proposed in the previous sections, and is shown in Algorithm 1.1. Again, Σ_Γ is initially empty, and α is first randomly initialized. The proposed optimizations do not affect soundness or completeness. The soundness and completeness of hbisat [5] allow establishing the following result.

Theorem 1. hinotos *is complete and sound for the satisfiability problem.*

4 Experimental Evaluation

4.1 Methodology

The empirical results presented in this section were obtained on servers running Red Hat Enterprise Linux WS release 4, with Intel Xeon 5160 Dual Core 3GHz processors and 4GB of RAM. For all experiments, the CPU time limit per instance was set to 1500 seconds.

hinotos[2] is implemented in C++ and can be configured to implement or not the optimizations proposed in Section 3. Hence, one possible configuration for hinotos corresponds to the actual hbisat algorithm [5]. However, the LS and the CDCL solvers used in hinotos are more efficient than the ones used in hbisat [5]. In hinotos the LS solver is adaptg2wsat+ [15] and the CDCL solver is Minisat2, whereas in the original hbisat the LS solver is

---

[2] hinotos complete documentation and binaries are publicly available on
http://satstore.ecs.soton.ac.uk/software/hinotos.

WalkSAT2004.11.15 [24] and the CDCL solver is Minisat1.14 [4]). Observe that other alternative CDCL and LS solvers could be considered for hinotos.

Moreover, hinotos can be configured to run the following configurations (the representative letter for each option is highlighted with parenthesis):

(i)nverse order pure incremental: Represents a purely incremental (and not hybrid) SAT solver. No LS solver is used in this version. Clauses are always sent in the same order, inverse from the order of appearance in the formula, according to the first criterion described in section 3.3;

(h)bisat implementation: Represents the original hbisat algorithm as described in [5] and presented in section 2.2;

(m)inimum amount hbisat-like: Implements option h, and in addition integrates the criterion for moving clauses described in section 3.3;

hi(n)otos method: Implements option h and integrates the optimizations proposed in section 3.2;

(r)emoving+minimum amount hinotos-like: Corresponds to the combination of the two previous options (i.e. m and n).

These five configurations for the hinotos solver were compared against Minisat versions 1.14 and 2. The main purpose of the first configuration was to evaluate the usefulness of the LS SAT solver for identifying sets of clauses to move to the CDCL SAT solvers. The second configuration allowed benchmarking the implementation of hbisat in hinotos, thus confirming similar performance on instances for which results for hbisat are known. Finally, the last three configurations evaluate whether the proposed optimizations are effective in practice.

4.2   Benchmarks

With the objective of conducting a comprehensive evaluation of the different algorithms, a total of 602 problem instances were selected from the following classes of instances:

IBM Formal Verification Benchmarks. Problem instances from Bounded Model Checking considering different numbers of computation steps [29];

Pimag Problem instances from pipelined-machine-verification problems [16]. All Pimag instances are unsatisfiable;

Formal Verification of Processors (fvp). Formal verification of buggy variants of an out-of-order super-scalar processor [27].

Calysto (csv). Benchmarks generated by the Calysto static checker [1] for software verification.

The results presented in this section extend and complete the preliminary experimental evaluation of [5], by considering 602 instances instead of the 24 studied in [5]. In order to reduce bias from too many instances from any of the classes considered, for classes ibm and csv only a subset of the available instances was considered, which was chosen arbitrarily. Nevertheless, for each of these classes a large number of instances was evaluated (respectively 198 and 152).

Table 1. Number of solved instances (and approximate average CPU time in seconds) for each configuration. CPU timeout is fixed to 1500 seconds.

CNF	Minisat		hinotos					#s/#t
	1.14	2	si	sh	sm	sn	sr	
c*	95(135)	**98(164)**	**98(194)**	97(189)	**98(202)**	96(188)	96(155)	102/126
f*	20(158)	**21(298)**	19(231)	18(161)	20(255)	**21(293)**	19(232)	21/42
g*	**42(139)**	41(144)	41(120)	38(150)	**42(186)**	41(199)	41(144)	42/42
pmg	157(139)	**160(177)**	158(179)	153(176)	**160(205)**	158(205)	156(161)	165/210
fvps	11(157)	**12(120)**	10(83)	**12(136)**	**12(150)**	10(135)	11(203)	14/20
fvpu	**19(98)**	18(39)	**19(83)**	18(75)	**19(100)**	18(44)	18(88)	20/22
fvp	30(119)	30(72)	29(83)	30(100)	**31(119)**	28(76)	29(131)	34/42
ibms	81(150)	84(192)	**87(185)**	78(234)	85(198)	83(222)	86(252)	93/93
ibmu	86(97)	85(109)	87(112)	83(176)	**88(138)**	85(164)	**88(170)**	88/88
ibm	167(97)	169(108)	**174(112)**	161(176)	173(137)	168(164)	**174(169)**	182/198
csvs	92(142)	95(127)	98(144)	117(197)	99(133)	**121(134)**	116(201)	129/129
csvu	8(85)	8(101)	8(87)	8(116)	8(115)	7(31)	**9(267)**	9/9
csv	100(138)	103(125)	106(140)	125(192)	107(132)	**128(128)**	125(206)	138/152
Alls	184(146)	191(155)	195(159)	207(207)	196(162)	**214(168)**	213(222)	237/242
Allu	270(105)	271(118)	272(125)	262(150)	**275(155)**	268(158)	271(136)	282/329
All	454(122)	462(133)	467(139)	469(175)	471(158)	482(163)	**484(174)**	519/602

4.3 Results

Table 1 summarizes the results obtained by all configurations of hinotos and the two versions of Minisat. Each configuration of hinotos is denoted by sX, where X is one of the possible configurations: i, h, m, n, and r. The first column contains the name of the class CNF formulas, where rows c*, f*, g* are problems in the Pimag category, rows fvps and fvpu (respectively ibms and ibmu, csvs and csvu) show the results on satisfiable and unsatisfiable instances of fvp (respectively ibm, csv) categories. The last column shows for each class of instances the total number of solved instances (by any solver) and the total number of instances. For each combination of tool configuration and instance category, the results shown are the number of solved instances followed (in parenthesis) by the average running time on *solved* instances. Analysis of table 1 allows concluding that no configuration, for either hinotos or Minisat, seems to give the best performance in terms of average run time for solved instances. Compared to Minisat, the performance improvements of hbisat are not clear, particularly for the first three classes of instances. For all classes of instances, the implementation of hbisat solves 7 more instances than Minisat2. Regarding the optimizations proposed in this paper, there are reasonable gains in terms of robustness, i.e. the number of instances solved. Indeed, configurations sn and sr solve significantly more instances than any other configuration, either for Minisat or for hinotos. A more detailed analysis for each class of problem instances indicates that Minisat and configuration sm perform better for the Pimag and fvp

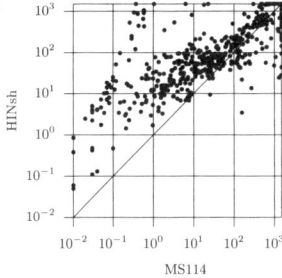 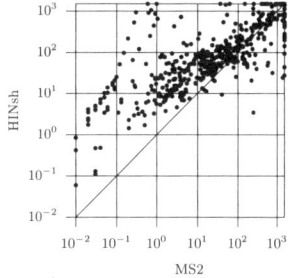

Fig. 2. Scatter plot: `hinotos-sh` vs. `Minisat1.14` (left hand side) and `Minisat2` (right hand side)

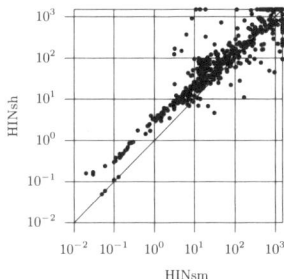 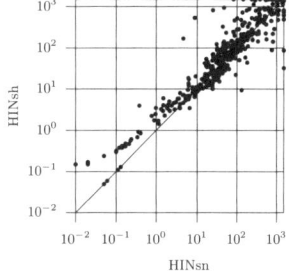

Fig. 3. Scatter plot: `hinotos-sh` vs. `hinotos-sm` (left hand side) and `hinotos-sn` (right hand side)

instances, whereas configurations `sn` and `sr` perform significantly better for the ibm and csv instances.

Figures 2 to 4 show scatter plots comparing different tool configurations. The names used are MS114, MS2, HINsh, HINsm, HINsn and HINsr, respectively for `Minisat1.14`, `Minisat2`, `hinotos-sh`, `hinotos-sm`, `hinotos-sn` and `hinotos-sr`. All run times are in seconds. Given the large number of configurations, only a subset of possible scatter plots is shown. Figure 2 evaluates `hinotos-sh` against `Minisat` versions 1.14 and 2. Albeit `hinotos-sh` aborts fewer instances than either version of `Minisat`, the plots indicate that this configuration is slower than either version of `Minisat` for most problem instances. Figure 3 shows the results for `hinotos-sh` against `hinotos-sm` and `hinotos--sn`. As can be concluded, for most instances, the run times for `hinotos-sm` and `hinotos-sn` are smaller than for `hinotos-sh`. This indicates that the optimizations proposed in this paper are in general effective, allowing smaller run times, besides being more robust. Finally, figure 4 shows scatter plots of `hinotos-sr` against `hinotos-sn` and `Minisat2`. As can be concluded, `hinotos-sr` in general has smaller run times than `hinotos-sn` (which has smaller run times than `hinotos-sh`), besides being more robust. The scatter plot on the right shows that

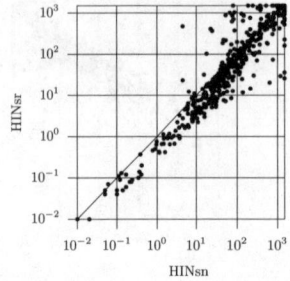
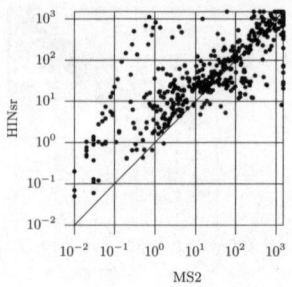

Fig. 4. Scatter plot: `hinotos-sr` vs. `hinotos-sn` (left hand side) and `Minisat2` (right hand side)

for most instances `Minisat2` has smaller run times than `hinotos-sr`. However, for a significant number of instances `hinotos-sr` has smaller run times than `Minisat2`. The plot also shows that several instances that `Minisat2` aborts, are solved by `hinotos-sr` in run times that do not exceed 100 seconds.

4.4 Analysis

The experimental results allow drawing a few conclusions. Existing versions of `Minisat` are less robust than most of the available configurations of `hinotos`. Out of 602 instances, the best version of `Minisat` (i.e. version 2) solves 462 instances. The more robust algorithm of `hinotos` solves 484 instances (almost 5% more instances solved than `Minisat2`). This result is quite significant in an industrial setting, and suggests that some of the configurations of `hinotos` should be considered for some classes of instances. On the other hand, a detailed analysis of the run times also suggests that for most instances, `Minisat2` has the smallest run time, and so should be the preferred solver. As a result, the experimental results given in this section suggest that an effective approach would be the following:

– Run `Minisat2` with a small CPU time limit, e.g. 100 seconds.
– If `Minisat2` does not solve the instance in the allowed CPU time limit, run one of the `hinotos` configurations (e.g. `hinotos-sr`) with a larger CPU time limit (e.g. 900 seconds).

The proposed configuration will be as robust as `hinotos-sr` and, for easier instances, run as effectively as `Minisat2`.

Moreover, from the results one might consider running the two solvers in parallel, without interaction between them. We conducted this experiment, and the results indicate that the LS solver is hardly ever useful. Moreover, the results show that running the two solvers in parallel is essentially equivalent to the original CDCL SAT solver. Finally, one should note that the algorithms proposed in `hinotos` complement the original SAT solvers, `Minisat2` and `Minisat1.14`. Together, `Minisat2`, `Minisat1.14` and `hinotos` can solve a significantly larger number of instances than any solver alone: 519 for all solvers compared with 484 for the best standalone solver, hinotos with configuration sr. These results suggest using a portfolio of SAT solvers similar to SATzilla [28] in industrial problems instances.

5   Related Work

Besides hbisat [5], other hybrid approaches have been proposed for SAT. A few are analyzed next. First, Mazure *at al.* [18] propose an approach essentially opposite to the one used in hbisat and hinotos. In this approach, a LS SAT solver is used to help a CDCL solver select the next decision variable. The motivation is that the use of LS SAT solver to guide the branching strategy can provide significant improvements. However, existing results are not promising [7]. Exploiting variable dependencies has been shown useful in local search algorithms for SAT. The approach of [11] proposes to extend the use of such dependencies by hybridizing WalkSAT, and the DPLL procedure Satz. At each node reached in the DPLL search tree to a fixed depth, the literal implication graph is constructed. Its strongly connected components are viewed as equivalency classes. Each one is substituted by a unique representative literal to reduce the constructed graph and the input formula. Finally, the implication dependencies are closed under transitivity. The resulting implications and equivalencies are exploited by WalkSAT at each node of the DPLL tree. The resulting algorithm [11] is an incomplete LS procedure that helps another LS SAT solver WalkSAT making better variable selections. Again, the approach, albeit efficient for some classes of satisfiable problem instances, is fundamentally different from hbisat and hinotos and unusable for unsatisfiable instances. Finally, Lardeux *at al.* [14] propose the GASAT algorithm, based on evolutionary algorithms and tabu search for SAT. The GASAT algorithm consists of a recombination stage based on a specific crossover and a tabu search stage. GASAT includes a crossover operator which relies on the structure of the clauses and a tabu search with specific mechanisms. The resulting algorithm is incomplete, and so not applicable to unsatisfiable problem instances.

6   Conclusions and Future Work

Hybrid incremental SAT solvers have recently been proposed as a possible approach for improving performance of CDCL SAT solvers [5]. Unfortunately, when an extended set of problem instances is considered, our experience is that the original hbisat algorithm is not effective when compared with different versions of Minisat. From the results, the conclusion is that for hbisat the CPU times increase, and the reduction in the number of aborted instances is negligible.

Motivated by these less positive results, this paper outlines a number of key optimizations to the original hybrid incremental SAT solver, hbisat. The experimental results, obtained on a wide range of problem instances, indicate that the proposed optimizations provide relevant performance improvements in terms of problem instances that can be solved, either with respect to Minisat (version 2) or hbisat. In terms of aborted instances, the best configuration of hinotos is significantly more effective than either version of Minisat, solving 5% more instances. In an industrial setting this is significant.

Despite the promising results, the experimental evaluation also suggests that no solver is the best option individually, and that three of the hinotos configurations should be considered as an alternative to Minisat2 for some classes of problem instances. The analysis of the results also indicates that Minisat2 usually performs better when a solution can be found in a reasonably small amount of time (e.g. 100 seconds). Hence, one strategy would be to consider running Minisat2 as the first option and, in case no solution is found, considering one of the hinotos configurations. Given the experimental results, the most promising configurations are hinotos-sr, hinotos-sn, and hinotos-sm, all of which include improvements proposed in this paper. A fairly orthogonal approach, that for some classes of instances yields promising results and so should be considered, is the hinotos-si configuration, which is also proposed in this paper.

Future work will address portfolios of configurations, based on the ideas used in SATzilla for different SAT algorithms [28]. Moreover, further tuning of the algorithm's components should be considered, e.g. improve variable lifting procedure. Finally, it might be interesting to compare results with different CDCL solvers and different LS algorithms (e.g. based on dynamic clause weighting like SAPS, PAWS or DLM [26]). Another line of work is to automatically select some of the heuristic numbers used by the hinotos algorithm.

Acknowledgement. This work is partially supported by EU grants IST/033709 and ICT/217069, and by EPSRC grant EP/E012973/1. The reviewers provided insightful comments to an earlier version of the paper.

References

1. Babić, D., Hu, A.J.: Structural Abstraction of Software Verification Conditions. Computer-Aided Verification, 371–383 (2007)
2. Biere, A., Cimatti, A., Clarke, E., Zhu, Y.: Symbolic model checking without BDDs. Tools and Algorithms for the Construction and Analysis of Systems, 193–207 (1999)
3. Davis, M., Putnam, H.: A computing procedure for quantification theory. Journal of the ACM 7, 201–215 (1960)
4. Een, N., Sörensson, N.: An extensible SAT solver. In: International Conference on Theory and Applications of Satisfiability Testing, pp. 502–518 (2003)
5. Fang, L., Hsiao, M.S.: A new hybrid solution to boost SAT solver performance. In: Design, Automation and Testing in Europe Conference, pp. 1307–1313 (2007)
6. Fang, L., Hsiao, M.S.: Private communications (2007)
7. Ferris, B., Froehlich, J.: WalkSAT as an Informed Heuristic to DPLL in SAT Solving. Technical report, CSE 573: Artificial Intelligence (2004)
8. Gent, I.P., Walsh, T.: Towards an understanding of hill-climbing procedures for SAT. In: National Conference on Artificial Intelligence, pp. 28–33 (1993)
9. Goldberg, E., Novikov, Y.: BerkMin: a fast and robust SAT-solver. In: Design, Automation and Testing in Europe Conference, pp. 142–149 (2002)

10. Gomes, C.P., Selman, B., Kautz, H.: Boosting combinatorial search through randomization. In: National Conference on Artificial Intelligence, pp. 431–437 (1998)
11. Habet, D., Li, C.M., Devendeville, L., Vasquez, M.: A hybrid approach for SAT. In: International Conference on Principles and Practice of Constraint Programming, pp. 172–184 (2002)
12. Hoos, H.: An adaptive noise mechanism for WalkSAT. In: National Conference on Artificial Intelligence, pp. 655–660 (2002)
13. Hoos, H., Stützle, T.: Stochastic Local Search: Foundations and Applications. Morgan Kaufmann, San Francisco (2004)
14. Lardeux, F., Saubion, F., Hao, J.-K.: GASAT: A genetic local search algorithm for the satisfiability problem. Evolutionary Computation 14(2), 223–253 (2006)
15. Li, C.M., Huang, W.Q., Zhang, H.: Combining adaptive noise and look-ahead in local search for SAT. In: International Conference on Theory and Applications of Satisfiability Testing, pp. 121–133 (2007)
16. Manolios, P., Srinivasan, S.K.: A parameterized benchmark suite of hard pipelined-machine-verification problems. In: Advanced Research Working Conference on Correct Hardware Design and Verification Methods, pp. 363–366 (2005)
17. Marques-Silva, J., Sakallah, K.: GRASP: A new search algorithm for satisfiability. In: International Conference on Computer-Aided Design, pp. 220–227 (1996)
18. Mazure, B., Sais, L., Grégoire, E.: Boosting complete techniques thanks to local search methods. Annals of Mathematics and Artificial Intelligence 22(3-4), 319–331 (1998)
19. McAllester, D., Selman, B., Kautz, H.: Evidence of invariants in local search. In: National Conference on Artificial Intelligence, pp. 321–326 (1997)
20. McMillan, K.L.: Applying SAT methods in unbounded symbolic model checking. Computer-Aided Verification, 250–264 (2002)
21. Moskewicz, M., Madigan, C., Zhao, Y., Zhang, L., Malik, S.: Engineering an efficient SAT solver. In: Design Automation Conference, pp. 530–535 (2001)
22. Prestwich, S., Lynce, I.: Local search for unsatisfiability. In: International Conference on Theory and Applications of Satisfiability Testing, pp. 283–296 (2006)
23. Ravi, K., Somenzi, F.: Minimal assignments for bounded model checking. Tools and Algorithms for the Construction and Analysis of Systems, 31–45 (2004)
24. Selman, B., Kautz, H., Cohen, B.: Noise strategies for improving local search. In: National Conference on Artificial Intelligence, pp. 337–343 (1994)
25. Selman, B., Levesque, H., Mitchell, D.: A new method for solving hard satisfiability problems. In: National Conference on Artificial Intelligence, pp. 440–446 (1992)
26. Thornton, J., Pham, D.N., Bain, S., Ferreira Jr., V.: Additive versus multiplicative clause weighting for SAT. In: National Conference on Artificial Intelligence, pp. 191–196 (2004)
27. Velev, M.N.: Using rewriting rules and positive equality to formally verify wide-issue out-of-order microprocessors with a reorder buffer. In: Design, Automation and Testing in Europe Conference, pp. 28–35 (2002)
28. Xu, L., Hutter, F., Hoos, H., Leyton-Brown, K.: SATzilla-07: The design and analysis of an algorithm portfolio for SAT. In: International Conference on Principles and Practice of Constraint Programming, pp. 712–727 (2007)
29. Zarpas, E.: Benchmarking SAT solvers for bounded model checking. In: International Conference on Theory and Applications of Satisfiability Testing, pp. 340–354 (2005)

Searching for Autarkies to Trim Unsatisfiable Clause Sets

Mark Liffiton and Karem Sakallah

Department of Electrical Engineering and Computer Science,
University of Michigan, Ann Arbor 48109-2121
{liffiton,karem}@eecs.umich.edu

Abstract. An *autarky* is a partial assignment to the variables of a
Boolean CNF formula that satisfies every clause containing an assigned
variable. For an unsatisfiable formula, an autarky provides information
about those clauses that are essentially independent from the infeasibility; clauses satisfied by an autarky are not contained in any minimal unsatisfiable subset (MUS) or minimal correction subset (MCS) of clauses.
This suggests a preprocessing step of detecting autarkies and trimming
such independent clauses from an instance prior to running an algorithm
for finding MUSes or MCSes. With little existing work on algorithms for
finding autarkies or experimental evaluations thereof, there is room for
further research in this area. Here, we present a novel algorithm that
searches for autarkies directly using a standard satisfiability solver. We
investigate the autarkies of several industrial benchmark suites, and experimental results show that our algorithm compares favorably to an
existing approach for discovering autarkies. Finally, we explore the potential of trimming autarkies in MCS- or MUS-extraction flows.

1 Introduction

Analysis of the infeasibility of unsatisfiable Boolean satisfiability problems has
recently received increasing attention, though still little when compared to the
efforts directed toward solutions to the problem of deciding the satisfiability of
a Boolean formula (SAT). In many cases, the answer returned by a SAT solver
given an infeasible formula, "UNSAT," is not sufficient information, and tools
for further analysis are necessary.

Two such tools are the related concepts of minimal unsatisfiable subsets
(MUSes) and minimal correction subsets (MCSes). Both MUSes and MCSes
are irreducible portions of a formula that contain information relevant to understanding and correcting the formula's infeasibility while ignoring unrelated
information. Several algorithms have been developed for computing MUSes and
MCSes, including algorithms for finding a single, often approximate MUS (e.g.,
[6,9,16,21]); finding a smallest MUS (SMUS, also called a minimum unsatisfiable
core) [14]; and finding both all MCSes and all MUSes [13]. The work in [13], a
focus of this paper, has found applications in finding all MCSes for circuit error
diagnosis [17] and all MUSes as part of an abstraction refinement flow [1].

H. Kleine Büning and X. Zhao (Eds.): SAT 2008, LNCS 4996, pp. 182–195, 2008.

Autarkies provide another tool for looking into the structure of an unsatisfiable formula; they essentially provide information about portions of the formula that can be considered independent of the infeasibility. Autarkies have recently been linked to MUSes in [12], where Kullmann, et al., develop a classification of clauses in Boolean formulas based on their involvement in MUSes, autarkies, and resolution refutations. They use CAMUS ("Compute All Minimal Unsatisfiable Subsets") [13], a tool for computing all MUSes of a Boolean formula, and the only existing full approach for finding autarkies of which we are aware (first introduced in [11]) to investigate the complete set of MUSes and the autarkies, respectively, of a set of industrial benchmarks. They do not report runtime results, and we are not aware of any other experimental research on algorithms for finding the largest, or maximum, autarky of an instance.

In [12], the authors suggest two directions of research that we have undertaken in this paper:

1. An algorithm that directly searches for autarkies could be developed and compared to their algorithm, which makes use of a "duality" between autarkies and resolution refutations to find autarkies indirectly.
2. As clauses involved in autarkies are never contained in any MUS, such clauses can be removed as a preprocessing step for computing MUSes of a formula. (This also holds for MCSes, as they are comprised of the same clauses as MUSes.)

We have developed a novel algorithm, named Sifter, that directly performs a complete search for maximum autarkies, and we compare it to the existing approach based on resolution proofs. We also investigate the use of this algorithm as a preprocessing step to trim autarkies from unsatisfiable instances before searching for MUSes or MCSes.

This paper is organized as follows. Section 2 lays out formal definitions and concepts used throughout the paper. We review previous work related to autarkies in Section 3, and in Section 4, we introduce our new algorithm for finding maximum autarkies, Sifter. Experimental results comparing Sifter to the previous approach and investigating its use as a preprocessing step for two algorithms that operate on unsatisfiable formulas are shown and discussed in Section 5. Finally, Section 6 concludes with a brief overview of the paper and potential future work.

2   Preliminaries

Boolean Satisfiability and Conjunctive Normal Form. Formally, a Boolean formula C in conjunctive normal form (CNF) is defined as follows:

$$C = \bigwedge_{i=1...n} C_i$$

$$C_i = \bigvee_{j=1...k_i} a_{ij}$$

where each *literal* a_{ij} is either a positive or negative instance of some Boolean variable (e.g., x_3 or $\neg x_3$, where the domain of x_j is $\{0, 1\}$), the value k_i is the number of literals in the *clause* C_i (a disjunction of literals), and n is the number of clauses in the formula. In more general terms, each clause is a *constraint* of the constraint system C. A CNF instance is said to be *satisfiable* (SAT) if there exists some assignment to its variables that makes the formula evaluate to 1 or TRUE; otherwise, we call it *unsatisfiable* (UNSAT). A *SAT solver* evaluates the satisfiability of a given CNF formula and returns a satisfying assignment of its variables if it is satisfiable, and some produce *resolution refutations* (or *resolution proofs*) for unsatisfiable instances, directed acyclic graphs containing the resolution steps used to prove unsatisfiability.

The following unsatisfiable CNF instance C will be used as an example in this paper. We will refer to individual clauses as C_i, where i refers to the position of the clause in the formula (e.g., $C_3 = (\neg x_1 \vee \neg x_2)$).

$$C = (x_1)(\neg x_1 \vee x_2)(\neg x_1 \vee \neg x_2)(\neg x_2 \vee x_3)(x_4 \vee x_5)(\neg x_4 \vee \neg x_5)$$

AtMost Constraints. Our algorithm employs *AtMost* constraints, a type of counting constraint that can be constructed from Boolean CNF constraints or added to a SAT solver with few modifications. Given a set of n literals $\{l_1, l_2, \ldots, l_n\}$ and a positive integer k, s.t. $k < n$, an AtMost constraint is defined as

$$\text{AtMost}(\{l_1, l_2, \ldots, l_n\}, k) \equiv \sum_{i=1}^{n} \text{val}(l_i) \leq k$$

where $\text{val}(l_i)$ is 1 if l_i is assigned TRUE and 0 otherwise. This constraint places an upper bound on the number of literals in the set assigned TRUE.

This constraint can be encoded into Boolean CNF using encodings such as in [18], or it can be implemented efficiently in a SAT solver that employs watched variables (such as MiniSAT [7], which we use in this work). An implementation of an AtMost constraint can watch the assignments to the variables in the constraint and immediately propagate the negation of each remaining literal once k of them have been assigned TRUE. On a closed SAT solver that does not allow for a built-in implementation of the AtMost constraint, the CNF encoding can still be used.

Minimal Unsatisfiable Subsets / Minimal Correction Subsets. The definitions of *Minimal Unsatisfiable Subsets* (MUSes) and *Minimal Correction Subsets* (MCSes) of clauses are important to this work, as we are looking at the use of autarkies in preprocessing steps for algorithms that find MUSes and MCSes. An MUS is a subset of the clauses of an unsatisfiable formula that is unsatisfiable and cannot be made smaller without becoming satisfiable. An MCS is a subset of the clauses of an unsatisfiable formula whose removal from that formula results in a satisfiable formula ("correcting" the infeasibility) and that is minimal in the same sense that any proper subset does not have that defining property. Any unsatisfiable formula can have multiple MUSes and MCSes, potentially exponential in the number of clauses.

As proven in [3], there is a duality between MUSes and MCSes such that for a given instance, the complete set of MUSes (resp. MCSes) can be generated by finding all minimal hitting sets of the complete set of MCSes (resp. MUSes). This fact is used in [13] as the foundation for CAMUS, a set of two algorithms that computes all MUSes of an instance by way of first computing all MCSes. A corollary of this is that the union of all MUSes is equivalent to the union of all MCSes.

Our example contains one MUS, $\{C_1, C_2, C_3\}$, and its MCSes are the single-clause sets $\{C_1\}$, $\{C_2\}$, and $\{C_3\}$.

Autarkies. An *autarky* (or *autark assignment*) is an assignment to a subset of a formula's variables that satisfies every clause containing one of the assigned variables. Following the meaning of the term in other fields, it is a *self-sufficient* partial assignment. Because we are interested in trimming clause sets in this work, we will refer to autarkies in terms of the clauses they satisfy. Thus, the maximum autarky for us is the largest set of clauses satisfiable by an autarky, as opposed to the largest partial assignment. The maximum autarky for our example formula C is $\{C_4, C_5, C_6\}$, which in this case is the complement of the one MUS, and it is satisfied by the partial assignment $\{x_3 = \text{TRUE}, x_4 = \text{TRUE}, x_5 = \text{FALSE}\}$.

As explained in [10], any clause satisfied by some autarky can not be contained in any MUS (nor in any MCS, as they are comprised of the same clauses). This motivates the idea of preprocessing unsatisfiable formulas by removing their maximum autarkies before searching for MUSes or MCSes.

Pure Literals. One simple form of autarky arises from *pure literals*. A pure literal is a variable that occurs in only one polarity (either always positive or always negated) in a CNF formula. In our example formula, x_3 is a pure literal, because $\neg x_3$ is not present. An assignment of TRUE to a pure literal will trivially satisfy any clause containing the corresponding variable, thus any such assignment is an autarky.

Pure literals can be found in a linear time scan of a formula. Removing the clauses satisfied by pure literals may cause other literals to become pure in the formula, so repeatedly detecting, recording, and removing pure literals is a simple first step for any algorithm that finds autarkies. The process terminates when the formula no longer contains pure literals.

3   Previous Work

Monien and Speckenmeyer [15] first introduced the concept of an autark assignment or autarky, using autarkies in a modification of the DPLL satisfiability algorithm [4,5] that reduced its complexity upper bound below 2^n splitting steps (for a formula with n variables). Autarkies were later used in another satisfiability algorithm by Van Gelder [20] named Modoc. Modoc integrates autarky pruning, removing those clauses satisfied by autarkies, into a resolution-based model elimination approach to satisfiability. Both Monien and Speckenmeyer's algorithm and Modoc find autarkies as side-effects of their operation, but neither is aiming to find the maximum autarky. Additionally, both find many more "conditional autarkies," i.e., autarkies that appear after propagating a partial assignment through the formula, than "top-level autarkies" for the entire formula.

More recently, Kullmann has investigated autarkies in several papers. In [10], he introduces the idea of *lean clause-sets*, sets of clauses that have no autarkies. The largest lean clause-set is the complement of the maximum autarky of a formula; all clauses can be partitioned into one or the other. Kullmann investigates a special case of autarky that can be found in polynomial time using linear programming, though this does not generalize to finding all autarkies. He also proves, with Theorem 3.16, that a set of clauses F is lean "if and only if every clause of F can be used by some resolution refutation of F." Conversely, a set of clauses $A \subseteq F$ is an autarky if and only if each clause in A can *not* be used in any resolution refutation of F. Later, in [11], Kullmann uses this fact to develop an algorithm for computing the maximum autarky. Using a SAT solver that provides a resolution refutation for unsatisfiable instances, one can iteratively remove the variables included in some resolution proof. When the reduced formula becomes satisfiable, the satisfying assignment is an autarky of the original formula. This is the algorithm to which we compare ours in Section 5.

Finally, Kullmann, et al. [12] use both autarkies and MUSes as tools to describe and examine unsatisfiable formulas. They characterize clauses in such formulas into several classes based on each clause's involvement in MUSes, resolution refutations, and autarkies. Clauses contained in every MUS are called "necessary"; those in any MUS are "potentially necessary"; "usable" indicates a clause is in some resolution refutation; and thus "unusable" refers to clauses in an autarky. Complements and intersections of these classes are defined as well. They experimentally evaluate a set of industrial benchmarks from an automotive product configuration domain [19], reporting on the MUSes and clauses in the different levels of "necessity" in each instance. To compute all MUSes of the instances, they use CAMUS [13], and they found maximum autarkies using the algorithm described in [11], implemented using the ZChaff SAT solver's ability to produce resolution refutations [21].

4   Searching for Autarkies

Our approach to the problem of finding the maximum autarky for a formula treats it as an optimization problem. We search for the largest partial assignment that satisfies the clauses it touches, i.e., the largest autarky, by explicitly searching in the space of all partial assignments and maximizing the size of the result (in terms of the number of satisfied clauses). Specifically, we "instrument" the formula to give a standard SAT solver the ability to enable and disable individual clauses and variables within its normal search, and we use AtMost constraints to perform a sliding objective maximization of the autarky size. This draws inspiration from a similar technique we employed in [13] that uses a less-involved instrumentation and the same optimization technique to allow a SAT solver to search for maximal satisfiable subsets of clauses. We directly exploit the efficiency gains made in SAT solvers in recent years by using an "off-the-shelf" solver; our algorithm works with any solver[1], so it can benefit from future improvements as well.

4.1 Instrumentation

To give a SAT solver the ability to search for autarkies, we instrument a formula C with the following modifications:

1. We replace every literal in the formula with a *literal-substitute*; x_j in the formula becomes x_j^1, while $\neg x_j$ is replaced with x_j^0.
2. Each clause C_i is augmented with a *clause-selector* y_i to form a new clause $C_i' = (y_i \rightarrow C_i) = (\neg y_i \vee C_i)$.
3. We create a *variable-selector* x_j^+ for every variable x_j. When x_j^+ is TRUE, x_j will be enabled, and it is disabled otherwise. For every variable x_j, we add clauses to relate its variable-selector x_j^+, its two literal-substitutes x_j^0 and x_j^1, and the value of the variable itself, x_j. In short, we want each literal-substitute to be TRUE when the variable is enabled (x_j^+ is TRUE) and x_j has the corresponding value. This leads to new clauses encoding the following: $(x_j^1 = x_j^+ \wedge x_j)$ and $(x_j^0 = x_j^+ \wedge \neg x_j)$.
4. Finally, we add clauses to require that a clause be enabled ($y_i =$ TRUE) if any one of its variables is enabled. Thus, for any x_j present in clause C_i, we add a clause $(x_j^+ \rightarrow y_i) = (\neg x_j^+ \vee y_i)$.

This is not the only option for instrumenting the formula; other encodings have the same effect. However, while preliminary experiments showed that similar encodings yield slightly different runtimes, the differences in efficiency were not substantial.

The complete instrumented formula for our example is too large to be useful here, but here we show the constraints produced from a single clause of the example, C_2:

$$C_2, (\neg x_1 \vee x_2) \implies \begin{cases} \text{1 \& 2: } (\neg y_2 \vee x_1^0 \vee x_2^1) \\ \quad\quad (x_1^1 = x_1^+ \wedge x_1)(x_1^0 = x_1^+ \wedge \neg x_1) \\ \text{3: } \\ \quad\quad (x_2^1 = x_2^+ \wedge x_2)(x_2^0 = x_2^+ \wedge \neg x_2) \\ \text{4: } (\neg x_1^+ \vee y_2)(\neg x_2^+ \vee y_2) \end{cases}$$

The clause derived from modifications 1 and 2 replaces the original clause, while the rest are additions. The clauses from modification 3 (presented in short-hand as equalities; each is three clauses in CNF) are specific to variables, and the complete formula will only contain each set once per variable. The final two clauses, resulting from modification 4, are specific to C_2.

---

[1] SAT solvers that implement AtMost constraints internally will likely perform better than those that require using a CNF encoding of them, but all will work.

With the formula instrumented in this way, any satisfying assignment will indicate an autarky of the original formula. The x_j^+ variables indicate which variables are "activated," i.e., included in the autarky; the original variables contain the autarky assignment; and the clauses satisfied by the autarky are represented by those y_i variables set to TRUE. One such assignment is the trivial solution in which all variables and all clauses are disabled. To find the maximum autarky, we must maximize the number of enabled clauses.

4.2   Our Algorithm

We maximize the number of enabled clauses (y_i variables assigned TRUE) by way of an iterative optimization approach. We use AtMost constraints to bound the number of disabled clauses, tightening the bound as solutions are found. If an autarky is found that leaves n clauses disabled, we start the search for a larger autarky by bounding the disabled clauses to $n - 1$. Eventually, if the instance is unsatisfiable, we will reach a bound k for which no solution can be found. At this point, we have proven that there exists an autarky of size $k - 1$ and none with size k, thus the previously found autarky is the maximum autarky.

Sifter(C)

---

1. $(C, \text{autarky}) \leftarrow \textbf{PureLits}(C)$
2. $C' \leftarrow \textbf{Instrument}(C)$
3. bound $\leftarrow |C| - 1$
4. **loop**
5. $C_b' \leftarrow C' \wedge \textbf{AtMost}(\{\neg y_1, \neg y_2, \ldots, \neg y_n\}, \text{bound})$
6. $(\text{isSAT}, \text{model}) \leftarrow \textbf{Solve}(C_b')$
7.   **if not** isSAT
8.   **return** autarky
9.   autarky $\leftarrow$ autarky $\cup$ **SatisfiedClauses**(model)
10.   bound $\leftarrow |C| - |\text{autarky}| - 1$

Fig. 1. Sifter finds the maximum autarky of a CNF formula C by "instrumenting" the instance and using a SAT solver to search for satisfying partial assignments

Figure 1 contains pseudocode for the complete algorithm, which we call Sifter. First, we repeatedly scan for pure literals, recording and removing them as described in Section 2: the call to **PureLits** returns 1) C with any clauses containing pure literals removed and 2) the set of such clauses as an initial autarky. We then instrument the formula and use the sliding objective method described above to find the rest of the maximum autarky or to prove that the pure literal approach found it in its entirety. The **Instrument** subroutine produces instrumented clauses via the modifications described in Section 4.1. The bound on the number of disabled clauses is set initially to $|C| - 1$ to begin the search by looking for an autarky that satisfies at least one clause, and the loop then proceeds by searching for a satisfying assignment, `model`, of the instrumented, bounded formula, C_b'. If none is found (`isSAT` is false), the algorithm returns `autarky`, which must be the maximum autarky. Otherwise, the satisfied clauses are added to `autarky`, the bound is set to search for an autarky that satisfies at least one more clause, and the loop repeats.

5   Experimental Results

Our two experimental goals are 1) to compare and contrast Sifter, our direct search-based approach for finding the maximum autarky, with the earlier iterative technique using resolution refutation trees [11], and 2) to investigate the value of trimming autarkies as a preprocessing step for finding MUSes and MCSes.

5.1   Comparing Search to an Iterated Resolution Proof Approach

We implemented Sifter in C++ using MiniSAT [7] version 1.12b (the last version containing support for AtMost constraints). We wrote the iterative approach [11], which we will call Scraper, as a Perl script. First, Scraper uses the pure literal elimination written for Sifter, making that phase equivalent in both implementations. Then, it employs the tools `zchaff` and `zverify_df` [21] from the ZChaff distribution `zchaff.64bit.2007.3.12` to repeatedly produce resolution refutations and eliminate the involved variables until the instance becomes satisfiable. We compiled all executables for the x86-64 instruction set using GCC 4.1.2 with standard optimizations, and all experiments were run under Linux (Fedora 7) on a 3.0GHz Intel Core 2 Duo E6850 with 4GB of RAM.

Figure 2 contains a log-log scatterplot comparing the runtimes of Sifter and Scraper on a variety of industrial benchmarks. Runtimes for Sifter are represented on the y-axis, so points lying below the diagonal indicate instances in which Sifter outperforms Scraper. A timeout of 600 seconds was used for every run, indicated by the dashed lines on the extremes of the chart; points on these lines indicate that a timeout was reached by the corresponding algorithms. The reported runtimes are processor time, which for Sifter are essentially equivalent to wall-clock time. Our implementation of Scraper, however, stores several intermediate results to disk; we ignore this I/O time in these results to estimate the runtime of a more efficient approach that retains everything in memory.

To provide a more complete understanding of these results, Table 1 lists some overall characteristics of each benchmark family. We list the minimum and maximum number of variables, number of clauses, and size of the maximum autarky (in clauses) for the instances in each family. The Benz benchmarks[2] are the automotive product configuration instances from [19] used in the experiments in [12]. The Miter family[3] contains equivalence checking instances from João Marques-Silva. The Dimacs instances are circuit benchmarks from the DIMACS set. The

---

[2] http://www-sr.informatik.uni-tuebingen.de/~sinz/DC/
[3] http://sat.inesc.pt/benchmarks/cnf/equiv-checking/instances/

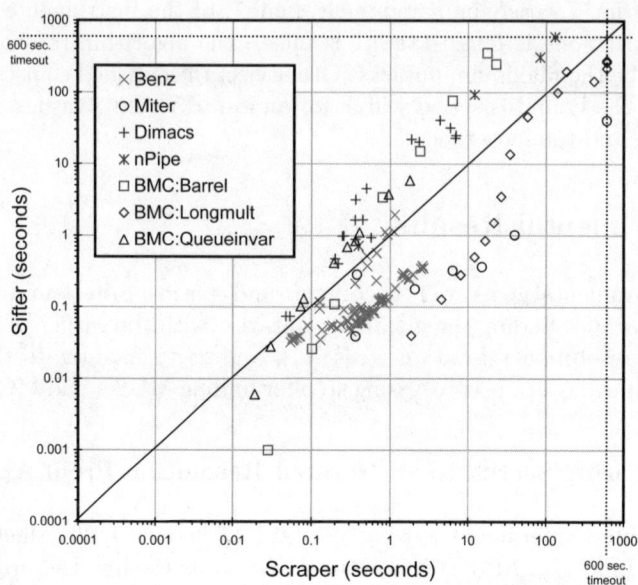

Fig. 2. Comparing the performance of Sifter and Scraper on a variety of benchmarks

nPipe instances are from Miroslav Velev's FVP-UNSAT-2.0 benchmarks[4], generated for the formal verification of microprocessors, with redundant variables removed. The BMC:[] instances[5] are formulas used in bounded model checking (BMC) as described in [2].

From Figure 2 and Table 1, we can draw several conclusions:

1. Across all of the benchmarks, neither Sifter nor Scraper dominates the other in terms of runtime. In some benchmarks, Scraper is faster, up to 20x, while in others, Sifter is faster, up to 46x.

2. In just those benchmarks with non-trivial autarkies, however, our Sifter algorithm is faster in nearly every instance. Specifically, looking at the Benz and Miter families (the autarkies covering 2 clauses in each BMC:Longmult instance are all found by pure-literal elimination alone), we see that Sifter outperforms Scraper by approximately one order of magnitude.

3. The presence and size of autarkies is fairly consistent *within* benchmark families. Each particular family in Dimacs, nPipe, and BMC:[] has either no autarkies in any instance or an autarky that covers 2 clauses in each. The Benz family consistently has autarkies that cover a large portion (between 32 and 98 percent) of each instance's clauses. Every instance in the Miter family has a non-empty autarky, though the autarky sizes vary more than they do in the Benz instances.

---

[4] http://www.miroslav-velev.com/sat_benchmarks.html

[5] http://www.cs.cmu.edu/~modelcheck/bmc/bmc-benchmarks.html

Table 1. Benchmark Characteristics

Family	Variables		Clauses		`autarky`	
	min	max	min	max	min	max
Benz	1,513	1,891	4,013	9,957	2,097	7,025
Miter	1,266	17,303	1,027	34,238	1	1,831
Dimacs	389	7,767	1,115	20,812	0	0
nPipe	861	15,469	6,695	394,739	0	0
BMC:Barrel	50	8,903	159	36,606	0	0
BMC:Longmult	437	7,807	1,206	24,351	2	2
BMC:Queueinvar	116	2,435	399	20,671	0	0

Overall, these conclusions imply a strategy for exploiting autarkies in practice. First, by searching for autarkies on a small representative set of instances from a particular application, one can determine whether the instances in that domain have autarkies at all. If none of the test set have autarkies of any appreciable size, then it is likely that none generated in the application will, in which case autarkies will be of no use. This is likely in applications such as bounded model checking, where performing a cone of influence reduction of the circuit will likely eliminate all autarkies. In these applications, checking for autarkies could be a simple test of the sanity of the CNF encoding. In the other case, in which instances do contain autarkies, it is probable that most if not all instances will have autarkies, and Sifter is likely the more efficient algorithm to use.

5.2   Trimming Autarkies to Boost Searching for MUSes and MCSes

Trimming autarkies holds the most promise for boosting algorithms that have a high complexity and are affected heavily by the number of clauses in an instance. An algorithm for finding any single unsatisfiable subformula, such as that developed in ZChaff [21], is unlikely to benefit from such boosting, as the time taken to find the maximum autarky will likely dwarf the runtime of the unboosted algorithm.

We identified two algorithms that *are* good candidates for this boosting. One is the first phase of CAMUS [13], which finds all MCSes of a formula as the first step of solving several related problems such as computing all MUSes or finding the smallest MUS (SMUS) of a CNF formula. In addition to computing MUSes, this algorithm has been applied (without the second phase) in a circuit error diagnosis system [17], in which the MCSes were used directly. A second algorithm with potential for boosting by trimming autarkies is that by Mneimneh, et. al. [14] for computing an SMUS directly, which we will refer to as SMUS. Both of these candidate algorithms use clause-selector variables (as used in Sifter and described in Section 4.1) and use a SAT solver to implicitly search through subsets of clauses. Therefore, both can benefit from the reduced search space produced by a reduction in the number of input clauses.

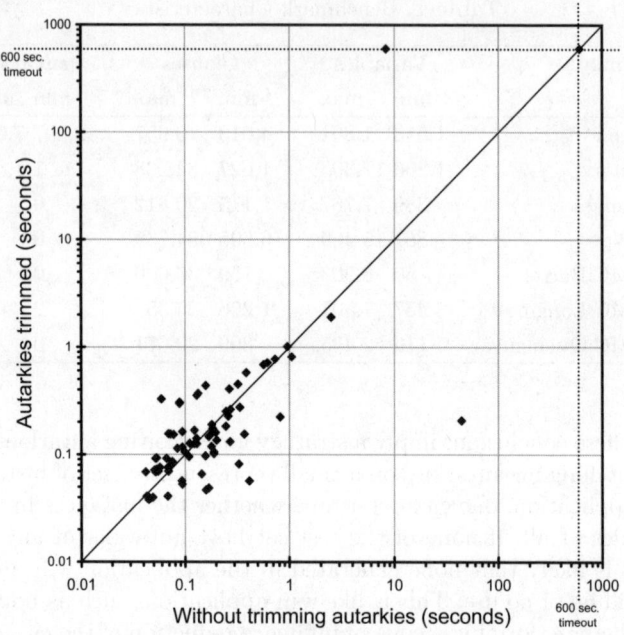

Fig. 3. Boosting SMUS by trimming autarkies for the Benz benchmarks

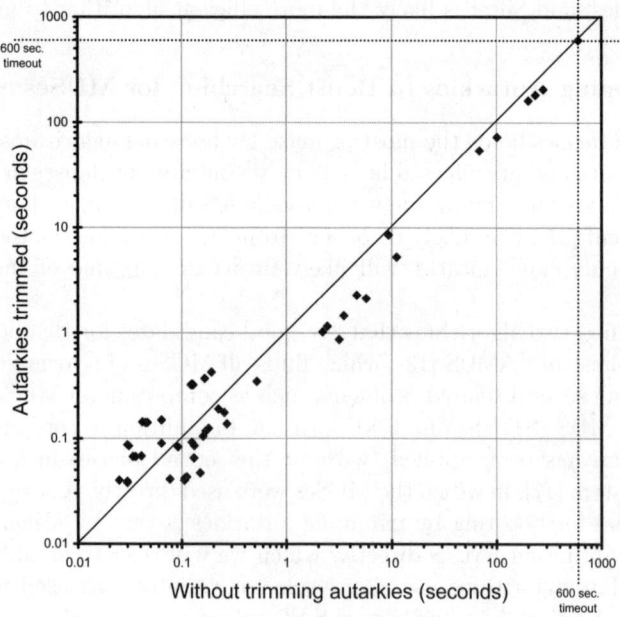

Fig. 4. Boosting CAMUS (first phase) by trimming autarkies for the Benz benchmarks

We investigated the impact of trimming autarkies on both of these algorithms for the Benz benchmarks, which have the largest autarkies, and the results are displayed in Figures 3 and 4. Each figure is a log-log scatterplot that charts the runtime of the specified algorithm alone on the x-axis against the runtime of the boosted version on the y-axis. The runtime reported for the boosted version is the sum of finding an instance's maximum autarky with Sifter and running the algorithm on the trimmed instance. A point below the diagonal indicates an instance for which the boosting produced a net decrease in runtime.

The results are mixed. In Figure 3, we see that the boosting does not produce markedly better or worse results overall for finding SMUSes with SMUS. While the runtimes for SMUS alone (not shown) do improve in nearly all cases when it is run on the trimmed instances, the runtime of Sifter outweighs this gain in many cases. There are two outliers: one in which SMUS's runtime improves by over two orders of magnitude when run on the trimmed instance, and another that takes less than 10 seconds on the untrimmed instance yet times out at 600 seconds on the trimmed version. These are artifacts of the susceptibility of combinatorial search algorithms like SMUS to variations in runtime due to minor ordering changes and similar effects.

The results for boosting the first phase of CAMUS, shown in Figure 4, show that the boosting does have value in some cases. For this algorithm, the runtime of Sifter can outweigh the decrease in runtime due to the boosting in cases with *small* runtimes (below 1 second in these benchmarks), but the boosted algorithm always outperforms the original algorithm in cases with longer runtimes. Taken as a whole, this is a net benefit, because the runtime increases in some "small" instances are far outweighed by the gains in the "large" instances. The total runtime, over all instances that did not time out in both techniques, decreased from 931 seconds on untrimmed instances to 704 seconds for the boosted algorithm, a 24% decrease in total runtime.

6   Conclusions and Future Work

We are aware of only one existing algorithm for computing maximum autarkies, presented in [11], and no experimental research investigating the runtime of finding maximum autarkies has been published prior to this work. Furthermore, little research has been conducted in the area of autarkies for Boolean satisfiability, and no "industrial" application has previously been identified for them.

In this paper, we have presented a new algorithm, Sifter, for finding maximum autarkies that searches for them directly with a standard SAT solver and an "instrumented" formula. We have evaluated it experimentally, comparing it to the existing approach based on iterated construction of resolution refutations, on a variety of industrial benchmarks. In our results, Sifter outperforms the other algorithm on benchmarks with autarkies, though the results are mixed on those with none.

We have also performed an initial exploration of the use of autarky trimming as a preprocessing step for complex algorithms for finding MUSes and MCSes.

We used Sifter to boost two different algorithms by trimming maximum autarkies from instances before searching for MUSes or MCSes. While the boosted version of an algorithm for finding the smallest MUS of a formula was not (overall) faster or slower than the normal version, the boosted version of the first phase of CAMUS [13], which finds all MCSes of an instance, was noticeably faster. The overhead of the trimming often outweighed runtime gains on instances that completed in under one second, but the trimming was beneficial on long-running instances; we obtained a total runtime reduction of 24% over all instances that did not time out.

As future work, we can look into improving the efficiency of Sifter, possibly by using a new encoding to instrument formulas or by employing a different optimization method. Also, more work can be done to explore the structure and characteristics of autarkies in real-world Boolean formulas; the results here show that there is much variation among benchmark families, with some containing no autarkies at all. The use of conditional autarkies (autarkies that arise following the assignment of some variables) in algorithms for analyzing unsatisfiable instances is worth investigating as well. In the area of boosting MUS algorithms, it would be interesting to compare autarky-trimming with the local search used to boost the first phase of CAMUS in [8] by quickly identifying candidate MCSes before the complete search begins.

Acknowledgments

This material is based upon work supported by the National Science Foundation under ITR Grant No. 0205288. Any opinions, findings and conclusions or recommendations expressed in this material are those of the authors and do not necessarily reflect the view of the National Science Foundation (NSF).

References

1. Andraus, Z.S., Liffiton, M.H., Sakallah, K.A.: Refinement strategies for verification methods based on datapath abstraction. In: Proceedings of the 2006 conference on Asia South Pacific design automation (ASP-DAC 2006), pp. 19–24 (2006)
2. Biere, A., Cimatti, A., Clarke, E.M., Zhu, Y.: Symbolic model checking without BDDs. In: Cleaveland, W.R. (ed.) ETAPS 1999 and TACAS 1999. LNCS, vol. 1579, pp. 193–207. Springer, Heidelberg (1999)
3. Birnbaum, E., Lozinskii, E.L.: Consistent subsets of inconsistent systems: structure and behaviour. Journal of Experimental and Theoretical Artificial Intelligence 15, 25–46 (2003)
4. Davis, M., Logemann, G., Loveland, D.: A machine program for theorem-proving. Communications of the ACM 5(7), 394–397 (1962)
5. Davis, M., Putnam, H.: A computing procedure for quantification theory. Journal of the ACM 7(3), 201–215 (1960)
6. Dershowitz, N., Hanna, Z., Nadel, A.: A scalable algorithm for minimal unsatisfiable core extraction. In: Sattar, A., Kang, B.-h. (eds.) AI 2006. LNCS (LNAI), vol. 4304, pp. 36–41. Springer, Heidelberg (2006)

7. Eén, N., Sörensson, N.: An extensible SAT-solver. In: Giunchiglia, E., Tacchella, A. (eds.) SAT 2003. LNCS, vol. 2919, pp. 502–518. Springer, Heidelberg (2004)
8. Grégoire, É., Mazure, B., Piette, C.: Boosting a complete technique to find MSSes and MUSes thanks to a local search oracle. In: Proceedings of the 20th International Joint Conference on Artificial Intelligence (IJCAI 2007), January 2007, vol. 2, pp. 2300–2305 (2007)
9. Grégoire, É., Mazure, B., Piette, C.: Local-search extraction of MUSes. Constraints 12(3), 325–344 (2007)
10. Kullmann, O.: Investigations on autark assignments. Discrete Applied Mathematics 107(1-3), 99–137 (2000)
11. Kullmann, O.: On the use of autarkies for satisfiability decision. In: LICS 2001 Workshop on Theory and Applications of Satisfiability Testing (SAT-2001). Electronic Notes in Discrete Mathematics, vol. 9, pp. 231–253 (2001)
12. Kullmann, O., Lynce, I., Marques-Silva, J.: Categorisation of clauses in conjunctive normal forms: Minimally unsatisfiable sub-clause-sets and the lean kernel. In: Biere, A., Gomes, C.P. (eds.) SAT 2006. LNCS, vol. 4121, pp. 22–35. Springer, Heidelberg (2006)
13. Liffiton, M.H., Sakallah, K.A.: Algorithms for computing minimal unsatisfiable subsets of constraints. Journal of Automated Reasoning 40(1), 1–33 (2008)
14. Mneimneh, M.N., Lynce, I., Andraus, Z.S., Silva, J.P.M., Sakallah, K.A.: A branch-and-bound algorithm for extracting smallest minimal unsatisfiable formulas. In: Bacchus, F., Walsh, T. (eds.) SAT 2005. LNCS, vol. 3569, pp. 467–474. Springer, Heidelberg (2005)
15. Monien, B., Speckenmeyer, E.: Solving satisfiability in less than 2^n steps. Discrete Applied Mathematics 10(3), 287–295 (1985)
16. Oh, Y., Mneimneh, M.N., Andraus, Z.S., Sakallah, K.A., Markov, I.L.: AMUSE: a minimally-unsatisfiable subformula extractor. In: Proceedings of the 41st Annual Conference on Design Automation (DAC 2004), pp. 518–523 (2004)
17. Safarpour, S., Liffiton, M., Mangassarian, H., Veneris, A., Sakallah, K.: Improved design debugging using maximum satisfiability. In: Proceedings of the International Conference on Formal Methods in Computer-Aided Design (FMCAD 2007), November 2007, pp. 13–19 (2007)
18. Sinz, C.: Towards an optimal CNF encoding of Boolean cardinality constraints. In: van Beek, P. (ed.) CP 2005. LNCS, vol. 3709, pp. 827–831. Springer, Heidelberg (2005)
19. Sinz, C., Kaiser, A., Küchlin, W.: Formal methods for the validation of automotive product configuration data. Artificial Intelligence for Engineering Design, Analysis and Manufacturing 17(1), 75–97 (2003)
20. Van Gelder, A.: Autarky pruning in propositional model elimination reduces failure redundancy. Journal of Automated Reasoning 23(2), 137–193 (1999)
21. Zhang, L., Malik, S.: Extracting small unsatisfiable cores from unsatisfiable Boolean formula. In: Giunchiglia, E., Tacchella, A. (eds.) SAT 2003. LNCS, vol. 2919, Springer, Heidelberg (2004)

Nenofex: Expanding NNF for QBF Solving

Florian Lonsing and Armin Biere

Johannes Kepler University

Abstract. The topic of this paper is Nenofex, a solver for quantified boolean formulae (QBF) in negation normal form (NNF), which relies on expansion as the core technique for eliminating variables. In contrast to eliminating existentially quantified variables by resolution on CNF, which causes the formula size to increase quadratically in the worst case, expansion on NNF is involved with only a linear increase of the formula size. This property motivates the use of NNF instead of CNF combined with expansion. In Nenofex, a formula in NNF is represented as a tree with structural restrictions in order to keep its size small and distances from nodes to the root short. Expansions of variables are scheduled based on estimated expansion cost. The variable with the smallest estimated cost is expanded first. In order to remove redundancy from the formula, limited versions of two approaches from the domain of circuit optimization have been integrated. Experimental results on latest benchmarks show that Nenofex indeed exceeds a given memory limit less frequently than a resolution-based QBF solver for CNF, but also that there is the need for runtime-related improvements.

1 Introduction

QBF [36, 19] is an important research area with many applications [28, 33, 6, 31, 21, 16, 24, 5, 30]. Progress has been impressive in recent years, but in practice QBF solvers lack the generic applicability on PSPACE hard problems as SAT solvers on NP hard problems.

We believe that one of the reasons is the usage of CNF as input *and* as internal format to reason about SAT problems. We argue that in QBF more general data structures are necessary. There is clear indication in the relevant literature [39, 37, 34] that supports this conjecture.

The most natural extension of CNF is NNF: NNF is still tree shaped, i.e. there is no sharing, and a formula in CNF is as well a formula in NNF. Before mentioned approaches [39, 34] and in essence all QBF solvers that learn solutions [29, 40, 20] use some kind of combination of DNF and CNF, which again can be considered NNF. As we argue in this paper, a tree representation has many advantages compared to a more general DAG representation.

We investigated the usage of NNF to reduce space usage of expansion based QBF solvers. Our experiments clearly indicate, that our prototype implementation Nenofex needs less space than the highly optimized expansion based solver Quantor [7], which works on CNF. Nenofex is also able to solve several instances, that could not be handled by Quantor.

H. Kleine Büning and X. Zhao (Eds.): SAT 2008, LNCS 4996, pp. 196–210, 2008.
© Springer-Verlag Berlin Heidelberg 2008

The success of Quantor is based on two techniques. First, it implements a fast scheduling algorithm for heuristically choosing the next variable to expand. The second contribution is a fast subsumption algorithm, that removes redundant clauses generated during expansions. Both techniques are also crucial for the efficiency of the SAT preprocessor SATeLite [17]. To maintain accurate expansion costs for NNF turns out to be much more difficult than for CNF. The same comment applies to redundancy removal for NNF. In Nenofex we considered both problems, but due to space constraints only present a solution to redundancy removal on NNF in this paper.

Another related application of a restricted form of NNF is knowledge compilation [12]. But it is unclear how to use it for QBF solving.

2   Motivation

In order to show that NNF is much more compact than CNF for representing the *result* of expansion, consider the following formula $F \equiv R \wedge X_0 \wedge X_1$ in CNF. The clause sets $X_0 = \{c_1, c_2, c_3\} = \{(\neg x \vee c \vee \neg d), (\neg x \vee d \vee \neg e), (\neg x \vee e \vee \neg c)\}$ and $X_1 = \{c_4, c_5, c_6\} = \{(x \vee f \vee \neg g), (x \vee g \vee \neg h), (x \vee h \vee \neg f)\}$ with $|X_0| = |X_1| = 3$ contain all clauses with negative and positive literals of variable x, respectively. Variable x has $n = 3$ negative and $p = 3$ positive literals. $R = \{(a \vee b)\}$ is the set of clauses which do not contain a literal of x (notation adopted from [14]).

Variable x may be expanded by copying F: $F \equiv F[x/0] \vee F[x/1]$ where expression $F[x/v]$ denotes the formula obtained from F by substituting value v for every literal of x. This yields

$$F \equiv ((R \wedge X_0 \wedge X_1)[x/0]) \vee ((R \wedge X_0 \wedge X_1)[x/1]) \tag{1}$$
$$F \equiv (R \wedge (X_0 \wedge X_1)[x/0]) \vee (R \wedge (X_0 \wedge X_1)[x/1]) \tag{2}$$
$$F \equiv R \wedge ((X_0 \wedge X_1)[x/0] \vee (X_0 \wedge X_1)[x/1]) \tag{3}$$
$$F \equiv R \wedge (X_0' \vee X_1') \tag{4}$$

which is $(a \vee b) \wedge (((c \vee \neg d) \wedge (d \vee \neg e) \wedge (e \vee \neg c)) \vee ((f \vee \neg g) \wedge (g \vee \neg h) \wedge (h \vee \neg f)))$. In the clause set X_0' (X_1') all negative (positive) literals of variable x have been deleted. Clauses in R have not been affected during expansion, hence this set can be factored out as shown in formula 3. Formulae 1 to 4 are not in CNF any more but in NNF.

If x is eliminated by resolution, then the set of resolvents $X_r = \{c_{i,j} \mid i = 1, \ldots, n, j = n+1, \ldots, n+p, c_{i,j} = (c_i \cup c_j) \setminus \{x, \neg x\}\}$ contains clauses $\{c_{1,4}, c_{1,5}, c_{1,6}, c_{2,4}, c_{2,5}, c_{2,6}, c_{3,4}, c_{3,5}, c_{3,6}\}$ where $|X_r| = n \cdot p = 3 \cdot 3 = 9$ clauses. After discarding sets X_0 and X_1 and adding X_r to F, we have $F \equiv (a \vee b) \wedge (c \vee \neg d \vee f \vee \neg g) \wedge (c \vee \neg d \vee g \vee \neg h) \wedge (c \vee \neg d \vee h \vee \neg f) \wedge (d \vee \neg e \vee f \vee \neg g) \wedge (d \vee \neg e \vee g \vee \neg h) \wedge (d \vee \neg e \vee h \vee \neg f) \wedge (e \vee \neg c \vee f \vee \neg g) \wedge (e \vee \neg c \vee g \vee \neg h) \wedge (e \vee \neg c \vee h \vee \neg f)$.

Expanding a variable in a formula can at most double its size. In particular, expanding some variable x on CNF as shown in formulae 1 to 4 will copy $n + p$ clauses (and introduce one logical connective), whereas eliminating the same variable by resolution will produce $n \cdot p$ clauses in the worst case. Expansion causes a formula to grow linearly in contrast to resolution, which is involved with a quadratic size increase. In the example, this is reflected in the sizes of the formulae resulting from expansion and resolution.

It is exactly this observation which motivates the use of NNF as formula representation in an expansion based QBF solver as Nenofex. We expect less size increase when eliminating existential variables by expansion on NNF than by resolution on CNF. When expanding universal variables, there is no advantage of expansion on NNF compared to CNF because in both cases the same set of clauses has to be copied. For an arbitrary formula in NNF, expansion of a variable will always yield a formula which is in NNF again. No transformations need to be carried out on the formula between expansions.

3   Preliminaries

For a set of variables V, a *literal* (or *occurrence*) is either a variable $v \in V$ or its negation $\neg v$. A *clause* is a disjunction over literals. A (propositional) formula is in *conjunctive normal form (CNF)* if it is a conjunction over clauses. A formula consisting of disjunctions, conjunctions and literals is in *negation normal form (NNF)* if the negation operator is applied to literals only.

A *quantified boolean formula (QBF)* $F \equiv S_1 S_2 \ldots S_n \phi$ consists of a propositional formula ϕ over a set of variables V and a *quantifier prefix* $S_1 S_2 \ldots S_n$. We assume that ϕ is in NNF. The quantifier prefix is an ordered set of *scopes* S_i, which forms a partition on the set of variables: $V = S_1 \cup S_2 \cup \ldots \cup S_n$ and $S_i \cap S_j = \emptyset$ for $1 \leq i, j \leq n$ and $i \neq j$. A scope S_i is *existential (universal)* if it is associated with an existential (universal) quantifier, written as $type(S_i) = \exists$ $(type(S_i) = \forall)$. A variable $x \in S_i$ where $type(S_i) = \exists$ $(type(S_i) = \forall)$ is existentially (universally) quantified. By convention, for two adjacent scopes S_i and S_{i+1}, where $1 \leq i < n$, $type(S_i) \neq type(S_{i+1})$.

The set of scopes is a linearly ordered set $S_1 < S_2 < \ldots < S_n$ which follows from the order of appearance of scopes S_i in the quantifier prefix. Scope S_1 is the outermost, scope S_n the innermost scope. Variables are ordered with respect to the order of scopes they belong to. For variables from the same scope, an arbitrary order may be chosen. Our definitions of QBF and scopes are similar to the ones in [7] except that formula ϕ is not in CNF but NNF.

There is strong indication that non-prefix orders are important for QBF reasoning [18,7,3,22]. Initially we experimented with non-prefix orders as well, but due to the complexity involved, we focus on non-CNF representation in this paper, except for on-the-fly miniscoping during expansions as in Quantor.

A *tree* $T = (N, E)$ consists of a set of nodes N and a set of directed edges $E \subseteq N \times N$ such that for exactly one node r called *root*, there is no $v \in N$ where $(v, r) \in E$, and for each node $w \in N \setminus \{r\}$, there exists exactly one $v \in N$, $v \neq w$, such that $(v, w) \in E$. If $(v, w) \in E$, then v is the *parent* of w and w is a *child* of v. The root is the only node in T which has no parent, any other node has exactly one parent. For nodes v and $w \in N$, a *path* of length k from v to w is a sequence of nodes $p_0, p_1, \ldots, p_k$ where $p_0 = v$, $p_k = w$ and $(p_i, p_{i+1}) \in E$ for

$0 \le i < k$. For any node v, there is a path of length 0 from v to v. The *level* of a node v is the length of the path from the root to v. For root r, $level(r) = 0$. If $(v, w) \in E$, then $level(w) = level(v) + 1$. For nodes v and w, if there is a path from v to w, then v is an *ancestor* of w and w is a *descendant* of v. Every node is ancestor and descendant of itself. The root is ancestor of any node in T.

A *common ancestor* of a pair of nodes v,w in T is a node which is ancestor of both v and w. The *least common ancestor (LCA)* of v,w, written as $lca(v, w)$ where $lca : N \times N \to N$, is their common ancestor with maximum level, that is, which is farthest away from the root. Commutativity and associativity of lca as an operator extend the definition from pairs to sets of nodes:

$$lca(n_1, n_2, \ldots, n_k) = \begin{cases} lca(lca(n_1, n_2), n_3, \ldots, n_k) & if \quad k \ge 3 \\ \text{least common ancestor of } n_1 \text{ and } n_2 & if \quad k = 2 \\ n_1 & if \quad k = 1 \end{cases}$$

4 Formula Representation

A formula ϕ in NNF is represented as a tree $T = (N, E)$, referred to as *NNF-tree*. The set of nodes N is partitioned into *operator nodes* N_O and *literal occurrence nodes* N_L (short: *literal nodes*), that is $N = N_O \cup N_L$ and $N_O \cap N_L = \emptyset$. The set N_O (N_L) comprises exactly the set of internal nodes (leaf nodes) of the tree.

The set N_O is partitioned into the sets $N_\vee$ and $N_\wedge$, that is $N_O = N_\vee \cup N_\wedge$ and $N_\vee \cap N_\wedge = \emptyset$. A node from the set $N_\vee$ ($N_\wedge$) is called *OR-node* (*AND-node*) and denotes the logical disjunction (conjunction) over its children. Operator nodes with n children, where $n \ge 2$, represent n-ary boolean functions.

A node $n_l \in N_L$ denotes one single (positive or negative) literal of some variable $x \in V$. Conversely, a literal of some variable x is represented by exactly one node $n_l \in N_L$. The *least common ancestor (LCA) of a variable $x \in V$*, written as $lca(x)$, is defined as the LCA over all of its literal nodes.

The structure of an NNF-tree is restricted as follows. Operator nodes may have an arbitrary number of children but must have at least two. For operator nodes $n_o \in N_\vee$ ($n_o \in N_\wedge$) and all its children c, either $c \in N_\wedge$ or $c \in N_L$ ($c \in N_\vee$ or $c \in N_L$), that is, the types of operator nodes and their children must be different. This corresponds to the application of associativity of disjunction and conjunction whenever possible. For operator nodes $n_o \in N_O$ and some variable $x \in V$, if n_o has a child $c_1 \in N_L$ which is a literal node of x, then n_o must not have another child $c_2 \in N_L$, $c_1 \ne c_2$, which is a literal node of x. Thus operator nodes must neither have multiple nor complementary literals of one and the same variable as children. The structural restrictions aim at keeping the NNF-tree small and node levels, that is distances between nodes and the root short. Fig. 1 shows a sample NNF-tree.

As alternative to trees, a representation related to *directed acyclic graphs (DAGs)* could have been used, which allow nodes to be *structurally shared* among several parents. A well-known, DAG-related formula representation are *And-Inverter Graphs (AIGs)* [26] where the set of operators is restricted to binary conjunction and negation. Methods for identifying structural sharing in AIGs

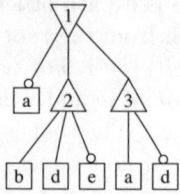

Fig. 1. NNF-tree for formula $\neg a \vee (b \wedge d \wedge \neg e) \vee (a \wedge \neg d)$. An AND-node (OR-node) is represented as a triangle $\triangle$ (inverted triangle $\bigtriangledown$) resembling the symbol for conjunction $\wedge$ (disjunction $\vee$), and a literal node as a box $\square$. A circle $\circ$ at a literal node denotes the negation operator. Labels of operator nodes in the figures are used as indices in the text, e.g. n_1 denotes the root of the NNF-tree in the example above.

have been studied in [9,8]. To our knowledge, structural sharing in combination with n-ary operators like in an NNF-tree has not been studied at a large extent, but obviously there is much more complexity involved. Furthermore, NNF-trees guarantee that a formula in CNF has a flat representation: each (non-unit) literal in the CNF corresponds to exactly one literal node with level 2, each clause to exactly one OR-node with level 1 and the conjunction over the clauses to one single AND-node at the root of the tree. It is impossible to achieve these correspondences with AIGs. DAGs complicate the implementation. For each node, the set of parents and children need to maintained under modifications of the graph. With trees, algorithms related to expansion (next section) and redundancy removal (section 6.1) are much easier to implement.

5   NNF Expansion

If a variable is expanded as shown in the introductory example (formulae 1 to 4), then parts of the formula might be copied unnecessarily and need to be factored out in order to reduce the size of the expanded formula.

We present *local expansion* for NNF, a method where only the relevant parts of a formula are copied and which does not require factoring out common subformulae in the expanded formula. Generally, our method can be regarded as *miniscoping* [2], which produces a non-prefix reduced scope through the application of standard quantifier rules, followed by expansion. In our approach (section 5.3), a minimal reduced scope is determined bottom up, starting from the literal occurrences of the expanded variable.

For a QBF $S_1 \ldots S_{n-1} S_n \phi$, we consider expansion of (1) existential or (2) universal variables from scope S_n (section 5.1), and expansion of (3) universal variables from scope S_{n-1} (section 5.2) only. Case (2) is irrelevant for formulae in CNF since *forall-reduction* [25,7] could remove all literals of universal variables in S_n instead: a universal literal can be removed from a clause if there is no existential literal in that clause whose variable belongs to a scope which is larger than the scope of the universal literal's variable. To our knowledge, it is not clear whether and how this operation can be applied to formulae in NNF. Replacing innermost universal literals by *false* is incorrect, because this would reduce the following formulae to *false* even though they are valid:

$$\forall x \ (x \vee \overline{x}) \qquad\qquad \forall x, y \ (xy \vee x\overline{y} \vee \overline{x}y \vee \overline{xy})$$

5.1 Innermost Expansion

Given a QBF $S_1 \ldots S_n \ \phi$ and some variable x in S_n where $type(S_n) = \exists$ or $type(S_n) = \forall$, let $ers(x)$ denote the *expansion relevant subformula* of variable x, which is the smallest subformula of ϕ which contains all literals of x.

Local expansion of variable x in ϕ is defined as follows:

$$S_1 \ldots S_n \ \phi \equiv S_1 \ldots (S_n \setminus \{x\}) \ \phi[ers(x) \ / \ (ers(x)[x/0] \otimes ers(x)[x/1])] \quad (5)$$

where operator $\otimes = \vee$ $(\otimes = \wedge)$ if $type(S_n) = \exists$ $(type(S_n) = \forall)$. In rule 5, ϕ is modified by replacing the expansion relevant subformula $ers(x)$ by a subformula consisting of two copies of $ers(x)$, where variable x is assigned *true* resp. *false*.

5.2 Non-innermost Expansion

Expansion of universal variables from scope S_{n-1} requires depending existential variables from S_n to be duplicated. Concerning CNF, methods for universal expansion and for identifying dependencies have been proposed in Quantor [7], sKizzo [4], quantifier trees [3] and bounded universal expansion [10]. For example, before some universal variable x from scope S_{n-1} is expanded in Quantor, the set of depending existential variables from scope S_n is computed via a connection relation. Then, all clauses which contain literals of x or of any depending variable are copied during expansion. This idea is generalized in [10,35].

Given a QBF $S_1 \ldots S_{n-1} S_n \ \phi$ with n scopes and some universal variable x in S_{n-1} where $type(S_{n-1}) = \forall$ and $type(S_n) = \exists$. Let $ers(x)$ be defined as in the previous section. Let D_x be the set of *depending existential variables* of x defined as follows (notation adopted from [10]):

$$D_x^{(0)} := \{y \in S_n \mid y \text{ has literals in } ers(x)\}$$
$$D_x^{(k+1)} := \{z \in S_n \mid z \text{ has literals in } ers(y') \text{ for some } y' \in D_x^k\}, k \geq 0$$
$$D_x := \bigcup_k D_x^k$$

Let $urs(x, D_x)$ denote the *expansion relevant subformula* of universal variable x with respect to D_x, which is the smallest subformula of ϕ which contains all literals of x and all literals of any existential variable $y \in D_x$. *Local expansion* of variable x in ϕ is defined as follows:

$$S_1 \ldots S_{n-1} S_n \ \phi \equiv \qquad\qquad\qquad\qquad\qquad (6)$$
$$S_1 \ldots (S_{n-1} \setminus \{x\})(S_n \cup D'_x) \ \phi[u \ / \ (u[x/0] \wedge u'[x/1])]$$

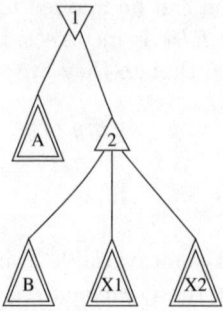

Fig. 2. NNF-tree for formula $A \vee (B \wedge X_1 \wedge X_2)$

where u stands for $urs(x, D_x)$ and $urs(x, D_x)'$ is obtained from $urs(x, D_x)$ by substituting y' for all literals of $y \in D_x$. D'_x is the set which contains duplicated variables y' for every $y \in D_x$. The definition of urs extends the one of ers from the previous section by taking the set of depending existential variables into account. In fact, the notion of $urs(x, D_x)$ is closely related to the CNF-based approaches in [7] and [10], where the set D_x is constructed via a connection relation between variables: v_i is locally connected to v_j if both occur in a common clause. In our NNF-based approach, the connection relation is generalized to subformulae.

5.3   Expansion Relevant LCAs

According to the definitions of expansion relevant subformulae ers and urs for some variable x in some formula ϕ, the *expansion relevant subtree* of x is defined to be the smallest subtree in the NNF-tree of ϕ which contains all literals of x.

In order to expand x in the NNF-tree, a correspondence between expansion relevant subformulae and subtrees as defined has to be established. The *expansion relevant LCA* of variable x is defined by node $lca(x)$ and the set of *LCA-children* of $lca(x)$. A child of node $lca(x)$ is an LCA-child if its subtree contains at least one literal of x. The subtree denoted by the expansion relevant LCA exactly corresponds to the expansion relevant subformula and vice versa.

In Fig. 2, subtrees X_1 and X_2 contain all literals of some variable x and node n_2 is $lca(x)$. The roots of subtrees X_1 and X_2 form the set of LCA-children and, together with node $lca(x)$, denote the expansion relevant subtree of x, which corresponds to $X_1 \wedge X_2$, the expansion relevant subformula of x. Generally, LCAs of variables without the notion of LCA-children are only an over-approximation for expansion relevant subtrees. In Fig. 2, the subtree of node $lca(x)$ contains subtree B as well, which does not contain literals of x.

In Nenofex, expansion relevant LCAs are computed incrementally in an upward directed search starting from each literal of the variable, where parents are successively visited and LCA-children are collected. Our approach requires $O(nm)$ time in the worst case, where n is the number of literals and m the maximum level of a literal which is expected to be small due to structural restrictions.

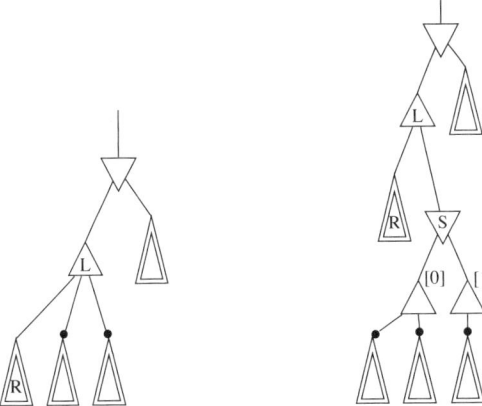

Fig. 3. Expansion template. Node n_L is $lca(x)$ for some existential variable x, subtree R does not contain literals of x and black dots indicate LCA-children. In the right NNF-tree, x has been expanded: OR-node n_S is parent of two copies of the expansion relevant subtree, where x is assigned *true* ([1]) and *false* ([0]).

In order to expand a variable in an NNF-tree, its expansion relevant subtree needs to be copied. Fig. 3 illustrates the situation for an existential variable whose LCA is an AND-node and which has two LCA-children. Expanding variable x in the formula in Fig. 2 yields the expanded formula $A \lor (B \land ((X_1 \land X_2)[x/0] \lor (X_1 \land X_2)[x/1]))$, as indicated by the right template in Fig. 3.

6   Implementation

The architecture of Nenofex is very similar to the one of Quantor. Variables are eliminated in cyclic fashion starting from the innermost scope, where scheduling is based on estimated elimination costs. Elimination of variables is interspersed with redundancy removal. If there is only one type of variables left, then the QBF is reduced to a SAT problem and forwarded to an internal SAT solver.

Fig. 4 shows the phases of the core algorithm in Nenofex. In either phase, the solver may return an answer if the NNF-tree has been deleted or the SAT solver has terminated. After an initialization phase (INIT in Fig. 4), where the problem instance is parsed and data structures are set up, *unit literals* and *pure literals* (or *unates*) [11] are eliminated until saturation.

6.1   Redundancy Removal

Local expansion avoids unnecessary copies of formula parts, but can not avoid redundancy in general. As in Quantor, which relies on subsumption checking, redundancy removal is crucial for Nenofex to achieve best performance. For this purpose, limited versions of two approaches from the domain of circuit optimization have been implemented where an NNF-tree is regarded as a circuit.

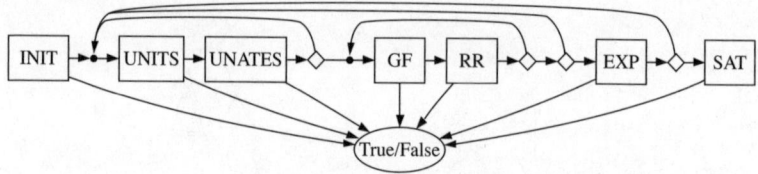

Fig. 4. Core algorithm of Nenofex. Parsing and initialization (INIT), elimination of units and unates (UNITS, UNATES), global flow (GF), redundancy removal (RR), expansion (EXP) and propositional SAT solving (SAT).

ATPG-Based Redundancy Removal. In *structural testing*, a test for a fault at a single line or gate in a circuit is a set of input values, called *test pattern*, by which wrong circuit outputs related to the faulty part can be detected. Test patterns can be generated algorithmically, which is the main purpose of *automatic test pattern generation (ATPG)* [1]. A fault which does not change the circuit's behaviour is redundant and the respective hardware may be removed.

A typical model for faults related to single lines (or signals) is the *stuck-at-fault model*. A line is stuck-at-1 (stuck-at-0), if it always carries *true* (*false*) regardless of the intended value. Detection and removal of redundant stuck-at faults can be used for circuit optimization. Testing a stuck-at fault in *ATPG-based redundancy removal* [1] comprises three steps. In *fault sensitization*, the fault is activated by assigning the corresponding signal the opposite value of the fault: for a stuck-at-1 fault, the signal is assigned *false*, otherwise *true*.

In *path sensitization*, the effect of the activated fault must be propagated *unambiguously* along a *fault path* to an output signal of the circuit. This can be achieved by assigning conservative values to all off-path inputs of gates along the fault path. Off-path inputs of OR-gates (AND-gates) must be assigned *false* (*true*). There might be exponentially many fault paths. If propagation on one path fails, then possibly all remaining paths have to be considered.

In *justification*, signal assignments made in the previous two steps must be justified by finding a set of circuit inputs which establish the configuration of internal signal assignments. Starting at an unjustified, assigned signal, its inputs are assigned recursively with justifying values. For example, an AND-gate which is assigned *false* may be justified by assigning *false* to one of its inputs. As in DPLL-based SAT solvers [13], justification involves making *decisions* which have to be undone during *backtracking* if conflicts between assignments occur.

If all fault paths and alternative assignments have been tried out but conflicts could not be resolved, then the fault is untestable: there is no set of input values such that the fault effect can be observed unambiguously at a circuit output. The corresponding hardware is redundant and may be removed, which can cause further faults to become redundant.

Global Flow. *Global flow* [27] is an approach for circuit minimization where implications are derived from signals which are then used to transform the circuit

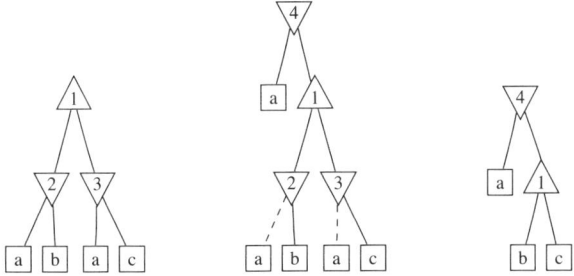

Fig. 5. Detecting distributivity by global flow and redundancy removal

without changing the logical flow of values. For any signal x in the circuit, there are four sets of implications defined as $F_{VW}(x) := \{s : x = V \rightarrow s = W\}$ where $V, W \in \{0, 1\}$ and s is a signal. Given the sets F_{VW} for some signal x, the following transformations are valid:

$$y \in F_{00}(x) : y \equiv x \wedge y \qquad y \in F_{10}(x) : y \equiv \neg x \wedge y$$
$$y \in F_{11}(x) : y \equiv x \vee y \qquad y \in F_{01}(x) : y \equiv \neg x \vee y$$

In order to optimize a circuit, first some signal x is chosen where subsets of implications in $F_{VW}(x)$ are computed, because full computation is complex. Next, an implication is chosen and the circuit is transformed according to the respective rule. Certain connections of x to other nodes may be removed, provided that the logical flow of the value from x to the implied node does not change. If circuit size is not decreased, then all modifications will be are reversed. These steps are carried out in cyclic fashion for all signals in the circuit.

Fig. 5 illustrates a typical situation where redundancy is detected by global flow *together with* ATPG-based redundancy removal: in the leftmost NNF-tree, literal a may be factored out by applying distributivity. This can not be detected by ATPG-based redundancy removal alone. When assigning literal a at n_2 *true* (and consequently variable a as well) in the leftmost NNF-tree, then n_1 (the root) will be assigned *true* as well, hence $a = true \rightarrow n_1 = true$ and n_1 may be replaced by $a \vee n_1$ which yields the second NNF-tree. Dashed edges indicate nodes with untestable stuck-at-0 faults. If the fault at literal a at n_2 is tested, then it must be assigned *true* in fault sensitization (that is, variable a will be assigned *true*), but this yields an unresolvable conflict at n_4, the only circuit output, where literal a will be assigned *true* as well. This is not a conservative value as required by path sensitization. The same argument applies for a at n_3, and the two literals may be removed which yields the NNF-tree on the right.

Limitations. The implementation of global flow (phase GF in Fig. 4) and ATPG-based redundancy removal (phase RR) is very limited. For GF, only implications from sets F_{00} and F_{11} are considered. GF alone is not capable of reducing the size of an NNF-tree but, together with RR (as shown in Fig. 5), can enable detection of redundancies which would remain undetected by RR. Modifications made by GF are never reversed, since they always produce additional redundant stuck-at-faults due to the tree shape of NNF.

General ATPG-based redundancy removal is NP-complete [1]. Our implementation runs in polynomial time, but is incomplete. We only use propagations and no decisions. It greatly benefits from the tree representation of NNF, because there is a single fault propagation path from the fault site to the root.

Phases GF and RR are carried out in cyclic fashion on a small subtree of the NNF-tree only. Generally, each optimization runs until saturation, but this can become problematic due to the amount of required runtime. Therefore, fixed limits are imposed on the size of the subtree and on the number of value propagations during GF and RR.

6.2 Expansion

Let S_i and S_{i-1} denote the current innermost and first non-innermost scope, respectively. Variables are selected for expansion depending on their estimated expansion costs (scores) and on the types of S_i and S_{i-1}. In Nenofex, generally a greedy strategy is applied: in order to keep the size of the NNF-tree small in each expansion, always the variable with minimum expansion cost is selected.

First, if $type(S_{i-1}) = \exists$ and $type(S_i) = \forall$, then only variables from S_i may be expanded. The variable with minimum score is expanded. Second, if $type(S_{i-1}) = \forall$ and $type(S_i) = \exists$, then variables from both scopes may be expanded. A variable from S_{i-1} is eligible for expansion iff the *preceding* expansion from S_i (1) caused the size of the NNF-tree to increase and (2) the size increase to exceed a heuristic *universal threshold* t_u. Initially, t_u is set to 10 nodes. If t_u is exceeded in an expansion, then expansion of *exactly one* variable from S_{i-1} will be *forced*, which causes t_u to increase by 10 and expansions from S_{i-1} to be disabled again. Expansions are forced because score computation for S_{i-1} likely becomes impractical when carried out each time before an expansion. This policy goes against the greedy expansion strategy because the minimum score variable from S_i may well be cheaper than the one in S_{i-1}.

The estimated expansion cost is calculated as a tight upper bound on the real expansion cost. It is measured in the number of nodes added to and removed from the NNF-tree. The costs are recalculated after every change to the NNF-tree, taking only the changed part into account. The details are complicated and due to space constraints have to be omitted. However, we clearly see a potential for speeding up this process by updating scores in a full incremental fashion.

6.3 SAT Solving

If the formula contains only one type of variables, then the QBF may be reduced to a SAT problem: a formula in CNF is generated from the NNF-tree which is forwarded to an internal SAT solver. If only existential variables are left, then a CNF will be generated which is satisfiable iff the formula denoted by the

Table 1. Comparison of Quantor and Nenofex in three different versions by number of instances where solvers succeeded, timed out (OOT) and ran out of memory (OOM). The number of instances solved by Nenofex decreases from (GF, RR) to (no GF, no RR), which indicates that GF and RR contribute positively to the solver's performance. On the other hand, time is traded for memory when enabling optimizations: values of OOT increase from (no GF, no RR) to (GF, RR), the opposite effect can be observed for memory. Note that (no GF, no RR) runs out of memory more often than Quantor, which applies subsumption checking, hence optimizations are crucial in combination with NNF as well. The last two lines show sums of actual maximum amounts of memory (in MB) consumed on each solved or unsolved instance ($MEM\cup$) and on instances solved by all four solvers ($MEM\cap$). The (unoptimized) node representation in Nenofex requires about twice as much memory than the pointer-based structures of clauses and literals in Quantor, which is reflected in the last line.

	Quantor	Nenofex		
		GF, RR	*no GF, RR*	*no GF, no RR*
Solved	**421**	361	352	313
OOT	**32**	124	103	83
OOM	683	**651**	681	740
MEM∪	1.10e6	1.15e6	1.17e6	1.23e6
MEM∩	10473	18472	19693	28422

Table 2. Number of instances where both or only one of Quantor and Nenofex (GF, RR) succeeded, timed out or ran out of memory. Nenofex solved 19 instances which Quantor could not solve. OOT and OOM indicate a similar tendency as Tab. 1.

	Quantor only	Both	Nenofex only	Sum
Solved	79	342	19	440
OOT	18	14	110	142
OOM	80	603	48	731

NNF-tree is *satisfiable*. Otherwise, a CNF will be generated which is satisfiable iff the formula denoted by the NNF-tree is *not a tautology*.

The algorithm for generating a CNF from an NNF-tree requires linear time in the number of nodes of the tree and is based on the Tseitin transformation [38]. Ideas from [15,32] are combined to reduce the number of clauses in the resulting CNF.

7   Experiments

Nenofex was compared to Quantor on the benchmark collection used for the competitive QBF evaluation in 2007 [23], which contains 1136 instances. Both solvers used the same version of PicoSAT as backend SAT solver. Tests were run on a cluster of Pentium IV 3 GHz workstations running Linux, where runtime and memory were limited by 900 seconds and 1.5 GB, respectively.

Concerning global flow (GF) and redundancy removal (RR), Nenofex was run in three versions: either both GF and RR are enabled (GF, RR), or only RR is enabled (no GF, RR) or both GF and RR are disabled (no GF, no RR). In either version, the size of the subtree subject to these optimizations was limited by 500 nodes. Table 1 shows an overall comparison of Quantor and Nenofex, and in Tab. 2, behaviour unique to each solver is indicated.

8   Conclusion

This paper showed that expansion on quantified NNF needs less space than CNF. However, it may be worthwhile to extend our algorithms to DAGs. So far we have only used NNF in an expansion based approach, which eliminates quantifiers from inside to the outside. As future work, one should also consider the combination of NNF with DPLL style algorithms. Finally we want to thank Ofer Strichman for very fruitful discussions on this subject.

References

1. Agrawal, V., Bushnell, M.: Essentials of Electronic Testing for Digital, Memory and Mixed-Signal VLSI Circuits. Kluwer, Dordrecht (2000)
2. Ayari, A., Basin, D.A.: QUBOS: Deciding quantified boolean logic using propositional satisfiability solvers. In: Aagaard, M.D., O'Leary, J.W. (eds.) FMCAD 2002. LNCS, vol. 2517, Springer, Heidelberg (2002)
3. Benedetti, M.: Quantifier Trees for QBFs. In: Bacchus, F., Walsh, T. (eds.) SAT 2005. LNCS, vol. 3569, pp. 378–385. Springer, Heidelberg (2005)
4. Benedetti, M.: sKizzo: A Suite to Evaluate and Certify QBFs. In: Nieuwenhuis, R. (ed.) CADE 2005. LNCS (LNAI), vol. 3632, pp. 369–376. Springer, Heidelberg (2005)
5. Benedetti, M.: Experimenting with QBF-based formal verification. In: Proc. CFV 2005 (2005)
6. Biere, A., Cimatti, A., Clarke, E., Zhu, Y.: Symbolic Model Checking without BDDs. In: Cleaveland, W.R. (ed.) ETAPS 1999 and TACAS 1999. LNCS, vol. 1579, Springer, Heidelberg (1999)
7. Biere, A.: Resolve and Expand. In: H. Hoos, H., Mitchell, D.G. (eds.) SAT 2004. LNCS, vol. 3542, pp. 59–70. Springer, Heidelberg (2005)
8. Bjesse, P., Borälv, A.: DAG-aware circuit compression for formal verification. In: Proc. ICCAD 2004 (2004)
9. Brummayer, R., Biere, A.: Local two-level and-inverter graph minimization without blowup. In: Proc. MEMICS 2006 (2006)
10. Bubeck, U., Kleine Büning, H.: Bounded Universal Expansion for Preprocessing QBF. In: Marques-Silva, J., Sakallah, K.A. (eds.) SAT 2007. LNCS, vol. 4501, pp. 244–257. Springer, Heidelberg (2007)
11. Cadoli, M., Giovanardi, A., Schaerf, M.: An algorithm to evaluate quantified boolean formulae. In: Proc. AAAI/IAAI 1998 (1998)
12. Darwiche, A.: Decomposable negation normal form. JACM 48(4) (2001)
13. Davis, M., Logemann, G., Loveland, D.W.: A machine program for theorem-proving. CACM 5(7) (1962)

14. Davis, M., Putnam, H.: A computing procedure for quantification theory. JACM 7(3) (1960)
15. Boy de la Tour, T.: An optimality result for clause form translation. Symb. Comput. 14(4) (1992)
16. Dershowitz, N., Hanna, Z., Katz, J.: Bounded Model Checking with QBF. In: Bacchus, F., Walsh, T. (eds.) SAT 2005. LNCS, vol. 3569, pp. 408–414. Springer, Heidelberg (2005)
17. Eén, N., Biere, A.: Effective Preprocessing in SAT Through Variable and Clause Elimination. In: Bacchus, F., Walsh, T. (eds.) SAT 2005. LNCS, vol. 3569, pp. 61–75. Springer, Heidelberg (2005)
18. Egly, U., Seidl, M., Tompits, H., Woltran, S., Zolda, M.: Comparing Different Prenexing Strategies for Quantified Boolean Formulas. In: Giunchiglia, E., Tacchella, A. (eds.) SAT 2003. LNCS, vol. 2919, pp. 214–228. Springer, Heidelberg (2004)
19. Garey, M., Johnson, D.: Computers and Intractability: A Guide to the Theory of NP-Completeness (1979)
20. Giunchiglia, E., Narizzano, M., Tacchella, A.: Learning for quantified boolean logic satisfiability. In: Proc. AAAI 2002 (2002)
21. Giunchiglia, E., Narizzano, M., Tacchella, A.: QBF Reasoning on Real-World Instances. In: H. Hoos, H., Mitchell, D.G. (eds.) SAT 2004. LNCS, vol. 3542, pp. 105–121. Springer, Heidelberg (2005)
22. Giunchiglia, E., Narizzano, M., Tacchella, A.: Quantifier structure in search based procedures for QBFs. In: Proc. DATE 2006 (2006)
23. Giunchiglia, E., Narizzano, M., Tacchella, A.: Quantified Boolean Formulas satisfiability library (QBFLIB) (2001), www.qbflib.org
24. Jussila, T., Biere, A.: Compressing BMC encodings with QBF. In: Proc. BMC 2006 (2006)
25. Kleine Büning, H., Karpinski, M., Flügel, A.: Resolution for quantified boolean formulas. Inf. Comput. 117(1) (1995)
26. Kuehlmann, A., Paruthi, V., Krohm, F., Ganai, M.K.: Robust boolean reasoning for equivalence checking and functional property verification. TCAD 21(12) (2002)
27. Kunz, W., Stoffel, D.: Reasoning in Boolean Networks: Logic Synthesis and Verification Using Testing Techniques. Kluwer, Dordrecht (1997)
28. Ladner, R.: The computational complexity of provability in systems of modal propositional logic. SIAM Journal on Computing 6(3) (1977)
29. Letz, R.: Lemma and Model Caching in Decision Procedures for Quantified Boolean Formulas. In: Egly, U., Fermüller, C. (eds.) TABLEAUX 2002. LNCS (LNAI), vol. 2381, Springer, Heidelberg (2002)
30. Mangassarian, H., Veneris, A., Safarpour, S., Benedetti, M., Smith, D.: A performance-driven QBF-based iterative logic array representation with applications to verification, debug and test. In: Proc. ICCAD 2007 (2007)
31. Otwell, C., Remshagen, A., Truemper, K.: An effective QBF solver for planning problems. In: MSV/AMCS (2004)
32. Plaisted, D., Greenbaum, S.: A structure-preserving clause form translation. Symb. Comput. 2(3) (1986)
33. Rintanen, J.: Constructing conditional plans by a theorem-prover. Journal of Artificial Intelligence Research 10 (1999)
34. Sabharwal, A., Ansótegui, C., Gomes, C., Hart, J., Selman, B.: QBF modeling: Exploiting player symmetry for simplicity and efficiency. In: Proc. SAT 2006 (2006)

35. Samer, M., Szeider, S.: Backdoor Sets of Quantified Boolean Formulas. In: Marques-Silva, J., Sakallah, K.A. (eds.) SAT 2007. LNCS, vol. 4501, pp. 230–243. Springer, Heidelberg (2007)
36. Stockmeyer, L.: The polynomial–time hierarchy. TCS 3(1) (1976)
37. Malik, S., Tang, D.: Solving Quantified Boolean Formulas with Circuit Observability Don't Cares. In: Biere, A., Gomes, C.P. (eds.) SAT 2006. LNCS, vol. 4121, pp. 368–381. Springer, Heidelberg (2006)
38. Tseitin, G.: On the complexity of derivation in propositional calculus. Studies in Constructive Mathematics and Mathematical Logic 2 (1968)
39. Zhang, L.: Solving QBF by combining conjunctive and disjunctive normal forms. In: Proc. AAAI 2006 (2006)
40. Zhang, L., Malik, S.: Towards a Symmetric Treatment of Satisfaction and Conflicts in Quantified Boolean Formula Evaluation. In: Van Hentenryck, P. (ed.) CP 2002. LNCS, vol. 2470, Springer, Heidelberg (2002)

SAT(ID): Satisfiability of Propositional Logic Extended with Inductive Definitions

Maarten Mariën, Johan Wittocx, Marc Denecker, and Maurice Bruynooghe

Department of Computer Science, Katholieke Universiteit Leuven, Belgium
{maartenm,johan,marcd,maurice}@cs.kuleuven.be

Abstract. We investigate the satisfiability problem, SAT(ID), of an extension of propositional logic with inductive definitions. We demonstrate how to extend existing SAT solvers to become SAT(ID) solvers, and provide an implementation on top of MiniSat. We also report on a performance study, in which our implementation exhibits the expected benefits: full use of the underlying SAT solver's potential.

1 Introduction

The SAT problem, deciding the satisfiability of propositional logic (PC) theories, is a major research theme. An important research direction is to develop SAT solvers for extensions of PC (e.g., SMT [22]). The use of extended languages leads to broader applicability of SAT-like systems, facilitates the modelling of applications, and may substantially reduce the size of encodings. All these benefits also hold for PC(ID), the extension of propositional logic with *inductive definitions (IDs)*, as we argue below. This paper presents a SAT solver for PC(ID). The satisfiability problem of this logic is called the *SAT(ID)* problem.

Inductive definitions occur very frequently in real-world problems. A familiar example of an inductive definition is that of reachability. Reachability has so many applications that, for example, the abstract software design analyzer Alloy supports it by a special purpose language construct [11]. FO(ID) [2,4] is an extension of first order logic with inductive definitions, it was proposed as the language of a general declarative problem solving framework based on model expansion [20,16]. PC(ID) is the propositional fragment of FO(ID). Current solvers reduce FO(ID) model expansion problems to SAT(ID) problems by grounding. Thus techniques and systems as those presented in this paper form the basic inference mechanism of this paradigm.

One approach to the SAT(ID) problem is by encoding the IDs into propositional logic formulas [23], but has the disadvantage that it may considerably increase theory size. The alternative approach is by supporting IDs in SAT solvers. In previous work, we developed algorithms for integrating inductive definitions in SAT, and implemented the SAT(ID) solver MidL [17,18]. However, these algorithms required changes to the unit propagation mechanism itself, so that we couldn't speak of a proper *extension* of SAT algorithms. In this work, by contrast, we show how SAT(ID) solvers can be built by extending SAT solvers with

H. Kleine Büning and X. Zhao (Eds.): SAT 2008, LNCS 4996, pp. 211–224, 2008.
© Springer-Verlag Berlin Heidelberg 2008

an additional propagation mechanism suitable for reasoning on inductive definitions. The obvious advantage from this approach is that the SAT(ID) solver benefits from any improvement made to the underlying SAT solver. In particular, it has the same performance on pure propositional problems as the underlying solver. This was not the case for MidL, which did not have all the fine tuning of a state of the art SAT solver. A further advantage is that by separating the two propagation mechanisms (one for propositional theories and one for IDs), we also strongly simplify the description of a SAT(ID) solver.

We have built an implementation that extends the popular solver MiniSat [6]. The resulting solver, MiniSat(ID), is more efficient than the systems described in [18] and [23]. We also compared with the best solvers of ASP [15], a formalism with similar expressivity, and found that MiniSat(ID) is competitive.

In summary, the contributions of this work are:

- An *extension* of a SAT solver for SAT(ID). This allows one to build SAT(ID) solvers that fully benefit from the best SAT solving technology.
- A *simplification* of the formalisation of a SAT(ID) solver.
- An *implementation* on top of MiniSat that performs substantially better than other SAT(ID) solvers and is at least competitive with ASP solvers.

After formally introducing SAT and SAT(ID) in Section 2, we present a high level overview of the requirements in Section 3, paving the way for the "recipe" for extending a SAT solver with ID propagations, detailed in Section 4. We evaluate the MiniSat(ID) solver in Section 5 and conclude in Section 6.

2   Preliminaries

2.1   Propositional Logic and SAT Solving

A vocabulary Σ is a set of atoms (propositions), a literal is an atom p or its negation $\neg p$. A propositional *formula* is built from literals using the binary connectives $\vee, \wedge, \equiv, \supset$. A propositional logic *theory* is a set of formulas. In this paper we assume, without loss of generality, that theories are in conjunctive normal form (CNF): all formulas are disjunctions of literals, called *clauses*.

A Σ-*assignment* A is a function $A : \Sigma \rightarrow \{t, f, u\}$ (u stands for "unknown"). An assignment is *total* when all elements are assigned either t of f and *partial* otherwise[1]. We leave out Σ when it is clear from the context. In this paper we assume that Σ always contains the atom $\bot$, which is not used in the theory, and is always assigned f. An assignment can be extended to literals: $A(\neg p) = \neg A(p)$ with $\neg t = f$, $\neg f = t$, $\neg u = u$, and to clauses and conjunctions of literals: $A(\bigvee C) = \bigvee_{c \in C} A(c)$ with $\bigvee V = t$ if the multiset of truth values V contains t, otherwise $\bigvee V = u$ if V contains u, $\bigvee V = f$ otherwise; likewise $A(\bigwedge C) = \bigwedge_{c \in C} A(c) = \neg \bigvee_{c \in C} \neg A(c)$. A theory T is *satisfiable* if it has a *model*: a total

---

[1] It is more convenient to use a three-valued formalization than to use a partial function that leaves elements undefined when assigned u in the three valued one.

assignment A that satisfies every clause of T, denoted $A \models T$. SAT is the problem of deciding whether a given theory is satisfiable.

SAT solving is the practice of answering the SAT problem. Although other solving techniques exist, the current state of the art solvers are based on DPLL [1] augmented with the two watched literal scheme (2WL) and with clause learning [21,24]. The propagation method that drives this search is *unit propagation*, whereby a literal is assigned true if it occurs as the only non-false literal in a clause. In Section 4.2 we introduce a highly abstracted version of the 2WL scheme; for more (implementation-oriented) details we refer the reader to [21].

2.2   Inductive Definitions

This section is based on work on FO(ID) [2,4]. In short, SAT(ID) is the satisfiability problem of the propositional fragment of FO(ID).

A *definition* Δ over vocabulary Σ is a set of *rules* of the form $p \leftarrow \varphi$, where $p \in \Sigma$, $\leftarrow$ is a new binary connective (*definitional implication*), and φ is a Σ-formula (the *body* of the rule). Propositional calculus (PC) extended with inductive definitions (IDs) is called PC(ID); a PC(ID) theory consists of a set of formulas and definitions. In this paper we assume, without loss of generality [26], that definitions are in *definitional normal form* (DefNF): for any $p \in \Sigma$, the definition contains at most one rule $p \leftarrow \varphi$, and either $\varphi = \bigvee B_p$ or $\varphi = \bigwedge B_p$, where B_p is a set of literals called the *body literals*. A PC(ID) theory is in DefNF if it contains only one definition, which is in DefNF, and its set of formulas is in CNF. In this paper we also assume that PC(ID) theories are in DefNF.

For a definition Δ, the atoms $Def(\Delta) = \{p \mid (p \leftarrow \varphi) \in \Delta\}$ are called *defined atoms* and the others the *non-defined* ones. Given an assignment A' of the *non*-defined atoms, we can simplify Δ by substituting these atoms by their truth values. The resulting definition contains only defined atoms and their negative literals; such a set of rules has a unique *well-founded model* [10], denoted $\mathrm{wfm}_\Delta(A')$. This well-founded model is a (possibly partial) assignment of the defined atoms. A definition Δ is *total* (with respect to theory T) if this well-founded model $\mathrm{wfm}_\Delta(A')$ is a total assignment (for all assignments A' that satisfy T). Definitions occurring in standard mathematical practice—when bug-free—are always total with respect to the rest of the theory in which they occur[2], and we restrict attention to these unless otherwise said. In Section 4.3 we will explain how to deal with non-total definitions.

A total assignment A satisfies a definition Δ, denoted $A \models \Delta$, if $A|_{Def(\Delta)} = \mathrm{wfm}_\Delta(A|_{\Sigma \setminus Def(\Delta)})$. We say that A extends $A|_{\Sigma \setminus Def(\Delta)}$ to a model of Δ. Two definitions Δ_1, Δ_2 are equivalent, denoted $\Delta_1 \equiv \Delta_2$, when for any total assignment A, $A \models \Delta_1$ iff $A \models \Delta_2$. A satisfies a PC(ID) theory if it satisfies every formula and every definition of the theory. The SAT(ID) problem is the satisfiability problem for PC(ID) theories.

---

[2] For many classes of definitions it is trivial to prove that they are total with respect to *any* (the empty) theory, e.g. for definitions that do not contain negation.

The well-founded semantics correctly formalizes the most common forms of inductive definitions in mathematics: both monotone induction, iterated induction, and non-monotone induction over a well-founded order [5].

Example 1. Consider irreflexive undirected graphs with nodes $\{a, b, c\}$. We represent the edge between nodes x and y by E_{xy}. Then R_x (for $x \in \{b, c\}$) in the following definition expresses the reachability of x from a (I_{xy} expresses the reachability of y from a via x): $\Delta_1 = \{R_b \leftarrow E_{ab} \vee I_{cb}, \quad R_c \leftarrow E_{ac} \vee I_{bc}, \quad I_{cb} \leftarrow R_c \wedge E_{bc}, \quad I_{bc} \leftarrow R_b \wedge E_{bc}\}$.

For instance, to determine reachability from a in a graph with only an edge between b and c, we start from the assignment A' with $A'(E_{ab}) = A'(E_{ac}) = \mathbf{f}$, and $A'(E_{bc}) = \mathbf{t}$. Simplifying the definition with A' yields $\Delta_1' = \{R_b \leftarrow I_{cb}, \quad R_c \leftarrow I_{bc}, \quad I_{cb} \leftarrow R_c, \quad I_{bc} \leftarrow R_b\}$. In wfm$_{\Delta_1'}(A')$, each of R_b, R_c, I_{cb} and I_c is false, hence the assignment $A_1 = \{E_{ab} \mapsto \mathbf{f}, E_{ac} \mapsto \mathbf{f}, E_{bc} \mapsto \mathbf{t}, R_b \mapsto \mathbf{f}, R_c \mapsto \mathbf{f}, I_{cb} \mapsto \mathbf{f}, I_{bc} \mapsto \mathbf{f}\}$ is a model of Δ_1.

For the well-founded semantics we refer to [10]. An important property, though, is that $A \models \Delta$ implies $A \models \Delta[\leftarrow/\equiv]$, where $\Delta[\leftarrow/\equiv]$ is the propositional theory obtained from Δ by substituting the symbol $\leftarrow$ by $\equiv$.[3] The converse is not true, Δ has fewer models than $\Delta[\leftarrow/\equiv]$. Hence Δ can cause extra constraints that result in more propagations than in $\Delta[\leftarrow/\equiv]$. Our contribution is to extend a SAT solver with these propagations.

There exist quadratic algorithms for computing a well-founded model in the context of a given assignment of the non-defined atoms, e.g. [9,14]. However, in the more general context of PC(ID) theories, such assignments are not given.

3 Requirements for a SAT(ID) Algorithm

In this section we simplify the formalisation presented earlier in [17]. We start by investigating on an example what are the differences between a definition Δ, and its propositional counterpart $\Delta[\leftarrow/\equiv]$.

Example 2. Example 1 continued. $\Delta_1[\leftarrow/\equiv]$ has two models extending A': one where all defined atoms are true, and one where they are all false. However, the first model is not a well-founded model. Indeed, analysing the reason for the truth of its defined atoms, one observes a circular reasoning as only possible explanation (R_b is true because I_{cb} is, ..., I_{bc} is true because R_b is).

We have to formalize the reasoning about the circularities in the models of $\Delta[\leftarrow/\equiv]$. We do so by using graphical structures called justifications [3].

Definition 1 (Justification). *A justification J for a definition Δ is a directed graph (N, E), where the nodes N are the literals that occur in Δ and E is a minimal set of edges satisfying: (1) for every rule $d \leftarrow \bigvee B_d \in \Delta$, E contains precisely one edge $(d, b), b \in B_d$; (2) for every rule $c \leftarrow \bigwedge B_c \in \Delta$, E contains*

---

[3] DefNF enables a trivial rewriting of formulas in the forms $\bigvee B_p$ and $\bigwedge B_p$ into CNF.

all edges $(c, b), b \in B_c$. The subjustification *of a literal l is the subgraph of J consisting of all paths starting in l.*

We denote the unique descendant in a justification J of a disjunctively defined atom d by $J(d)$. Hence a justification is uniquely characterized by specifying $J(d)$ for each disjunctively defined atom d. Note that the edges are directed from defined atoms to body literals.

Example 3. Consider Example 1. For the disjunctive rule defining R_b one can choose $J(R_b)$ as either E_{ab} or I_{cb}; similarly $J(R_c)$ is either E_{ac} or I_{bc}. This results in $2 * 2 = 4$ possible justifications for Δ_1, one of which is shown on the right.

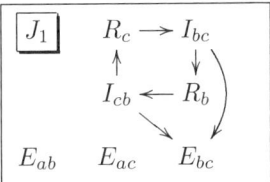

A justification reflects a possible reason for the truth of a defined atom: for a disjunctively defined atom, it suffices that one disjunct is true; for a conjunctively defined atom, all conjuncts are needed.

To formalize the notion that a defined atom is assigned a truth value in accordance with a justification, we introduce the notion of support.

Definition 2 (Support and witness). *A justification J supports a (partial) assignment A iff, for each defined atom p, $A(p) = A(\bigwedge\{b | (p, b) \in J\})$ when $A(\bigwedge\{b | (p, b) \in J\})$ differs from $\boldsymbol{u}$. A supporting justification that contains no cycles (is cycle-free) is a witness for A. The subjustification of a defined atom d is a witness of d if it is cycle-free and part of a supporting justification.*

Observe that for a total assignment the condition simplifies into: for each defined atom p, $A(p) = A(\bigwedge\{b | (p, b) \in J\})$.

Some definitions contain atoms whose truth can never be justified in a well-founded way (their falsity can); they have to be false in all models. Define $\circlearrowright = \{p \mid$ any subjustification for p contains a cycle $\}$. These atoms have to be false in any model. Defining $\Delta^{\circlearrowright}$ as the definition obtained by replacing in Δ the rule $p \leftarrow \varphi$ by $p \leftarrow \perp$ for each $p \in \circlearrowright$ enforces this. It has the advantage of reducing the size of the definition and of simplifying the solver.

Example 4. Let $\Delta_4 = \{p \leftarrow q \vee r, \quad q \leftarrow p, \quad r \leftarrow p, \quad s \leftarrow t \vee a, \quad t \leftarrow s\}$. Then p, q, r are defined in terms of each other, so they are part of a cycle in any justification; hence: $\Delta_4^{\circlearrowright} = \{p \leftarrow \perp, \quad q \leftarrow \perp, \quad r \leftarrow \perp, \quad s \leftarrow t \vee a, \quad t \leftarrow s\}$.

Lemma 1. $\Delta \equiv \Delta^{\circlearrowright}$.

A witness for a (partial) assignment A reflects a well-founded reasoning for the truths in A and the following theorem holds:

Theorem 1. *Let Δ be a total definition, A a total assignment. $A \models \Delta$ iff $A \models \Delta[\leftarrow/\equiv]$ and there is a justification for $\Delta^{\circlearrowright}$ that is a witness for A.*

Example 5. Continuing from Examples 1, 2, 3, we have $\Delta_1^\emptyset = \Delta_1$. We search a model A_4 of Δ_1 that extends A'. The justification J_1 with $J_1(R_b) = I_{cb}$, $J_1(R_c) = I_{bc}$ contains a cycle, hence it cannot be a witness for A_4. The other justifications have either $J(R_b) = E_{ab}$, or $J(R_c) = E_{ac}$, or both. For any of them to be a witness of A_4, they should support A_4, hence we have respectively either $A_4(R_b) = \boldsymbol{f}$, or $A_4(R_c) = \boldsymbol{f}$, or both. Then because also $A_4 \models \Delta_1[\leftarrow/\equiv]$ is required, $A_4(R_b) = A_4(R_c) = A_4(I_{cb}) = A_4(I_{bc}) = \boldsymbol{f}$, i.e. $A_4 = A_1$.

In other words, if we have the partial assignment A', we can strengthen it (propagate) by making R_b and R_c false, based on the above reasoning.

The above theorem suggests the following structure for a SAT(ID) algorithm:

- initialize, to find $\Delta^\emptyset$,
- apply SAT on $\Delta^\emptyset[\leftarrow/\equiv]$ (and on the propositional part of the theory),
- maintain a witness for the assignment found by the SAT solver.

4   Recipe for a SAT(ID) Solver

4.1   Construction of $\Delta^\emptyset$ and of a Cycle-Free Justification

The purpose of the initialisation step is to identify the atoms of $\circlearrowright$ and to construct $\Delta^\emptyset$. We do so by marking literals for which a witness exists. Applying the folowing rules until a fixpoint is reached marks literals with a stratification level.

- Mark all non-defined atoms and all negative literals with 0.[4]
- Mark an unmarked disjunctively defined atom d with level $l + 1$ when one of its body atoms is marked with level l
- Mark an unmarked conjunctively defined atom d with level $l + 1$ when all of its body atoms are marked with levels $\leq l$.

Atoms that remain unmarked cannot have a witness and belong to $\circlearrowright$. A cycle-free justification of $\Delta^\emptyset$ can now be derived by assigning to $J(d)$ a literal in the body with a smaller stratification level (for all disjunctively defined atoms).

4.2   Main Search Procedure

Recall from Section 3 that the goal is to apply SAT on $\Delta^\emptyset[\leftarrow/\equiv]$, and to maintain during the search a witness for the current partial assignment. We start with a high level description of the procedure and then elaborate on the details.

The main procedure iterates over the following steps until either a solution is found (a total assignment and a witness for it) or the search space is exhausted.

---

[4] We can mark all negative literals with level 0 because we assume all definitions are total. We come back to this issue in Section 4.3.

1. Select a cycle-free justification J_{cf}.
2. Use the SAT solver to update the current assignment by performing unit propagation on $\Delta^{\emptyset}[\leftarrow/\equiv]$ (and on the propositional part of the theory). (If no propagation is possible, first an undefined atom is assigned t or f.)
3. Use the state of the SAT solver to construct a supporting justification J_s.
4. If J_s is not a witness then:

   4a Use J_{cf} to adjust J_s so that it becomes a witness. In case this fails, a set $Cycle$ is obtained of defined atoms that cannot have a witness under the current assignment (in every supporting justification it holds that the subjustification of the atom has a cycle).

   4b If J_s is still not a witness, make sure all atoms in $Cycle$ will be set to false in the next iteration, e.g. by extending the theory with *reason clauses*.

Step 4a, finding a witness justification if there is one, is the step on which the efficiency of the whole procedure hinges most. A straightforward algorithm for this step can be derived from the algorithm from Section 4.1: in the marking phase, also truth values should be taken into account to ensure a supporting justification. However, this results in a bottom-up algorithm, which works globally. Instead, a top-down algorithm which works as locally as possible is much more efficient for most definitions; we describe such an algorithm below.

Example 6. Let $\Delta_6 = \{p \leftarrow q \vee r, \quad q \leftarrow p\}$, and let the current assignment have r false, and p and q undefined. A cycle-free justification for Δ_6 has $J_{cf}(p) = r$ but is not supporting. A supporting justification for Δ_6 has $J_s(p) = q$ but has a cycle. Adjusting $J_s(p)$ such that it becomes a witness is not possible; the adjustment fails and the $Cycle$-set $\{p, q\}$ is returned. In the next iteration, unit propagation will set p and q to false; the cycle-free justification J_{cf} becomes a supporting one and a solution is obtained.

Step 1: Selection of a cycle-free justification. In the first iteration, the cycle-free justification J_{cf} is constructed with the procedure described in Section 4.1 (note that it is a witness as the initial assignment assigns u to all atoms). In later iterations, the most recent witness is used as cycle-free justification, i.e., the supporting justification J_s of the previous iteration becomes the cycle-free justification when it is a witness. Taking the most recent witness keeps the difference with the supporting justification to be constructed in step 3 to a minimum and, as will become clear, this reduces the amount of work in step 4a.

Step 2: SAT solving. This is a propagation step by the underlying SAT-solver. Note that this includes clause learning and backtracking when propagation leads to the detection of a conflict.

Step 3: Construct a Supporting Justification. We assume here that the SAT solver we are extending implements unit propagation using the 2WL scheme. This scheme keeps clauses satisfiable by maintaining for each clause an invariant on its two watched literals. Let $W_1(c)$ and $W_2(c)$ be the watched literals of clause c and A the current assignment, then the invariant is as follows: either $A(W_1(c)) = t$ or $A(W_1(c)) = u \land A(W_2(c)) \neq f$.[5]

We use the watching literals to construct a supporting justification J_s. This requires to set $J_s(d)$ for every disjunctively defined atom d. Let $d \leftarrow \bigvee B_d$ be the rule defining d. $\Delta[\leftarrow/\equiv]$ contains the clause $\neg d \lor \bigvee B_d$; let W_1 and W_2 be the watched literals of this clause. If $W_1 = \neg d$ then $J_s(d) = W_2$, otherwise $J_s(d) = W_1$. Knowing that the current assignment is the result of a propagation step, one can easily verify that J_s is a supporting justification.

Step 4a: Find a witness. The supporting justification J_s can have cycles. Because J_{cf} is cycle-free, it must be the case that each cycle in J_s contains at least one disjunctively defined atom d with $J_s(d) \neq J_{cf}(d)$. Let us call such atoms *cycle sources*. Cycle sources belong to the set of atoms on which both justifications disagree: $DS = \{d \mid J_s(d) \neq J_{cf}(d)\}$. The overall strategy is to check for each element cs in DS whether it is a cycle source, and, if so, to perform local adjustments on the supporting justification so that cs is no longer part of a cycle ("justifying cs"). Obviously, the smaller DS, the less work this step requires.

These local adjustments may or may not update $J_s(cs)$ and may update elements in which J_s does not differ from J_{cf}. The point is that updates break up the cycles passing through cs. One may wonder whether these updates may introduce new cycles for which no element is present in DS. When we explain the details of the adjustments we will argue (inside the description of Analyse(cs)) that this is not possible. Hence, if all elements of DS can be successfully processed, J_s becomes cycle-free, i.e., a witness of the current assignment.

For what concerns the elements d in DS that are false under the current assignment, one can observe that, due to the propagation, all body literals of the rule defining d must be false as well. Hence for such d, J_s remains supporting when setting $J_s(d) = J_{cf}(d)$ and these elements can be removed from DS.

The further processing then consists of justifying each element cs in DS until either DS is empty and hence J_s is a witness of the current assignment or some cs could not be justified, in which case a set $Cycle$ as described in the high level algorithm is returned.

Analyse(cs) (A first version of such an algorithm was in [17])

The disjunctively defined atom cs is not false in the current assignment and possibly belongs to a cycle in the supporting justification J_s.

In an initialisation step, the procedure marks all atoms as *unsafe* that are on a path in J_s that leads to cs. This means that all atoms that belong to an eventual cycle are marked as unsafe; however, also other atoms can be marked as unsafe. If $J_s(cs)$ is not marked, then there is no cycle passing through cs and we are done. Otherwise, cs is a cycle source and J_s has to be adjusted.

By "to justify a disjunctively defined atom d" we mean setting $J_s(d)$ such that d is no longer part of a cycle through cs; "to justify a conjunctively defined

---

[5] For other schemes, one first has to use the current assignment to determine for each clause two literals that satisfy the two watching literals invariant.

atom" means showing that all its body literals are justified. The purpose of the algorithm is to justify cs.

A disjunctively defined atom d can be justified either by setting $J_s(d)$ to a literal that is not marked unsafe, or by setting it to an atom that in turn can be similarly justified. To this end, a working queue Q, initialised with cs, is maintained: Q contains atoms that can still be tried to be justified.

The algorithm also maintains a set $Cycle$, initialised with cs, of atoms that are waiting to be justified. If the algorithm fails, then the elements of $Cycle$ have not been justified.

Atoms are popped from Q and processesed, until either it is ensured that cs is no longer part of a cycle or Q is empty, in which case $Cycle$ is returned. Let d be the popped element. If it is no longer marked as unsafe, it has already been justified and the next element can be popped. Otherwise, two cases are distinguished. They rely on a procedure Justify(q) described afterwards.

d **is disjunctively defined.** Let B_d be the body of the defining rule. If B_d has a literal b that is neither marked nor false, then set $J_s(d) = b$ (d is not false under the current assignment, hence, to preserve support, b has to be non false as well) and perform Justify(d). Note that d is now justified: b is not marked and hence has no path to cs. Furthermore, it has no path to d either, because all atoms with a path to d in J_s are marked. Therefore adding the edge $d \rightarrow b$ cannot create a new cycle.[6]

If all non false literals in B_d are marked (they are atoms as negative literals cannot be marked), the ones that are not yet in $Cycle$ are all pushed on Q and added to $Cycle$: justifying any of them suffices to justify d (Justify will take care of doing that).

d **is conjunctively defined.** Let B_d be the body of the defining rule. If B_d has no marked literal, d is justified (not part of a cycle through cs), so Justify(d) is performed. Otherwise, a marked atom q is selected from B_d (preferably one already in $Cycle$) and, if not yet in $Cycle$, added to it and pushed on Q. This atom q is called the *guard* of d. Adding only this guard to Q (or none if already in $Cycle$), instead of all marked body atoms, has the advantage that no computation time will be lost on other body atoms in case this guard cannot be justified. In case it can, Justify(q) will add d to Q again for reconsideration.

Justify(q). The "unsafe" mark is removed from the atom q and the atom is removed from $Cycle$. Moreover, if it is an element from DS then it can be removed from DS as well, as it can no longer be a cycle source. Finally, if it is cs itself, we are done and can start with processing the next element in DS. If it is not cs, we have to continue:

– For every disjunctively defined atom $d \in Cycle$ with $q \in B_d$, set $J_s(d) = q$ and perform Justify(d). Indeed, if q is no longer part of the cycle through cs, then so is d in the changed J_s.

---

[6] We do not neccessarily have a witness for b (and d) as b could be part of another cycle with cycle source in DS.

– For every conjunctively defined atom d that has q as guard, d is (again) pushed on $\mathcal{Q}$.

If $\mathcal{Q}$ becomes empty before cs could be justified, some atoms (at least cs itself) are still in $Cycle$, and all possible supporting subjustifications for them have been exhaustively searched; none has been found that does not cycle through cs. This implies that these atoms cannot have a witness (a cycle-free subjustification of a justification supporting the current assignment), hence their truth can never be justified in a well-founded way and it is correct to add the learning clauses in step 4b.

The use of a queue as datastructure, and not a stack, means that we search for a subjustification of cs that does not cycle through cs by making possible modifications to the supporting justification breadth first, i.e., starting with edges most close to cs.

Step 4b: Learning clauses from the Cycle-set [7]. When the supporting justification cannot be adjusted into a cycle-free supporting justification, the set $Cycle$ of defined atoms that cannot have a witness subjustification is returned. These atoms have to be set to false in the current assignment as they necessarily have to be false in each well-founded model extending the current assignment. To properly integrate this with the SAT solver and its backtracking search, this is achieved by extending the theory with an appropriate learned clause for each of these atoms.

Define $Ante = \left(\bigcup_{d \in Cycle, d \text{ disjunctively defined}} B_d \right) \setminus Cycle$, i.e.: all body atoms of the disjunctively defined atoms in the cycle set except the cycle atoms themselves. The falsity of those literals forces the falsity of the atoms in the cycle set. A so-called *loop formula* $\bigvee Ante \vee \neg (\bigvee Cycle)$ captures this (adapted from [13]); its CNF contains one reason clause for each atom in $Cycle$.

4.3   Checking Totality of Definitions

When the search terminates with a SAT model A, the existence of the witness J_s guarantees (Th. 1) that A is also a SAT(ID) model (if the definition was total).

Deciding whether a definition is total is as hard as the SAT(ID) problem itself, i.e., NP-hard, hence there is no cheap test that could be used to inform the user if he has written an non-total definition. However, when a model is obtained as above, one can do some verification. One can calculate $\mathrm{wfm}_\Delta(A')$, where A' is the restriction of A to the non-defined atoms, and check whether this well-founded model coincides with the original assignment. If it does not, some atoms will be undefined in the well-founded model; it can be pointed out to the user that the definition is not total (the undefined atoms and assignment A' can help the user in localising the error). Note that coincidence for A' does not imply that the definition is total. The well-founded model could be three valued for a different consistent assignment to the non-defined atoms.

---

[7] For solvers that do not implement clause learning, the negation of each atom in $Cycle$ has to be added to the propagation queue.

5   Implementation on Top of **MiniSat**

5.1   Introduction

We have implemented the above algorithms as an extension to the popular SAT solver MiniSat [6]. The resulting program is called *MiniSat(ID)*, and its code is available on [19]. It contains several efficiency improvements compared to the abstracted version as presented here:

- Instead of constructing a supporting justification J_s in step 3 and later updating it, we only record the *changes* with respect to the cycle-free justification J_{cf}. When J_s eventually becomes a witness, then these changes are applied on J_{cf} in step 1.
- In the initialization, MiniSat(ID) computes strongly connected components (SCCs) in the graph determined by the edges $\{(q, b) \mid b$ is an atom of $B_q\}$. This information can later be used: e.g. to mark as unsafe only atoms in the same SCC as the cycle source.
- When there have been no backtracks since last iteration, finding DS is done by inspecting, using the propagation trail, only those atoms d whose $J_s(d)$ has changed.

MiniSat's implementation of 2WL is particularly suited for finding J_s: a clause is stored as an array of literals, with the watched literals always being the first two in the array, and the first watched literal never false.

Apart from obvious changes such as command-line options processing, the only change we did to the code of MiniSat proper, was in the "propagate()" method, where we call code to perform steps 3, 4 and 1 of our algorithm before completing the method.

5.2   Evaluation

SAT(ID) is a new research domain; no organized benchmarking suites exist for it yet. We know of 3 SAT(ID) solvers: idsat [23], MidL [18], and MiniSat(ID). The first works by translating IDs into propositional logic, the second is our earlier work mentioned in the introduction, and the last is a contribution of this paper.

To improve the evaluation, we include also ASP solvers in our comparisons. These can solve the same problems as SAT(ID) solvers, using similar encodings. We have included clasp [7] (which was the winner of the ASP competition in 2007 [8]), SModels$_{cc}$ [25], and cmodels [12]. cmodels uses MiniSat as an underlying SAT solver: it iteratively calls the solver and adds loop formulas until the found SAT model is also an ASP model, or the search space is exhausted.

We use 2 sets of problem instances with different characteristics: Hamiltonian Cycle problems and Sokoban problems. The full set of encodings and instances that we used and results obtained is available on [19]. Here we provide aggregated results in Tables 1 and 2.[8]

---

[8] Version numbers of all the solvers used: MidL-2.2.1, idsat-0.9.5, clasp-1.0.5, SModels$_{cc}$-1.08 (using option "nolookahead" as advised by the authors), cmodels-3.75, MiniSat-2.0b. All experiments were run on a P4 2.8GHz with 1GB memory.

We observe that MiniSat(ID) always performs very similarly to clasp, mostly slightly better. cmodels, on the other hand, shows very variable results. On the Hamiltonian Cycles, it shows its real drawback: whenever there are many cycles in the problem, timings get out of hand, because many iterations are needed. But on Sokoban instances, where reachability is required to express that the agent can reach the square from where his pushing move starts, most squares are reachable most of the time. Consequently, the first model found by the SAT solver underlying cmodels is usually already an ASP model. On these problems, most of the testing for cycles done by solvers as MiniSat(ID) and clasp is redundant.

It is possible to make PC encodings of both Hamiltonian Cycle and Sokoban problems (i.e., encodings that do not require IDs), as well as to automatically translate the PC(ID) instances to PC using idsat. However, both of these approaches yield PC instances that are orders of magnitude bigger than their corresponding PC(ID) instances. Indeed, using MiniSat on these PC instances yields only timeouts; we have not included these in the tables.

Table 1. Evaluation on Hamiltonian Cycle problems: avg. time in *sec*, (no. of runs > 300*sec*). Averaged over 10 instances.

Size	MiniSat(ID)	clasp	MidL	SModels$_{cc}$	cmodels
200 × 1800	**1.23**	2.08	6.90	16.94	22.74
200 × 2600	**1.94**	3.22	6.41	38.86	35.09
250 × 1800	6.41	**1.44**	28.06 (1)	18.48	94.14
250 × 2600	**2.19**	2.40	24.36	30.30	92.13
300 × 3600	**3.28**	5.61	23.81	50.60	229.54 (4)
300 × 4800	**4.90**	6.51	35.96	79.52	178.98 (1)
350 × 3600	**3.62**	5.09	38.51 (2)	n/a (10)	195.91 (9)
350 × 4800	**5.53**	6.66	50.30 (3)	n/a (10)	167.94 (7)

Table 2. Evaluation on Sokoban problems: avg. time in *sec*, (no. of runs > 300*sec*). Averaged over 12 instances.

MiniSat(ID)	clasp	MidL	SModels$_{cc}$	cmodels
7.60	7.57	n/a (12)	38.26	**5.16**
10.85	13.37	n/a (12)	63.08	**8.02**
20.33	20.59	n/a (12)	100.59 (1)	**14.35**
56.02	57.59	223.53 (11)	127.40 (3)	**31.73**
16.16	24.40	n/a (12)	69.66 (2)	**6.55**
16.71	**13.70**	33.2 (11)	82.42 (2)	24.43
9.65	9.68	139.65 (6)	64.69	**5.75**
34.19	34.20	219.9 (9)	93.76 (4)	**25.63**

6   Conclusions and Related Work

This work is based on earlier work from [17]. In particular, the proof of Theorem 1 in Section 3 is based on a similar theorem there, and the Analyse algorithm from

Section 4.2 is in essence the same algorithm as first published there, though the presentation differs considerably. In that paper, the definition of justification was more complex (it involved also justifications for negative literals), and also the concepts $\Delta^\varnothing$, J_s, and J_{cf} are new to this paper: these changes have enabled us to simplify substantially from [17].

The main contribution of this work, however, is to show how a SAT(ID) solver can be built *by extending an existing SAT solver*. To do so, we have made the ID propagation mechanism (steps 1, 3 and 4 of our algorithm) as independent as efficiently possible from the unit propagation mechanism: apart from adding literals on the propagation queue or reason clauses to the theory for ID-related propagations, it only requires *inspecting* the SAT solver's state. By contrast, our earlier solver MidL [18], which was a native SAT(ID) implementation, performed ID related changes to the solver's state during unit propagations.

We have implemented these ideas by extending MiniSat: our experiments show that the resulting solver, MiniSat(ID), exhibits the expected benefits. These experiments included also ASP solvers: ASP is a related formalism with the same expressivity as PC(ID). However, while PC(ID) is simply an extension of propositional logic with the natural concept of inductive definitions, ASP has a more complex relationship with propositional logic [13].

References

1. Davis, M., Logemann, G., Loveland, D.W.: A machine program for theorem-proving. Commun. ACM 5(7), 394–397 (1962)
2. Denecker, M.: Extending Classical Logic with Inductive Definitions. In: Palamidessi, C., Moniz Pereira, L., Lloyd, J.W., Dahl, V., Furbach, U., Kerber, M., Lau, K.-K., Sagiv, Y., Stuckey, P.J. (eds.) CL 2000. LNCS (LNAI), vol. 1861, pp. 703–717. Springer, Heidelberg (2000)
3. Denecker, M., De Schreye, D.: Justification semantics: A unifying framework for the semantics of logic programs. In: LPNMR 1993, pp. 365–379. MIT Press, Cambridge (1993)
4. Denecker, M., Ternovska, E.: A Logic of Non-monotone Inductive Definitions and Its Modularity Properties. In: Lifschitz, V., Niemelä, I. (eds.) LPNMR 2004. LNCS (LNAI), vol. 2923, pp. 47–60. Springer, Heidelberg (2003)
5. Denecker, M., Vennekens, J.: Well-Founded Semantics and the Algebraic Theory of Non-monotone Inductive Definitions. In: Baral, C., Brewka, G., Schlipf, J. (eds.) LPNMR 2007. LNCS (LNAI), vol. 4483, pp. 84–96. Springer, Heidelberg (2007)
6. Eén, N., Sörensson, N.: An extensible SAT-solver. In: Giunchiglia, E., Tacchella, A. (eds.) SAT 2003. LNCS, vol. 2919, pp. 502–518. Springer, Heidelberg (2004)
7. Gebser, M., Kaufmann, B., Neumann, A., Schaub, T.: clasp: A Conflict-Driven Answer Set Solver. In: Baral, C., Brewka, G., Schlipf, J. (eds.) LPNMR 2007. LNCS (LNAI), vol. 4483, pp. 260–265. Springer, Heidelberg (2007)
8. Gebser, M., Liu, L., Namasivayam, G., Neumann, A., Schaub, T., Truszczynski, M.: The First Answer Set Programming System Competition. In: Baral, C., Brewka, G., Schlipf, J. (eds.) LPNMR 2007. LNCS (LNAI), vol. 4483, pp. 3–17. Springer, Heidelberg (2007)
9. Van Gelder, A.: The alternating fixpoint of logic programs with negation. Journal of Computer and System Sciences 47(1), 185–221 (1993)

10. Van Gelder, A., Ross, K.A., Schlipf, J.S.: The well-founded semantics for general logic programs. J. ACM 38(3), 620–650 (1991)
11. Jackson, D.: Software Abstractions: Logic, Language, and Analysis. MIT Press, Cambridge (2006)
12. Lierler, Y.: cmodels – SAT-Based Disjunctive Answer Set Solver. In: Baral, C., Greco, G., Leone, N., Terracina, G. (eds.) LPNMR 2005. LNCS (LNAI), vol. 3662, pp. 447–451. Springer, Heidelberg (2005)
13. Lin, F., Zhao, Y.: ASSAT: Computing answer sets of a logic program by sat solvers. Artif. Intell. 157(1-2), 115–137 (2004)
14. Lonc, Z., Truszczyński, M.: On the Problem of Computing the Well-Founded Semantics. In: Palamidessi, C., Moniz Pereira, L., Lloyd, J.W., Dahl, V., Furbach, U., Kerber, M., Lau, K.-K., Sagiv, Y., Stuckey, P.J. (eds.) CL 2000. LNCS (LNAI), vol. 1861, pp. 673–687. Springer, Heidelberg (2000)
15. Marek, V.W., Truszczyński, M.: Stable models and an alternative logic programming paradigm. In: The Logic Programming Paradigm: a 25-Year Perspective, pp. 375–398. Springer, Heidelberg (1999)
16. Mariën, M., Wittocx, J., Denecker, M.: The IDP framework for declarative problem solving. In: Search and Logic: Answer Set Programming and SAT, pp. 19–34 (2006)
17. Mariën, M., Wittocx, J., Denecker, M.: Integrating Inductive Definitions in SAT. In: Dershowitz, N., Voronkov, A. (eds.) LPAR 2007. LNCS (LNAI), vol. 4790, pp. 378–392. Springer, Heidelberg (2007)
18. Mariën, M., Wittocx, J., Denecker, M.: MidL: a SAT(ID) solver. In: 4th Workshop on Answer Set Programming: Advances in Theory and Implementation, pp. 303–308 (2007)
19. MiniSat(ID), http://www.cs.kuleuven.be/~dtai/krr/software/minisatid.html
20. Mitchell, D.G., Ternovska, E.: A framework for representing and solving NP search problems. In: AAAI 2005, pp. 430–435. AAAI Press / The MIT Press (2005)
21. Moskewicz, M., Madigan, C., Zhao, Y., Zhang, L., Malik, S.: Chaff: Engineering an efficient SAT solver. In: DAC 2001, pp. 530–535. ACM Press, New York (2001)
22. Nieuwenhuis, R., Oliveras, A., Tinelli, C.: Solving SAT and SAT modulo theories: From an abstract Davis–Putnam–Logemann–Loveland procedure to DPLL(t). J. ACM 53(6), 937–977 (2006)
23. Pelov, N., Ternovska, E.: Reducing Inductive Definitions to Propositional Satisfiability. In: Gabbrielli, M., Gupta, G. (eds.) ICLP 2005. LNCS, vol. 3668, pp. 221–234. Springer, Heidelberg (2005)
24. Silva, J.P.M., Sakallah, K.A.: GRASP: A search algorithm for propositional satisfiability. IEEE Trans. Computers 48(5), 506–521 (1999)
25. Ward, J., Schlipf, J.S.: Answer Set Programming with Clause Learning. In: Lifschitz, V., Niemelä, I. (eds.) LPNMR 2004. LNCS (LNAI), vol. 2923, pp. 302–313. Springer, Heidelberg (2003)
26. Wittocx, J., Vennekens, J., Mariën, M., Denecker, M., Bruynooghe, M.: Predicate Introduction Under Stable and Well-Founded Semantics. In: Etalle, S., Truszczyński, M. (eds.) ICLP 2006. LNCS, vol. 4079, pp. 242–256. Springer, Heidelberg (2006)

Towards More Effective Unsatisfiability-Based Maximum Satisfiability Algorithms

Joao Marques-Silva and Vasco Manquinho

School of Electronics and Computer Science, University of Southampton, UK
IST/INESC-ID, Technical University of Lisbon, Portugal
jpms@ecs.soton.ac.uk, vmm@sat.inesc-id.pt

Abstract. The MaxSAT problem and some of its well-known variants find an increasing number of practical applications in a wide range of areas. Examples include different optimization problems in system design and verification. However, most MaxSAT problem instances from these practical applications are too hard for existing branch and bound algorithms. One recent alternative to branch and bound MaxSAT algorithms is based on unsatisfiable subformula identification. A number of different unsatisfiability-based MaxSAT algorithms have been developed, which are effective at solving different classes of problem instances. All MaxSAT algorithms based on unsatisfiable subformula identification require using additional Boolean variables, either to allow relaxing some of the clauses or to encode cardinality constraints used by these algorithms. As a result, these algorithms may require using a significant number of additional Boolean variables, that can exceed the original number of variables for some problem instances. This paper proposes techniques for effectively reducing the number of auxiliary variables that must be used in unsatisfiability-based MaxSAT algorithms. Experimental results indicate that the techniques for reducing the number of auxiliary variables are effective, and contribute to more efficient MaxSAT algorithms.

1 Introduction

Maximum Satisfiability (MaxSAT) and variants allow modeling an increasingly large number of optimization problems in an also growing number of practical settings. The recent application of MaxSAT and variants in design debugging and verification of complex designs [11, 4, 5] motivated the development of new MaxSAT algorithms, capable of solving large structured problem instances common to these application domains. Despite the significant improvements made in recent years to standard branch and bound MaxSAT algorithms, in practice existing branch and bound algorithms are unable to solve the vast majority of problem instances from practical applications.

One recent promising line of research is the development of MaxSAT solvers based on the identification of unsatisfiable subformulas (or cores) [4, 8, 9]. These MaxSAT algorithms are built on top of SAT solvers, and so can exploit the most effective SAT techniques [2]. Moreover, these algorithms rely extensively on the ability of modern SAT solvers for producing certificates of unsatisfiability [12]. Even though the organization of existing unsatisfiability-based MaxSAT algorithms is fairly different, these algorithms also share a number of key common characteristics. For example, all unsatisfiability-based MaxSAT algorithms iteratively identify and relax unsatisfiable subformulas. The approach for relaxing unsatisfiable subformulas is well-known

H. Kleine Büning and X. Zhao (Eds.): SAT 2008, LNCS 4996, pp. 225–230, 2008.
© Springer-Verlag Berlin Heidelberg 2008

(e.g. see [8] for an overview), and consists of adding *relaxing* (or *blocking*) variables to each clause in each identified unsatisfiable subformula. Even though existing experimental results suggest great promise for unsatisfiability-based MaxSAT algorithms, many problem instances are still too complex even for the most effective algorithms. One clear potential drawback of unsatisfiability-based MaxSAT algorithms is the iterated addition of auxiliary variables. For many problem instances, it is possible that the number of additional variables becomes far larger than the original number of variables. As a result, besides the increase of search space, the much larger number of variables can often have a negative effect on SAT solvers. This paper proposes techniques for reducing the number of additional variables used in unsatisfiability-based MaxSAT algorithms. The first technique addresses the encoding of the cardinality constraints relating blocking variables. The second technique addresses the reduction of the actual number of blocking variables. Experimental results, obtained on a wide range of practical problem instances, indicates that the reduction of additional variables can often contribute to significantly reduce run times. The paper is organized as follows. The next two sections introduce MaxSAT and variants, existing branch and bound algorithms, and recent unsatisfiability-based algorithms for MaxSAT. Afterwards, Section 4 proposes the new techniques for reducing the number of variables. The new MaxSAT solvers are evaluated in Section 5 and the paper concludes in Section 6.

2   Maximum Satisfiability

This section provides definitions and background knowledge for the MaxSAT problem; familiarity with SAT and related topics is assumed [2]. The maximum satisfiability (MaxSAT) problem can be stated as follows: given a SAT instance represented in Conjunctive Normal Form (CNF), compute an assignment to the variables that maximizes the number of satisfied clauses. Variants of the MaxSAT problem include the partial MaxSAT, the weighted MaxSAT problem and the partial weighted MaxSAT problem. In the partial MaxSAT problem some clauses (i.e. the *hard* clauses) must be satisfied, whereas others (i.e. the *soft* clauses) may not be satisfied. Weighted variants are addressed elsewhere (e.g. see [5]). MaxSAT algorithms have been subject to significant improvements over the last decade (see for example [6, 5] for a review of past work). Despite the clear relationship with the SAT problem, most modern SAT techniques cannot be applied directly to the MaxSAT problem [6, 5]. As a result, the most successful MaxSAT algorithms implement branch and bound search, and integrate sophisticated lower bounding and inference techniques [5, 6]. Effective lower bounding techniques are based on unit propagation, whereas effective inference techniques can be viewed as based on specific resolution patterns. One alternative approach for solving the MaxSAT problem is to use Pseudo-Boolean Optimization (PBO). An overview is provided in [8].

3   Unsatisfiability-Based MaxSAT Algorithms

As mentioned in the previous section, one of the major drawbacks of the PBO model for MaxSAT is the large number of blocking variables that must be considered. The ability to reduce the number of required blocking variables is expected to improve significantly the ability of SAT/PBO based solvers for tackling instances of MaxSAT. Moreover, any solution to the MaxSAT problem will be unable to satisfy clauses that *must* be part of an unsatisfiable subformula. Consequently, one approach for reducing the number of blocking variables is to associate blocking variables only with clauses that are part of unsatisfiable subformulas. However, it is not simple to identify all clauses that are part of unsatisfiable subformulas. One alternative is the identification and relaxation of unsatisfiable subformulas. A number of unsatisfiability-based MaxSAT algorithms have been proposed in recent years [4, 8, 9]. The first algorithm [4] (referred to as msu1) iteratively finds unsatisfiable cores, adds new blocking variables to the non-auxiliary clauses in the unsatisfiable core, and requires that exactly one of the new blocking variables must be assigned value 1. The algorithm terminates whenever the CNF formula is satisfiable, and the number of assigned blocking variables is used for computing the solution to the MaxSAT problem instance. The clauses used for implementing the cardinality constraints are declared auxiliary; all other clauses are non-auxiliary. Observe that each non-auxiliary clause may receive more than one blocking variable, and the total number of blocking variables a clause receives corresponds to the number of times the clause is part of an unsatisfiable core. In the msu1 algorithm [4] the pairwise encoding is used for encoding AtMost1 constraints. In contrast, msu1.1 [8][1] proposes different linear encodings. Also, msu1.1 uses AtMost1 constraints on the blocking variables associated with each clause. Alternative unsatisfiability-based MaxSAT algorithms, msu3 and msu4, were proposed recently [8, 9]. Both msu3 and msu4 use a *single* cardinality constraint to constrain the number of blocking variables that can be assigned value 1, and so ensure that at most one blocking variable is required for each clause. msu4 iterates between lower and upper bounds on the number of blocking variables. In contrast, msu3 resembles msu1 and variants, where only a lower bound on the number of blocking variables is updated. Existing experimental results indicate that msu1.1 is often more efficient than either for msu3 or msu4 for most problem instances.

4   Reducing the Number of Additional Variables

This section describes two techniques for reducing the number of variables. The first one addresses the encoding of cardinality constraints, while the second one extends the same ideas to the way blocking variables are used. The original unsatisfiability-based (partial) MaxSAT algorithm [4] used the pairwise encoding for the AtMost 1 cardinality constraints (this algorithm will be referred to as msu1). A more effective approach is to use a linear encoding for the AtMost1 cardinality constraint (e.g. [8] compares a number of alternative linear encodings). For problem instances with large unsatisfiable cores, the linear encodings are significantly more effective. The linear encodings use a linear number of additional variables, the auxiliary variables, and a linear number of clauses. The number of additional variables, albeit linear, can be a potential drawback for some problem instances.

One approach to reduce the number of additional variables is to use the recently proposed *bitwise encoding* [10]. Consider an AtMost1 constraint $\sum_{i=0}^{k-1} x_i \le 1$. Create r auxiliary variables, where $r = 1$ if $k = 1$ and $r = \lceil \log k \rceil$ if $k > 1$. Let $v_0, \ldots, v_{r-1}$

---

[1] For consistency, the algorithm msu2 in [8] is renamed to msu1.1.

be the auxiliary variables. Now associate with each x_i the binary representation of i. Finally, for each i create the clauses: $(\neg x_i \vee p_j), j = 0, \ldots, r - 1$, where $p_j = v_j$ if the binary representation of i has value 1 in position j, and $p_j = \neg v_j$ otherwise. For an AtMost1 constraint with k variables, the bitwise encoding requires $\mathcal{O}(\log k)$ variables and $\mathcal{O}(k \log k)$ clauses, i.e. $\mathcal{O}(\log k)$ for each variable in the AtMost1 constraint. Observe that linear encodings (e.g. [8]) require a linear number of auxiliary variables and a linear number of clauses. Hence, the bitwise encoding trades off variables for clauses.

Given that the number of iterations of msu1 and msu1.1 is $\mathcal{O}(m)$ [8], where m is the number of clauses in the original formula, the number of additional variables for these algorithms is in $\mathcal{O}(m^2)$. By using the bitwise encoding, this asymptotic complexity remains unchanged, but the actual constant is considerably smaller. Also, observe that the number of blocking variables will be the same; only the number of auxiliary variables used for encoding the AtMost1 constraint is reduced. One should also observe that the created binary clauses only selects which variable of the AtMost1 constraint *can* be assigned value 1; all the other variables are required to be assigned value 0. Hence, the encoding effectively represents an AtMost1 constraint. In order to encode the constraint Exactly1, it would suffice to simply add a clause to capture the AtLeast1 constraint (e.g. [8]). The modified algorithm, using a logarithmic number of auxiliary variables for representing AtMost1 constraints, is referred to as msu1.2.

The use of the bitwise encoding [10] above for reducing the number of auxiliary variables, motivates using the same ideas for actually reducing the number of blocking variables. Instead of selecting at most one blocking variable out of set of k blocking variables, the bitwise encoding now operates on the clauses of each identified unsatisfiable core. This essentially allows eliminating the blocking variables by working directly with the auxiliary variables used in the bitwise encoding. For an unsatisfiable core with k clauses, the proposed encoding will require r auxiliary variables, where $r = 1$ if $k = 1$ and $r = \lceil \log k \rceil$ if $k > 1$. Moreover, the encoding will require $\mathcal{O}(\log k)$ variables and $\mathcal{O}(k \log k)$ new clauses, i.e. $\mathcal{O}(\log k)$ new clauses for each original clause in the unsatisfiable core. Hence each original clause ω_i in an unsatisfied core *generates* $\mathcal{O}(\log k)$ new clauses. The proposed approach needs to take into consideration when a clause has already been relaxed. Assume clause ω_{ij}, relaxed from an original clause ω_i, is included in an identified unsatisfiable core. Then *all* clauses generated from ω_i need to be re-relaxed. The modified algorithm, using a logarithmic number of blocking variables for each unsatisfiable core, is referred to as msu2.

5   Results

This section summarizes results on MaxSAT and partial MaxSAT instances from practical applications. A number of classes of instances were considered. Class DEBUG [11] represents design debugging MaxSAT instances. FIR [7] represents filter design partial MaxSAT instances. Class SYN [7] represents logic synthesis partial MaxSAT instances. Finally, class MTG [7] represents minimum size test pattern generation partial MaxSAT instances. All partial MaxSAT instances are obtained by translating restricted pseudo-Boolean problem instances into partial MaxSAT (e.g. using a recently proposed translation [5]). A number of MaxSAT solvers was considered, namely: maxsatz [6],

Table 1. Number of aborted instances, with a 1000 seconds timeout

Class	#I	maxsatz	minimaxsat	minisat+	msu1	msu1.1	msu1.2	msu2	msu3	msu4
DEBUG	65	62	65	63	22	14	**8**	11	24	23
FIR	59	–	45	37	15	14	**9**	12	32	44
SYN	74	–	46	44	48	44	**42**	**42**	47	51
MTG	215	–	7	**0**	44	44	52	60	11	16
Total	413	–	163	144	129	116	**111**	125	114	134

the best performing solver in the MaxSAT 2007 evaluation [1]; minimaxsat [5], a recent competitive MaxSAT solver; and the PBO formulation of the MaxSAT problem solved with minisat+ [3], one of the best performing PBO solvers [7]. Moreover, the unsatisfiability-based MaxSAT solvers considered were msu1 [4]; msu1.1 [8] (renamed from msu2 for naming consistency); msu3 [8]; msu4 [9]; and the new algorithms described in this paper msu1.2 and msu2. All msu algorithms are built on top of the same unsatisfiable core extractor, implemented with minisat 1.14 [2]. Other alternative MaxSAT algorithms (see [8, 9] for an overview) are known not to be efficient for these instances. Moreover, minisat+ was run with its best configuration for these classes of instances and, for the partial MaxSAT instances, the original PBO instances were considered. The results for all MaxSAT solvers on all problem instances were obtained on a Linux server running RHE Linux, with a Xeon 5160 3.0 GHz dual-core processor. For the experiments, the available physical memory of the server was 2 GByte. The time limit was set to 1000 seconds per instance. Table 1 summarizes the number of aborted instances for the MaxSAT solvers considered. Overall, CPU times correlate well with the number of aborted instances and are not shown due to lack of space (see [8, 9] for plots for the other algorithms). For maxsatz, only MaxSAT results are shown, since maxsatz cannot be used with partial MaxSAT instances. The solvers exhibiting the best performance are highlighted in bold. The results indicate that the new algorithms are more efficient than previous MaxSAT algorithms. With the exception of class MTG, MaxSAT algorithms are now vastly superior to minisat+, one of the best performing PBO solvers [3]. For classes DEBUG and FIR, the instances aborted by either msu1.2 and msu2 are a fraction of the instances aborted by minisat+. Similarly, the two branch and bound algorithms considered, maxsatz and minimaxsat, perform much worse than *any* of the unsatisfiability-based MaxSAT algorithms. The results indicate that msu1.2 is the overall best performing algorithm. msu2 does not perform as well, especially for class MTG. This is in part explained by the growth in the number of clauses that msu2 often requires. The improvements introduced by msu1.2 allow significantly reducing the number of aborted instances for classes DEBUG and FIR, and also reducing the number of aborted instances for class SYN. For class MTG the results for msu1.2 are worse than for msu1.1, and also worse than for msu3 and msu4. Moreover, the results also suggest that msu1.2 is the preferred algorithm for classes DEBUG, FIR and SYN, and that minisat+ is the preferred algorithm for class MTG. Together, msu1.2 and minisat+ abort only 50 instances, thus motivating a portfolio of algorithms for MaxSAT. The experimental results confirm that techniques for reducing the number of additional variables are often effective. The performance of msu2 is affected by the significant increase in the number of clauses that can often take place. As a result, a mixed approach, involving msu1.2 and msu2 should be considered.

6   Conclusions

Despite the significant improvements in MaxSAT algorithms over the last few years [6,5], current state of the art MaxSAT solvers are ineffective on many problem instances obtained from practical applications [11,5,9]. This paper continues recent work on developing MaxSAT algorithms based on identification of unsatisfiable sub-formulas [4, 8,9], by proposing effective techniques for reducing the number of additional variables that must be used. Two different algorithms are developed msu1.2 and msu2. The experimental results indicate that the proposed techniques are effective, and that msu1.2 is the most effective algorithm. The results also suggest that a mixed approach between msu1.2 and msu2 is expected to provide the most efficient approach.

Acknowledgement. This work is partially supported by EPSRC grant EP/E012973/1, by EU grants IST/033709 and ICT/217069, and by FCT grants POSC/EIA/61852/2004 and PTDC/EIA/76572/2006.

References

1. Argelich, J., Li, C.M., Manyà, F., Planes, J.: MaxSAT evaluation,
   http://www.maxsat07.udl.es
2. Een, N., Sörensson, N.: An extensible SAT solver. In: International Conference on Theory and Applications of Satisfiability Testing, pp. 502–518 (May 2003)
3. Een, N., Sörensson, N.: Translating pseudo-Boolean constraints into SAT. Journal on Satisfiability, Boolean Modeling and Computation 2 (March 2006)
4. Fu, Z., Malik, S.: On solving the partial MAX-SAT problem. In: International Conference on Theory and Applications of Satisfiability Testing, pp. 252–265 (August 2006)
5. Heras, F., Larrosa, J., Oliveras, A.: MiniMaxSat: a new weighted Max-SAT solver. In: International Conference on Theory and Applications of Satisfiability Testing, pp. 41–55 (May 2007)
6. Li, C.M., Manyà, F., Planes, J.: New inference rules for Max-SAT. Journal of Artificial Intelligence Research 30, 321–359 (2007)
7. Manquinho, V., Roussel, O.: Pseudo-Boolean evaluation,
   http://www.cril.univ-artois.fr/PB07
8. Marques-Silva, J., Planes, J.: On using unsatisfiability for solving maximum satisfiability. Computing Research Repository, abs/0712.0097 (December 2007)
9. Marques-Silva, J., Planes, J.: Algorithms for maximum satisfiability using unsatisfiable cores. In: Design, Automation and Testing in Europe Conference (March 2008)
10. Prestwich, S.D.: Variable dependency in local search: Prevention is better than cure. In: International Conference on Theory and Applications of Satisfiability Testing, pp. 107–120 (May 2007)
11. Safarpour, S., Mangassarian, H., Veneris, A., Liffiton, M.H., Sakallah, K.A.: Improved design debugging using maximum satisfiability. In: Formal Methods in Computer-Aided Design, pp. 13–19 (November 2007)
12. Zhang, L., Malik, S.: Validating SAT solvers using an independent resolution-based checker: Practical implementations and other applications. In: Design, Automation and Testing in Europe Conference, pp. 10880–10885 (March 2003)

A CNF Class Generalizing Exact Linear Formulas

Stefan Porschen and Ewald Speckenmeyer

Institut für Informatik, Universität zu Köln,
Pohligstr. 1, D-50969 Köln, Germany
{porschen,esp}@informatik.uni-koeln.de

Abstract. The *fibre view* on clause sets, previously introduced in [12], is used in the present paper to define and investigate subclasses of CNF that appear to be polynomial time solvable w.r.t. SAT. The most interesting of these classes is a generalization of exact linear formulas, namely formulas such that each pair of distinct clauses has all variables in common or exactly one. By definition, in an exact linear formula each pair of distinct clauses has exactly one variable in common. SAT-solving for exact linear formulas was shown to be easy in [14]. Here we provide an algorithm solving the decision and counting variants of SAT for the generalized class in polynomial time. Moreover we study some other structurally defined formula classes on the basis of the fibre view. We show that these classes have the property that their members all are satisfiable or all are unsatisfiable.

Keywords: CNF satisfiability, exact linear formula, hypergraph, fibre-transversal.

1 Introduction

Exploiting the *fibre view* on clause sets recently introduced in [12] we consider some structurally defined subclasses of CNF regarding their behaviour w.r.t. SAT. The most interesting of these classes yields a generalization of *exact linear formulas*. The class of linear CNF formulas has recently been studied in [14] revealing its general NP-completeness w.r.t. SAT. Each pair of distinct clauses of a linear formula by definition has at most one variable in common. And requiring that there should be *exactly one* variable in the intersection of the variable sets of each pair of distinct clauses, one arrives at the subclass of exact linear formulas. SAT-solving was shown to be easy for exact linear formulas in [14]. Here we consider the class of formulas where each pair of distinct clauses has all variables in common or exactly one, thus extending exact linear formulas. The members of this class are called *exact linearly-based* formulas. We design an algorithm providing polynomial-time SAT-decidability for this class. Moreover, by a slight modification we are also able to show that the counting variant #-SAT for exact linearly-based formulas belongs to P. Furthermore, we study some other structurally defined formula classes using the fibre view on clause sets, and show that

H. Kleine Büning and X. Zhao (Eds.): SAT 2008, LNCS 4996, pp. 231–245, 2008.

they behave trivially w.r.t. SAT meaning that its members always are satisfiable. There are known several classes, for which SAT can be tested in polynomial time, such as quadratic formulas, (extended and q-)Horn formulas, matching formulas etc. [1,3,4,5,6,8,9,10,16,17]. The classes studied in this paper appear not to belong to one of these classes. On the other hand, mixing polynomial-time classes, in general, yields classes for which SAT becomes NP-complete, as already is the case for Horn and quadratic formulas [13], cf. also [8]. Closely related to some of these classes is the classic theorem of Schaefer in [15]. That theorem classifies satisfiability problems w.r.t. their complexity. The theorem does not automatically apply if restrictions on the number of occurrences of variables in CNF formulas are valid explicitly or implicitly. E.g. in [7] it is shown that whereas unrestricted k-SAT is NP-complete, for $k \geq 3$, it behaves trivially (i.e. all formulas are satisfiable) if each clause has length exactly k and no variable occurs in more than $f(k)$ clauses; it gets NP-complete then if variables are allowed to occur at most $f(k)+1$ times. Here $f(k)$ asymptotically grows as $\lfloor 2^k/(e \cdot k) \rfloor$; this bound has meanwhile been improved by other authors. However, it seems to be unknown whether one can expect a dichotomy result like Schaefer's regarding the occurrence number. The classes studied in the present paper do not meet the requirement that all clauses have a constant equal length, but seem to have a hidden or implicit bound on the maximal number of occurences along with additional structure.

The paper is organized as follows: The next section collects the notation and some preliminaries. Section 3 describes the fibre view on clause sets, and elaborates a subclass only containing unsatisfiable members. Section 4 contains the main part of the paper, namely showing that the decision, search and counting variants of SAT for exact linearly-based formulas can be solved in polynomial time. In Section 5 we apply the fibre view concept on some further CNF classes showing that they behave trivially w.r.t. SAT. Finally, in Section 6 we collect some conclusions and open problems.

2   Notation and Preliminaries

To fix the notation, let CNF denote the set of duplicate-free conjunctive normal form formulas over propositional variables $x \in \{0, 1\}$. A *positive (negative)* literal is a (negated) variable. The *negation (complement)* of a literal l is $\bar{l}$. Each formula $C \in$ CNF is considered as a clause set, and each clause $c \in C$ is represented as a literal set which in addition is assumed to be free of complemented pairs $\{x, \bar{x}\}$. For formula C, clause c, literal l, by $V(C), V(c), V(l)$ we denote the variables contained (neglecting negations), correspondingly. $L(C)$ is the set of all literals in C. The length of C is denoted as $\|C\|$. For $U \subset L(C)$, let $C(U) := \{c \in C | c \cap U \neq \varnothing\}$; we simply write $C(l)$, if $U = \{l\}$. The satisfiability problem (SAT) asks, whether input $C \in$ CNF has a *model*, which is a truth assignment $t : V(C) \to \{0, 1\}$ assigning at least one literal in each clause of C to 1. For $C \in$ CNF, let $M(C)$ be the space of all models of C and let UNSAT $:=$ CNF $-$ SAT. It is convenient to identify truth assignments with $|V|$-clauses in the following simple way: Let $x^0 := \bar{x}$, $x^1 := x$. Then we can identify $t : V \to \{0,1\}$ with the literal set $\{x^{t(x)} | x \in V\}$, and, for $b \subset V$, the *restriction* $t|_b$ is identified with the literal set $\{x^{t(x)} | x \in b\}$. The collection of all literal sets obtained as just described by running through all total truth assignments $V \to \{0,1\}$ is denoted as W_V. We call W_V the *hypercube (hc) formula (over V)*, since its clauses correspond 1:1 to the vertices of a hypercube of dimension $|V|$. E.g., for $V = \{x,y\}$, we have $W_V = \{xy, \bar{x}y, x\bar{y}, \bar{x}\bar{y}\}$ writing clauses as literal strings. For a clause c, we denote by c^γ the clause in which all its literals are complemented. Similarly, let $t^\gamma = 1 - t : V \to \{0,1\} \in W_V$, $C^\gamma := \{c^\gamma | c \in C\}$, and for $\mathcal{C} \subseteq \mathrm{CNF}$, let $\mathcal{C}^\gamma := \{C^\gamma | C \in \mathcal{C}\}$. We call C *asymmetric* if for each $c \in C$ we have $c^\gamma \notin C$. Asym $\subset \mathrm{CNF}$, denotes the set of all asymmetric formulas.

3   The Fibre View on Clause Sets

The *fibre view*, as introduced in [12], regards a clause set C as composed of *fibres* over a hypergraph: All clauses c of C *projecting* onto the same variable set $b = V(c)$, when negations are eliminated, form the *fibre* C_b *over* b, namely $C_b = \{c \in C | V(c) = b\}$. The collection of these *base elements* b forms a hypergraph, the *base hypergraph* $\mathcal{H}(C) = (V(C), B(C))$ of C, where $B(C) = \{b := V(c) | c \in C\}$. Hence, C is the disjoint union of all its fibres: $C = \bigcup_{b \in B(C)} C_b$. Conversely, we can also start with a given arbitrary hypergraph $\mathcal{H} = (V, B)$ serving as a base hypergraph if its vertices $x \in V$ are regarded as Boolean variables such that for each $x \in V$ there is a (hyper)edge $b \in B$ containing x. Recall that, for any $b \in B$, W_b is the hypercube formula over the set of variables in b. Then the set of all clauses over $\mathcal{H}$ is $K_\mathcal{H} := \bigcup_{b \in B} W_b$, also called the *total clause set over $\mathcal{H}$*. W_b is the fibre of $K_\mathcal{H}$ over b. For example, given $V = \{x_1, x_2, x_3\}$, and $B = \{b_1 := x_1, b_2 := x_1x_2, b_3 := x_1x_3\}$, we have $K_\mathcal{H} = W_{b_1} \cup W_{b_2} \cup W_{b_3}$, where $W_{b_1} = \{x_1, \bar{x}_1\}$, $W_{b_2} = \{x_1x_2, \bar{x}_1x_2, x_1\bar{x}_2, \bar{x}_1\bar{x}_2\}$, and $W_{b_3} = \{x_1x_3, \bar{x}_1x_3, x_1\bar{x}_3, \bar{x}_1\bar{x}_3\}$ are the hc formulas over b_1, b_2, and b_3, respectively.

A *formula over $\mathcal{H}$* (or a *$\mathcal{H}$-based*) formula is a subset $C \subseteq K_\mathcal{H}$ such that $C_b := C \cap W_b \neq \varnothing$, for each $b \in B$. Given a $\mathcal{H}$-based formula $C \subseteq K_\mathcal{H}$ with the additional property that $\bar{C}_b := W_b - C_b \neq \varnothing$ holds, for each $b \in B$, then we can define its *$\mathcal{H}$-based complement formula* $\bar{C} := \bigcup_{b \in B} \bar{C}_b = K_\mathcal{H} - C$ with fibres $\bar{C}_b$. For example, given $\mathcal{H} = (V, B)$ with $V = \{x_1, x_2, x_3\}$, and $B = \{x_1x_2, x_1x_3\}$, let $C = \{x_1\bar{x}_2, x_1x_2, x_1\bar{x}_3, \bar{x}_1\bar{x}_3\}$ then $K_\mathcal{H} = C \cup \bar{C}$ where $\bar{C} = \{\bar{x}_1x_2, \bar{x}_1\bar{x}_2, x_1x_3, \bar{x}_1x_3\}$. A *fibre-transversal (f-transversal)* of $K_\mathcal{H}$ (not to be confused with a hitting set) is a $\mathcal{H}$-based formula $F \subset K_\mathcal{H}$ such that $|F \cap W_b| = 1$, for each $b \in B$. Hence F is a formula containing exactly one clause of each fibre W_b of $K_\mathcal{H}$; let that clause be refered to as $F(b)$. For convenience let $\mathcal{F}(K_\mathcal{H})$ be the set of all f-transversals of $K_\mathcal{H}$. An important type of f-transversals F are those containing each variable of V as a pure literal, that is, occurring in F with a single polarity only. Such f-transversals are called *compatible* and have the property that $\bigcup_{b \in B} F(b) \in W_V$. Let $\mathcal{F}_{\mathrm{comp}}(K_\mathcal{H})$ be the collection of all compatible f-transversals of $K_\mathcal{H}$. As a simple example for a compatible f-transversal, consider the base hypergraph $\mathcal{H} = (V, B)$ with variable set $V := \{x_1, x_2, x_3\}$ and $B :=$

$\{b_1 := x_1 x_2,\ b_2 := x_1 x_3,\ b_3 := x_2 x_3\}$. Then, e.g., the clauses $c_1 := \bar{x}_1 x_2 \in W_{b_1}$, $c_2 := \bar{x}_1 \bar{x}_3 \in W_{b_2}$ and $c_3 := x_2 \bar{x}_3 \in W_{b_3}$, denoted as literal strings, form a compatible f-transversal of the corresponding $K_{\mathcal{H}}$, because $c_1 \cup c_2 \cup c_3 = \bar{x}_1 x_2 \bar{x}_3 \in W_V$. In a certain sense orthogonal to compatible f-transversals are the *diagonal* f-transversals. By definition, a diagonal f-transversal F meets each compatible f-transversal F' of $K_{\mathcal{H}}$ in at least one clause; formally: for each $F' \in \mathcal{F}_{\mathrm{comp}}(K_{\mathcal{H}})$ we have $F \cap F' \neq \varnothing$. Let $\mathcal{F}_{\mathrm{diag}}(K_{\mathcal{H}})$ be the collection of all diagonal f-transversals of $K_{\mathcal{H}}$.

As for the total clause set $K_{\mathcal{H}}$ we can define f-transversals for a $\mathcal{H}$-based formula $C \subset K_{\mathcal{H}}$: An f-transversal F of C contains exactly one clause of each fibre C_b of C. The collection of all f-transversals of C is denoted as $\mathcal{F}(C)$. We also define compatible and diagonal f-transversals of C via $\mathcal{F}_{\mathrm{comp}}(C) := \mathcal{F}(C) \cap \mathcal{F}_{\mathrm{comp}}(K_{\mathcal{H}})$, and $\mathcal{F}_{\mathrm{diag}}(C) := \mathcal{F}(C) \cap \mathcal{F}_{\mathrm{diag}}(K_{\mathcal{H}})$.

The following result characterizes satisfiability of a formula C in terms of compatible f-transversals in its based complement formula $\bar{C}$ (cf. [12]):

Theorem 1. *For $\mathcal{H} = (V, B)$, let $C \subset K_{\mathcal{H}}$ be a $\mathcal{H}$-based formula such that $\bar{C}$ is $\mathcal{H}$-based, too. Then C is satisfiable if and only if $\bar{C}$ admits a compatible f-transversal, i.e. $\mathcal{F}_{\mathrm{comp}}(\bar{C}) \neq \varnothing$.*

PROOF. Suppose C is satisfiable and let $t \in W_V$ be one of its models. Then for each base point $b \in B = B(C) = B(\bar{C})$, the restriction $t|_b$ of t to b satisfies all clauses of the fibre W_b of $K_{\mathcal{H}}$ except for the clause $t^\gamma|_b$ obtained from $t|_b$ via complementing all literals. Hence $F_t(b) := t^\gamma|_b$ is a member of $\bar{C}$ and therefore $F_t := \{F_t(b)|b \in B\}$ is a compatible f-transversal of $\bar{C}$ because $\bar{C}$ is $\mathcal{H}$-based and $\bigcup F_t = \bigcup_{b \in B} t^\gamma|_b = t^\gamma$.

Conversely, let F be a compatible f-transversal of $\bar{C}$. Then, by definition of compatibility and because V has no isolated variables $t := \bigcup_{b \in B} F(b) \in W_V$ is a truth assignment. And we claim that via complementing all assignments we obtain a model t^γ of C. Indeed, suppose the contrary, meaning that there is a base point b and a clause c over b belonging to C that is not satisfied by t^γ. Then this clause must have the form $c = t|_b \in C$, but this means a contradiction to $F(b) = t|_b \in \bar{C}$ as F was assumed to be an f-transversal of $\bar{C}$. $\square$

Whereas compatible f-transversals always exist, it is not clear whether diagonal transversals exist in $K_{\mathcal{H}}$. However, if there are diagonal transversals, then each fixed compatible transversal meets all diagonal transversals in $K_{\mathcal{H}}$. We have some more easy observations regarding f-transversals.

Proposition 1. *(1) $\mathcal{F}_{\mathrm{comp}}(K_{\mathcal{H}}) \cong W_V$ (means isomorphism),*
(2) $\mathcal{F}_{\mathrm{diag}}(K_{\mathcal{H}}) = \{F \in \mathcal{F}(K_{\mathcal{H}})|\forall t \in W_V \exists b \in B : F(b) = t|_b\}$,
(3) $\mathcal{F}_{\mathrm{comp}}(K_{\mathcal{H}})^\gamma = \mathcal{F}_{\mathrm{comp}}(K_{\mathcal{H}})$,
(4) $\mathcal{F}_{\mathrm{diag}}(K_{\mathcal{H}})^\gamma = \mathcal{F}_{\mathrm{diag}}(K_{\mathcal{H}})$,
(5) $\mathcal{F}_{\mathrm{diag}}(K_{\mathcal{H}}) \cap \mathcal{F}_{\mathrm{comp}}(K_{\mathcal{H}}) = \varnothing$,
(6) $F \in \mathcal{F}_{\mathrm{diag}}(K_{\mathcal{H}}) \Leftrightarrow F \in \mathrm{UNSAT}$,
(7) $F \in \mathcal{F}_{\mathrm{comp}}(K_{\mathcal{H}}) \Rightarrow F \in \mathrm{SAT}$.

PROOF. Assertion (1) follows from Theorem 1. Assertion (2) says that each diagonal f-transversal meets each clause in W_V and thus follows from (1) immediately.

Assertion (3) is obvious. Let $F \in \mathcal{F}_{\mathrm{diag}}(K_{\mathcal{H}})$ and assume there is $F' \in \mathcal{F}_{\mathrm{comp}}(K_{\mathcal{H}})$ such that $F'(b) \neq F^{\gamma}(b)$, for all $b \in B$, equivalent to $F'^{\gamma}(b) \neq F(b)$, for all $b \in B$, contradicting that F is diagonal, so yielding (4). Assume there is $F \in \mathcal{F}_{\mathrm{diag}}(K_{\mathcal{H}}) \cap \mathcal{F}_{\mathrm{comp}}(K_{\mathcal{H}})$, then with (3) also $F^{\gamma} \in \mathcal{F}_{\mathrm{comp}}(K_{\mathcal{H}})$ holds, but we have $F^{\gamma}(b) \neq F(b)$, for each $b \in B$, therefore $F \notin \mathcal{F}_{\mathrm{diag}}(K_{\mathcal{H}})$ yielding a contradiction implying (5). According to Theorem 1, we have $F \in \mathrm{UNSAT}$ iff $\mathcal{F}_{\mathrm{comp}}(\bar{F}) = \varnothing$ iff, $\forall F' \in \mathcal{F}_{\mathrm{comp}}(K_{\mathcal{H}})$, there is $b \in B$ such that $F'(b) = F(b) \in F$ iff $F \in \mathcal{F}_{\mathrm{diag}}(K_{\mathcal{H}})$, hence (6). (7) is implied by (6) according to (5); moreover for $F \in \mathcal{F}_{\mathrm{comp}}(K_{\mathcal{H}})$, $\bigcup F \in W_V$ specifically satisfies F. $\qquad\square$

Thus, we have three types of f-transversals composing $\mathcal{F}(K_{\mathcal{H}})$, namely compatible f-transversals which are always satisfiable, diagonal ones which (which do not exist in each case but) are always unsatisfiable, and finally, f-transversals that neither are compatible nor diagonal but are always satisfiable.

Definition 1. *A formula $D \subseteq K_{\mathcal{H}}$ is called a* diagonal *formula if, for each $F \in \mathcal{F}_{\mathrm{comp}}(K_{\mathcal{H}})$, $F \cap D \neq \varnothing$ holds.*

Obviously each $F \in \mathcal{F}_{\mathrm{diag}}(K_{\mathcal{H}})$ (if existing) is a diagonal formula. Since a diagonal formula D contains a member of each compatible f-transversal the complement formula $\bar{D}$ cannot have a compatible f-transversal. Therefore $D \in \mathrm{UNSAT}$ according to Theorem 1, and we have:

Proposition 2. *A formula is unsatisfiable iff it contains a diagonal subformula.*

Consider a simple application of the concepts above: Recall that a hypergraph is called *Sperner* if no hyperedge is contained in another hyperedge [2]. Similarly, we call a formula $C \in \mathrm{CNF}$ *simple* if no clause is contained in another one. A non-simple formula C can easily be turned into a SAT-equivalent simple one by removing each clause $c' \in C$ that properly contains another clause. If C is simple its base hypergraph $\mathcal{H}(C) = (V(C), B(C))$ can either be Sperner or non-Sperner. Assuming $\mathcal{H}(C) = \mathcal{H}(\bar{C})$ we have that $\mathcal{H}(C)$ Sperner implies that $\bar{C}$ is simple. The case that $\mathcal{H}(C) = \mathcal{H}(\bar{C})$ is non-Sperner, but both C and $\bar{C}$ are simple is illustrated by the following example (where clauses and edges are represented as strings):

$$C = \{xy, x\bar{y}z, \bar{x}yz, \bar{x}\bar{y}z, x\bar{y}\bar{z}, \bar{x}y\bar{z}, \bar{x}\bar{y}\bar{z}\}$$
$$\bar{C} = \{x\bar{y}, \bar{x}y, \bar{x}\bar{y}, xyz, xy\bar{z}\}$$
$$B(C) = \{xy, xyz\}$$

Theorem 2. *Let $C \in \mathrm{CNF}$ be such that $\mathcal{H}(C) = \mathcal{H}(\bar{C})$ is non-Sperner, but both C and $\bar{C}$ are simple. Then C and $\bar{C}$ are unsatisfiable.*

PROOF. For proving that C is unsatisfiable, it is sufficient to show that $\bar{C}$ cannot have a compatible f-transversal according to Theorem 1. Since $\mathcal{H}(C)$ is non-Sperner there are $b, b' \in B(C)$ with $b \subset b'$. Now, for each f-transversal $F \in \mathcal{F}(\bar{C})$, we have $F(b) \neq F(b')$, and $F(b) \not\subseteq F(b')$ because $\bar{C}$ is assumed to be simple. That means there is $x \in b$ such that $x \in F(b)$, $\bar{x} \in F(b')$ or vice versa, hence $F(b) \cup F(b') \supset \{x, \bar{x}\}$ is not compatible implying that $C \in \mathrm{UNSAT}$. By exchanging the roles of C and $\bar{C}$ we also obtain that $\bar{C}$ is unsatisfiable. $\qquad\square$

Corollary 1. *If C is simple and satisfiable then either*
(i) $\mathcal{H}(C)$ is Sperner and $\bar{C}$ is simple or
(ii) neither $\mathcal{H}(C)$ is Sperner nor $\bar{C}$ is simple and, for each pair $b_1 \subset b_2 \in B(C)$,
there are $c_1 \subset c_2 \in \bar{C}$ such that $V(c_i) = b_i$, $i = 1, 2$.

The criterion in (ii) above is not sufficient for satisfiability of C: Let $b_1 \subset b \in B(C)$ such that $c_1 \subset c \in \bar{C}$ and moreover let $b'_1 \subset b' \in B(C)$ such that $c'_1 \subset c' \in \bar{C}$, where $V(c) = b$, $V(c') = b'$, $V(c_1) = b_1$, and $V(c'_1) = b'_1$. Now assume that $b \cap b' \neq \varnothing$, and that c, c' are the only clauses over b, b' in $\bar{C}$. Clearly, if $c|_{b \cap b'} \neq c'|_{b \cap b'}$ then there is no compatible f-transversal of $\bar{C}$, so C has no model.

4 Formulas over Exact Linear Base Hypergraphs

Returning to the general discussion, let $\mathcal{H} = (V, B)$ be a non-empty base hypergraph, then clearly $\mathcal{F}_{\text{comp}}(K_{\mathcal{H}})$ is non-empty we even have $|\mathcal{F}_{\text{comp}}(K_{\mathcal{H}})| = 2^{|V|}$ according to Prop. 1 (1). However, a priori it is not clear whether in general $\mathcal{F}_{\text{diag}}(K_{\mathcal{H}}) \neq \varnothing$ holds, too. Actually, this depends on the structure of the base hypergraph $\mathcal{H}$. To that end, let us consider an interesting and guiding example regarding satisfiability of certain formulas over an *(exact) linear* base hypergraph $\mathcal{H} = (V, B)$. By definition, $\mathcal{H}$ linear has the property $|b \cap b'| \leq 1$, for all distinct $b, b' \in B$, and the exact linear case is defined replacing $\leq$ with $=$. Recall that a hypergraph $\mathcal{H} = (V, B)$ is called *loopless* iff $|b| \geq 2$, for all $b \in B$. In [14] *(exact) linear* formulas are discussed in more detail. In an (exact) linear formula the variable sets of distinct clauses have at most (resp. exactly) one member in common.

Lemma 1. *[14] Each exact linear formula without unit clauses is satisfiable.*

From the Lemma we conclude that if the base hypergraph $\mathcal{H} = (V, B)$ is exact linear and loopless then for the corresponding total clause set $\mathcal{F}_{\text{diag}}(K_{\mathcal{H}})$ $= \varnothing$ holds. Indeed, no unsatisfiable f-transversal can exist then, because each is exact linear and we are done by contraposing Proposition 1 (6). So indeed there are hypergraphs admitting no diagonal f-transversal. The reverse question, namely are there hypergraphs at all such that the total clause sets have diagonal f-transversals, also is answered positive: Each *unsatisfiable* linear formula obviously is an f-transversal of the total clause set over the underlying linear base hypergraph, so it is diagonal.

Fact 1. *The notion of (diagonal) f-transversals generalizes the notion of (unsatisfiable) linear formulas.*

We now address the class of all CNF formulas over exact linear base hypergraphs, called *exact linearly-based formulas*, for short. It is easy to see that this class corresponds to the class of CNF formulas such that the variable sets of *each pair* of clauses have exactly one or all members in common. Clearly, each exact linear formula also is exact linearly-based. In the following we investigate some aspects of SAT regarding the class of exact linearly-based formulas.

To a $\mathcal{H}$-based formula C, for arbitrary $\mathcal{H}$, we can assign its *fibre graph* $G(C)$ as follows: Each clause c of C corresponds to a vertex. And vertices $c, c' \in C$ form an edge iff (1) they belong to distinct fibres of C, i.e. there are $b, b' \in B$, $b \neq b'$, such that $c \in C_b$, $c' \in C_{b'}$; and (2) $c \cap c' \neq \varnothing$. In terms of the fibre graph we obtain the following characterization of satisfiability in the case of exact linear bases. Recall that a clique is a (sub)graph such that each pair of its vertices are joined by an edge.

Proposition 3. *Let C be exact linearly-based such that $\mathcal{H}(C) = \mathcal{H}(\bar{C}) =: \mathcal{H} = (V, B)$. Then C is satisfiable iff $G(\bar{C})$ admits a clique of size $|B|$.*

PROOF. In view of Theorem 1 we prove that $\bar{C}$ admits a compatible f-transversal if and only if $G(\bar{C})$ has a clique of size $|B|$. Assume $G(\bar{C})$ has such a clique F, then it contains exactly one member of each of the fibres of $\bar{C}$, hence the clauses corresponding to the vertices in F form an f-transversal of $\bar{C}$, also denoted as F. Suppose that F is not a compatible f-transversal. Then the union of all clauses in F contains a variable x and its negation $\bar{x}$, i.e. a complemented pair. As by definition no clause of C has a complemented pair, there are two clauses c and c' in F such that $x \in c$ and $\bar{x} \in c'$. Since $\mathcal{H}$ is linear and F is an f-transversal $V(c)$ and $V(c')$ do not have another variable in common than x meaning $c \cap c' = \varnothing$. Because c and c' form an edge in $G(\bar{C})$ we have $c \cap c' \neq \varnothing$ yielding a contradiction, so F is a compatible f-transversal.

Conversely, assume that $\bar{C}$ has a compatible f-transversal F. Then it has size $|B|$ because $\bar{C}$ is a $\mathcal{H}$-based formula by assumption. Suppose there are two members c, c' in F whose vertices do not form an edge in $G(\bar{C})$. Then $c \cap c' = \varnothing$ implying $V(c) \cap V(c') = \varnothing$ contradicting exact linearity of $\mathcal{H}$. In conclusion, the members of F correspond to a $|B|$-clique in $G(\bar{C})$.   □

Note that if $\bar{C}$ is not a $\mathcal{H}$-based formula then C is unsatisfiable trivially because it contains a complete W_b as fibre subformula. Moreover notice that the proof above is not valid for arbitrary base hypergraphs. The next observation concerns rather specific exact linearly-based formulas:

Lemma 2. *Let $\mathcal{H} = (V, B)$ be an exact linear base hypergraph such that there is a vertex $x \in V$ occuring in each $b \in B$. Let $C \subset K_{\mathcal{H}}$ be a $\mathcal{H}$-based formula such that $\bar{C}$ also is $\mathcal{H}$-based. Then we have: $C \in \mathrm{UNSAT}$ if and only if*

$$(*) \quad |\{b \in B | \forall c \in \bar{C}_b : x \in c\}| > 0 \quad and \quad |\{b \in B | \forall c \in \bar{C}_b : \bar{x} \in c\}| > 0$$

Moreover, assuming that C is given in terms of its fibre subformulas we can check it for satisfiability in linear time $O(\|C\|)$.

PROOF. In view of Prop. 3 it is sufficient to show that $G(\bar{C})$ admits no $|B|$-clique iff $(*)$ in the assertion is true. But this is obvious because $(*)$ holds for $\bar{C}$ iff there is a fibre subformula $\bar{C}_b$ containing x as pure literal, and (at least) one other $\bar{C}_{b'}$ containing $\bar{x}$ as pure literal. Hence no clause contained in $\bar{C}_b$ can be joined in $G(\bar{C})$ to a clause in $\bar{C}_{b'}$ and we proved the first assertion.

For verifying the second assertion, we describe a simple algorithm checking $(*)$, for $\bar{C}$, via inspecting C in linear time. To that end initialize two counters

$N_x \leftarrow 0$ and $N_{\bar{x}} \leftarrow 0$. Then check each fibre subformula $C_b \subset C$, $b \in B$, whether it contains exactly $2^{|b|-1}$ clauses containing literal x, respectively literal $\bar{x}$. If the first holds true then increase counter $N_{\bar{x}}$ by one; if the second is true then increase counter N_x by one; otherwise do not modify the counters, respectively. Proceed in this way until either $N_{\bar{x}} > 0$, and $N_x > 0$ then stop, and return $C \in$ UNSAT. Or the whole formula has been inspected and $N_{\bar{x}} = 0 = N_x$ then return $C \in$ SAT. The algorithm works correctly, as it is easy to see that $(*)$ holds iff $N_{\bar{x}} > 0$ and $N_x > 0$. $\qquad \square$

The next algorithm, instead of searching $G(\bar{C})$, works via inspecting C itself:

Theorem 3. *For C exact linearly-based and represented in terms of its fibre subformulas,* SAT *can be decided in polynomial time.*

PROOF. For C exact linearly-based, by definition, there is an exact linear base hypergraph $\mathcal{H} = (V, B)$ such that $C \subseteq K_{\mathcal{H}}$. If $\mathcal{H}$ is not loopless, we are done by Lemma 2. If C is exact linear and loopless, we are done according to Lemma 1 in linear time, by inspection of the input. If $\bar{C}$ is not $\mathcal{H}$-based then C cannot be satisfiable because it contains at least one complete fibre W_b of $K_{\mathcal{H}}$. So assume that $\bar{C}$ is $\mathcal{H}$-based. The basic idea is as follows: We say that a clause c of the input formula C over n variables *meets* a truth assignment $t \in W_V$ iff $c \subseteq t$. A single clause c obviously meets exactly $2^{n-|c|}$ truth assignments. Suppose we are able to calculate fast the total number N of all distinct truth assignments met by all clauses of the input formula C. Then we have $C \in$ SAT iff $N < 2^n$. Indeed, only in that case C does not contain a diagonal formula since the $\mathcal{H}$-based complement formula $\bar{C}$ of C admits a compatible f-transversal corresponding to a truth assignment not met by any $c \in C$ according to Prop. 1. So, we are done refering to Theorem 1.

To that end, recall that $C_b \subset W_b$ denotes the fibre subformula of C over $b \in B$. Let l be an arbitrary literal occurring in the input formula C, and recall that $C_b(l) \subset C_b$ is the subformula of C_b of all clauses containing literal l, where $V(l) \in b$. The clauses in $C_b(l)$ in total meet exactly $\mu(l, b) := |C_b(l)| \cdot 2^{n-|b|}$ of all 2^{n-1} truth assignments containing l. The determination of the number of truth assignments met by the clauses in the input formula C is organized as follows: First we compute a variable x called a *maximum variable* and the corresponding edges $b_x, b_{\bar{x}} \in B$ (smallest indices if ambiguous) satisfying

$$\mu(x, b_x) + \mu(\bar{x}, b_{\bar{x}}) = \max\{\mu(y, b) + \mu(\bar{y}, b')| y \in V, b, b' \in B\}$$

Note that $|C_b(l)| = 0$ if $l \notin L(C_b)$, specifically if $V(l) \notin b$ meaning $b \notin B(x)$. (Here, we have $B(x) = \{b \in B | x \in b\}$ regarding B as a positive monotone clause set.) It is possible that $b_x = b_{\bar{x}}$. Next we perform at most two independent runs of a Procedure ComputeCoverNumber(l, p). The first one for $l = x$ and a second one for $l = \bar{x}$; returning in p the number of all l-containing truth assignments, that are met by the clauses of C containing l. Each of these executions of Procedure ComputeCoverNumber(l, p) is initiated only if $\mu(l, b_l) < 2^{n-1}$ meaning that the fibre subformulas corresponding to the maximum variable x do not meet all 2^{n-1} possible l-containing truth assignments. Clearly, the runs of the procedure for x and $\bar{x}$ can be processed independently. Finally, the numbers of truth assignments

met that are returned in p are added, and the algorithm returns `unsatisfiable`
iff the total value equals 2^n.

Procedure `ComputeCoverNumber`(l, p) consists of two subprocedures. The first
one is entered only if there is at least one fibre subformula C_b with $x \in b$ besides
C_{b_l} in which literal l occurs. The subprocedure then computes all additional l-
containing truth assignments met by the clauses in all these fibre subformulas. The
second subprocedure is devoted to determine all further truth assignments containing l met by the remaining fibre subformulas C_b with $x = V(l) \notin b$. Similarly,
it is entered only if there is at least one fibre subformula C_b with $b \in B - B(x)$.

The first subprocedure of `ComputeCoverNumber` proceeds as follows: W.l.o.g.
let $\mathcal{C}_l := \{C_{b_1}, C_{b_2}, \ldots, C_{b_s}\}$, for $s \geq 1$, be the collection of all remaining fibre
subformulas with $x = V(l) \in b_i$, and $|C_{b_i}(l)| > 0$, for all $1 \leq i \leq s$, where
$b_1 := b_l$. Obviously, a fibre formula C_b with $b \in B(x)$ but $C_b(l) = \varnothing$ cannot
contribute to the set of met l-containing truth assignments. Let $m_j := |C_{b_j}(l)|$ and
$m'_j := |W_{b_j}(l) - C_{b_j}(l)| = 2^{|b_j|-1} - m_j$, for $1 \leq j \leq s$. Note that by assumption
we have $m'_j > 0$, for $1 \leq j \leq s$. Now we claim that the number α_l of l-containing
truth assignments met by the clauses of the subformulas in $\mathcal{C}_l$ is given by:

$$(*) \quad \alpha_l(s) := \sum_{j=1}^{s} \left[m_j \cdot 2^{n+(j-1)-\sum_{q=1}^{j} |b_q|} \cdot \prod_{k=1}^{j-1} m'_k \right]$$

where as usual $\prod_{k=1}^{j-1} m'_k := 1$ if $j = 1$. We prove the claim by induction on
$s := |\mathcal{C}_l| \geq 1$. If $s = 1$ then clearly $\alpha_l(1) = \mu(l, b_l)$ and $(*)$ can easily be verified
to be correct. So, let $s \geq 1$ and assume that $(*)$ is true for all values not greater
than s. Let $\mathcal{C}_l$ have cardinality $s + 1$, then the first s members of $\mathcal{C}_l$ meet $\alpha_l(s)$
distinct l-containing truth assignments by the induction hypothesis.

All clauses in $C_{b_{s+1}}(l)$ contain literal l and literals over the same $|b_{s+1}| - 1$
variables that do not occur in any other b_j, $1 \leq j \leq s$, because $\mathcal{H}$ is exact
linear. Let $\Delta_l(s + 1)$ be the number of additional l-containing truth assignments
met by the clauses in $C_{b_{s+1}}(l)$ but not by those in $\mathcal{C}_l - \{C_{b_{s+1}}(l)\}$. Further, each
clause of $C_{b_{s+1}}(l)$ contributes the same number R of additional l-containing truth
assignments. This is true because these clauses meet pairwise distinct parts over
the range $b_{s+1} - \{x\}$ of truth assignments not already met by the clauses in
$\mathcal{C}_l - \{C_{b_{s+1}}(l)\}$; we thus have $\Delta_l(s + 1) = m_{s+1} \cdot R$.

To determine R we need the number of all l-containing truth assignments
not already met by the clauses in $\mathcal{C}_l - \{C_{b_{s+1}}(l)\}$. Consider arbitrary clauses
$c_i \in W_{b_i}(l) - C_{b_i}(l)$, $1 \leq i \leq s$, i.e. the complements of the fibre subformulas
in $\mathcal{C}_l - \{C_{b_{s+1}}(l)\}$. The union $d := \bigcup_{i=1}^{s} c_i$ yields a literal set of cardinality
$|d| = 1 + \sum_{i=1}^{s} (|b_i| - 1)$ which clearly cannot be contained in any of the l-
containing truth assignments met by the clauses in $\mathcal{C}_l - \{C_{b_{s+1}}(l)\}$. Hence a
clause $c \in C_{b_{s+1}}(l)$ meets each l-containing truth assignment enlarging a literal
string d as constructed above with c. Of such a truth assignment consequently
then $r := 1 + \sum_{i=1}^{s} (|b_i| - 1) + |b_{s+1}| - 1$ positions are already fixed yielding 2^{n-r}
distinct such truth assignments containing that literal string d and c. Clearly,
we can construct $m'_1 \cdot m'_2 \cdots m'_s$ distinct literal strings d as above. Each yielding

2^{n-r} truth assignments met by each fixed clause of $C_{b_{s+1}}(l)$. So we obtain $R = m'_1 \cdot m'_2 \cdots m'_s \cdot 2^{n-r}$ and therefore

$$\Delta_l(s+1) = m_{s+1} \cdot 2^{n+s-\sum_{q=1}^{s+1}|b_q|} \cdot \prod_{k=1}^{s} m'_k$$

additional l-containing truth assignments. Now we conclude by induction

$$\alpha_l(s+1) = \Delta_l(s+1) + \alpha_l(s)$$

$$= m_{s+1} \cdot 2^{n+s-\sum_{q=1}^{s+1}|b_q|} \cdot \prod_{k=1}^{s} m'_k +$$

$$+ \sum_{j=1}^{s} \left[m_j \cdot 2^{n+(j-1)-\sum_{q=1}^{j}|b_q|} \cdot \prod_{k=1}^{j-1} m'_k \right]$$

$$= \sum_{j=1}^{s+1} \left[m_j \cdot 2^{n+(j-1)-\sum_{q=1}^{j}|b_q|} \cdot \prod_{k=1}^{j-1} m'_k \right]$$

in hamony with $(*)$, for $s + 1$, finishing the proof of the claim.

Clearly, number α_l can be determined performing a simple loop recalling that by assumption $m_j > 0$, for all $1 \le j \le s$:

$z \leftarrow m_1 \cdot 2^{n-|b_1|}$

$p \leftarrow z$

for $j = 1$ **to** $s - 1$ **do**

    $z \leftarrow z \cdot m'_j \cdot \frac{m_{j+1}}{m_j} \cdot 2^{1-|b_{j+1}|}$

    $p \leftarrow p + z$

do

So finally, we have to check whether $p = 2^{n-1}$. In order to avoid calculations with possibly large number 2^n it is sufficient instead to compute $p' := p/2^n$. That means to start with $z \leftarrow m_1 \cdot 2^{-|b_1|}$ and finally to check whether $p' = 1/2$.

Recall that the second subprocedure of ComputeCoverNumber(p, l) is devoted to determine all additional l-containing truth assignments met by all clauses in fibre subformulas C_b with $b \in B - B(x)$ meaning $x = V(l) \notin b$. Clearly this subprocedure needs to be started only if $p' < 1/2$, because otherwise all l-containing truth assignments are met already. For explaining the second subprocedure, let c' be a clause of a fibre subformula $C_{b'}$ with $b' \in B - B(x)$. We claim that c' meets a *not yet encountered* l-containing truth assignment if and only if for each $b \in B(x)$ there is $c \in W_b(l) - C_b(l)$ with $c' \cap c \neq \emptyset$. To prove the claim, recall that by assumption we have $m'_b = |W_b(l) - C_b(l)| > 0$ for all $b \in B(x)$ since otherwise $\mu(l, b_l) = 2^{n-1}$ and Procedure ComputeCoverNumber(p, l) would not have been entered at all. So, there always is at least one selection $S := \{c \in W_b(l) - C_b(l)| b \in B(x)\}$. As above we build the literal string $d = \bigcup S$ satisfying $|d| = 1 + \sum_{b \in B(x)}(|b| - 1)$ and $|V(c') \cap V(d)| = |S|$, for the chosen clause c', because of exact linearity. None of the clauses in C_l can meet any l-containing truth assignment enlarging d. Clearly c' meets such a truth assignment iff $|c' \cap d| = |S|$ which is equivalent to $c' \cap c \neq \emptyset$ for all $c \in S$ proving the claim.

W.l.o.g. let $\mathcal{C} := \{C_{b_{s+1}}, \ldots, C_{b_{s+r}}\}$ with $r := |\mathcal{C}| \geq 1$, be the collection of all fibre subformulas of C neither containing x nor $\bar{x}$. For $c' \in C_{b_{s+1}}$, let $\{y_i\} = V(c') \cap b_i$ which, for all $1 \leq i \leq s$, are uniquely determined and pairwise distinct because of exact linearity, hence $|c'| \geq s$. Assume that $l_i \in c'$ is the corresponding literal with $V(l_i) = y_i \neq x$. Let $n_l := \sum_{q=1}^{s} |b_q| - (s-1)$ be the number of distinct variables in $V(\mathcal{C}_l)$. Let $\lambda_i(c')$ be the number of clauses in $W_{b_i}(l) - C_{b_i}(l)$ containing literal l_i, $1 \leq i \leq s$. Clearly, l_i is contained in exactly $2^{|b_i|-2}$ clauses of $W_{b_i}(l)$. So, if l_i occurs t_i times in $C_{b_i}(l)$ we have $\lambda_i(c') = 2^{|b_i|-2} - t_i$, for $1 \leq i \leq s$. Since each clause in $C_{b_{s+1}}$ already fixes $n_l + (|b_{s+1}| - s)$ of n positions, we conclude with the claim above that all clauses in $C_{b_{s+1}}$ together meet exactly

$$2^{n - n_l - (|b_{s+1}| - s)} \sum_{c \in C_{b_{s+1}}} \prod_{j=1}^{s} \lambda_j(c)$$

additional l-containing truth assignments.

On that basis we obtain via induction on $r = |\mathcal{C}| \geq 1$ analogous to the argumentation for the first subprocedure, that all members of $\mathcal{C}$ together meet

$$\sum_{k=1}^{r} \left[2^{n - n_l - \sum_{j=1}^{k} f(j)} \sum_{c \in C_{b_{s+k}}} \left(\prod_{j=1}^{s+k-1} \lambda_j(c) \right) \right]$$

many additional distinct l-containing truth assignments. Here, for $1 \leq j \leq r$,

$$f(j) := |b_{s+j}| - \left| \bigcup_{i=1}^{s+j-1} (b_{s+j} \cap b_i) \right| \in \{0, \ldots, |b_{s+j}| - s\}$$

is the number of variables remaining from b_{s+j} if each variable contained in one of $\{b_1, \ldots, b_{s+j-1}\}$ is removed. Hence, $n_l + \sum_{j=1}^{k} f(j) = |V(\{b_1, \ldots, b_{s+k}\})|$.

Since the formula is represented through its fibres $C_b, b \in B$, and $|B| \leq V(C)$ because of exact linearity [14], all needed values can be collected in polynomial time. □

The method above can easily be adapted to solve the counting problem #-SAT for exact linearly-based formulas in polynomial time: execute both subprocedures yielding the counts N_1, N_2 respectively, then return $2^n - (N_1 + N_2)$. Further, a poly-time algorithm for the search problem can be provided by self-reduction: Iteratively set a variable and check by the algorithm above, whether the resulting formula remains satisfiable. In the negative case, fix the selected variable complementary, etc. Though our class is not stable under partial assignments, the algorithm above still works since the underlying hypergraph shrinks.

Moreover notice that checking whether the subformula $\{C_b | b \in B(x)\}$ of C already is unsatisfiable could be done fast according to Lemma 2, instead of running the first subprocedure twice. However if that does not yield unsatisfiability, in general we cannot simply proceed with the second subprocedure because we do not know how many compatible f-transversals have been met by the clauses in the fibre subformulas over $B(x)$.

5 The Fibre View Further Exploited

Next we investigate some other formula classes using the fibre view concept, besides exact linearly-based formulas. For $\mathcal{H} = (V, B)$, let $C \subset K_{\mathcal{H}}$ such that $B(C) = B = B(\bar{C})$. If $C \in$ SAT then according to Prop. 1 (1) for each $t \in M(C)$ there is a unique $F \in \mathcal{F}_{\text{comp}}(K_{\mathcal{H}})$ with $\bigcup F = t$. We now address the question in which case each model t of C corresponds to an $F \in \mathcal{F}_{\text{comp}}(C)$, i.e. corresponds to a compatible f-transversal of the *formula itself*.

Lemma 3. *If $C \in$ CNF $\cap$ SAT, $B(C) = B(\bar{C})$, such that for each $t \in M(C)$ there is $F \in \mathcal{F}_{\text{comp}}(C)$ with $\bigcup F = t$. Then $\bar{C} \in$ SAT and $\bigcup F' \in M(\bar{C})$ for each $F' \in \mathcal{F}_{\text{comp}}(\bar{C})$; and vice versa.*

PROOF. Let $F \in \mathcal{F}_{\text{comp}}(\bar{C})$ and $t := \bigcup F \in W_V$, then according to the proof of Theorem 1, t^γ is a model of C. By assumption there is $F' \in \mathcal{F}_{\text{comp}}(C)$ such that $t^\gamma = \bigcup F'$. Hence, again by Theorem 1, t is a model of $\bar{C}$ as claimed, specifically $\bar{C} \in$ SAT. For the vice versa assertion exchange the roles of C and $\bar{C}$.   □

Next we provide a formula class satisfying the assumption of the last lemma. Recall that an asymmetric formula C has the property that $c \in C$ implies $c^\gamma \notin C$ and that Asym $\subseteq$ CNF denotes the class of all asymmetric formulas.

Lemma 4. *Let $C \in$ CNF be such that $B(C) = B(\bar{C})$ and $\bar{C} \in$ Asym. Then $C \in$ SAT implies $\bar{C} \in$ SAT and each $t \in M(C)$ corresponds to a compatible f-transversal in $\mathcal{F}_{\text{comp}}(C)$; and vice versa.*

PROOF. Let $t \in M(C) \neq \varnothing$ then for each $b \in B(C)$ $t|_b$ satisfies all of W_b except for $(t|_b)^\gamma$ which thus must be a clause of $\bar{C}$. And $\bar{C} \in$ Asym implies that $t|_b \in C$ for each $b \in B(C)$. Hence $\{t|_b | b \in B(C)\}$ is a compatible f-transversal of C. It follows that $\bar{C} \in$ SAT and that for each $t \in M(C)$ there is $F \in \mathcal{F}_{\text{comp}}(C)$ with $\bigcup F = t$. For the vice versa assertion exchange the roles of C and $\bar{C}$.   □

Corollary 2. *Let $C \in$ Asym such that also $\bar{C} \in$ Asym and $B(C) = B(\bar{C})$. Then $C \in$ SAT if and only if $\bar{C} \in$ SAT.*   □

In view of Theorem 1 we know that a formula is satisfiable if and only if the complement formula admits a compatible f-transversal. Therefore the specific class of formulas C such that every f-transversal of C is compatible is of interest, because any f-transversal gives rise to a model of $\bar{C}$ then, and vice versa. A characterization of that very specific class can be provided as follows: Let $\mathcal{H} = (V, B)$ be a base hypergraph. Then in general there is a 2-partition of V given by sets V_I and V_U. Here V_U contains all vertices occurring in only one edge $b \in B$, and V_I contains all remaining vertices occurring in at least one edge intersection.

Lemma 5. *Let $C \in$ CNF such that $\mathcal{H}(C) = \mathcal{H}(\bar{C}) =: \mathcal{H} := (V, B)$. Then $\mathcal{F}_{\mathrm{comp}}(C) = \mathcal{F}(C)$ iff $(*)$: each variable in $V_I \subseteq V$ is a pure literal in C.*

PROOF. Assume that $(*)$ is true. Then each $y \in b \cap b'$ for all $b, b' \in B, b \neq b'$ is a pure literal in C and therefore each $F \in \mathcal{F}(C)$ is compatible. Conversely, $F \in \mathcal{F}(C)$ is compatible iff all variables are pure literals in F. Hence if each $F \in \mathcal{F}(C)$ is compatible it is easy to see that all variables occurring in edge intersections must be pure literals in C. □

Let $\mathrm{CNF}_{\mathrm{comp}}$ denote the class of all formulas $C \in$ CNF with $B(C) = B(\bar{C})$, and such that $\mathcal{F}(\bar{C}) = \mathcal{F}_{\mathrm{comp}}(\bar{C})$. As an example for $C \in \mathrm{CNF}_{\mathrm{comp}}$, let $\mathcal{H} = (V, B)$ with $V = \{q, r, s, t, u, v, x, y\}$, $B = \{b_1 = xy, b_2 = yuv, b_3 = vxr, b_4 = rst, b_5 = txq\}$, then the following formula $\bar{C}$

$$\bar{C} = x\bar{y} \quad \bar{y}uv \quad vxr \quad rs\bar{t} \quad \bar{t}xq$$
$$\bar{y}\bar{u}v \qquad\qquad r\bar{s}\bar{t} \quad \bar{t}x\bar{q}$$

where clauses are arranged fibrewise obviously has the property that $\mathcal{F}(\bar{C}) = \mathcal{F}_{\mathrm{comp}}(\bar{C})$, hence $C = K_{\mathcal{H}} - \bar{C} \in \mathrm{CNF}_{\mathrm{comp}}$ implying $C \in$ SAT.

Theorem 4. *We can check in polynomial time whether an input formula $C \in$ CNF belongs to $\mathrm{CNF}_{\mathrm{comp}}$. In the positive case a model can be provided in polynomial time assuming for both that C is represented through its fibre subformulas.*

PROOF. Let $\mathcal{H} = (V, B) = \mathcal{H}(C) = \mathcal{H}(\bar{C})$. In linear time we check whether C contains a whole W_b as fibre subformula, in which case $C \notin \mathrm{CNF}_{\mathrm{comp}}$. Otherwise, in view of Lemma 5 we have to check whether each variable in V_I is a pure literal in $\bar{C}$. First, we check in B which variables occur uniquely yielding V_U and set $V_I = V - V_U$. For each $y \in V_I$ we check whether it occurs in different polarities in $\bar{C}[B(y)] := \{\bar{C}_b | b \in B(y)\}$ simultaneously. Recall that $B(y) = \{b \in B | y \in b\}$. This test can be performed in linear time $O(\|C[B(y)]\|)$ similar to the procedure presented in the second part of the proof of Lemma 2: Initialize two counters $N_y \leftarrow 0$ and $N_{\bar{y}} \leftarrow 0$. Then check each fibre subformula C_b, $b \in B(y)$, whether it contains exactly $2^{|b|-1}$ clauses containing literal y, respectively literal $\bar{y}$. If the first holds true then increase counter $N_{\bar{y}}$ by one; if the second is true then increase counter N_y by one; otherwise do not modify the counters, respectively. Proceed in this way until either $N_{\bar{x}} > 0$ and $N_x > 0$ then stop, and return $C \notin \mathrm{CNF}_{\mathrm{comp}}$, because y occurs in both polarities (in distinct fibre subformulas). Or the whole formula has been inspected. Then we set a flag to $C \in \mathrm{CNF}_{\mathrm{comp}}$ if and only if $(*)$: $N_{\bar{y}} = 0$ and $N_y = |B(y)|$, or vice versa. The test works correctly, as it is easy to see that $(*)$ holds iff the clauses in the fibre subformulas in $\bar{C}[B(y)]$ either all contain y or $\bar{y}$. If in that way the algorithm does not return $C \notin \mathrm{CNF}_{\mathrm{comp}}$ we have checked all $y \in V_I$, and we know that $C \in \mathrm{CNF}_{\mathrm{comp}}$.

   Finally, to provide a model of $C \in \mathrm{CNF}_{\mathrm{comp}}$ we simply need to select one clause $c_b \in \bar{C}_b = W_b - C_b$ for each $b \in B$ ensuring that $\bigcup_{b \in B} c_b^\gamma \in M(C)$ according to Theorem 1. For fixed $b = \{b_{i_1}, \ldots, b_{i_{|b|}}\}$, the selection can be performed, e.g., by ordering the members $c = \{b_{i_1}^{\varepsilon_{i_1}(c)}, \ldots, b_{i_{|b|}}^{\varepsilon_{i_{|b|}}(c)}\}$ in C_b according to the lexicographic order of the vectors $(\varepsilon_{i_1}(c), \ldots, \varepsilon_{i_{|b|}}(c)) \in \{0,1\}^{|b|}$, and taking clause in the first gap w.r.t. all of W_b, for each $b \in B$. $\square$

6   Concluding Remarks and Open Problems

We provided several CNF subclasses that are polynomial time decidable resp. behave trivially regarding SAT using the fibre view on clause sets. The most interesting are exact linearly-based formulas. We showed that the decision, search and counting variants of SAT restricted to this class all are polynomial time solvable. We leave it as an open problem to design a more direct algorithm for the decision, resp. search variants of SAT for exact linearly-based formulas. A future work perspective is to elaborate more deeply the relationships between the polynomial time classes studied here to other such classes. Finally, it might be of interest to relate Theorem 1 to the characterization of satisfiability of CNF formulas in the framework of binary decision diagrams.

Acknowledgement. We would like to thank the anonymous reviewers for their valuable comments.

References

1. Aspvall, B., Plass, M.R., Tarjan, R.E.: A linear-time algorithm for testing the truth of certain quantified Boolean formulas. Inform. Process. Lett. 8, 121–123 (1979)
2. Berge, C.: Hypergraphs. North-Holland, Amsterdam (1989)
3. Boros, E., Crama, Y., Hammer, P.L.: Polynomial time inference of all valid implications for Horn and related formulae. Annals Math. Artif. Int. 1, 21–32 (1990)
4. Boros, E., Hammer, P.L., Sun, X.: Recognition of q-Horn formulae in linear time. Discrete Appl. Math. 55, 1–13 (1994)
5. Franco, J., Gelder, A.v.: A perspective on certain polynomial-time solvable classes of satisfiability. Discrete Appl. Math. 125, 177–214 (2003)
6. Kleine Büning, H., Lettman, T.: Propositional logic, deduction and algorithms. Cambridge University Press, Cambridge (1999)
7. Kratochvil, J., Savicky, P., Tusa, Z.: One more occurrence of variables makes satisfiability jump from trivial to NP-complete. SIAM J. Comput. 22, 203–210 (1993)
8. Knuth, D.E.: Nested satisfiability. Acta Informatica 28, 1–6 (1990)
9. Lewis, H.R.: Renaming a Set of Clauses as a Horn Set. J. ACM 25, 134–135 (1978)
10. Minoux, M.: LTUR: A Simplified Linear-Time Unit Resolution Algorithm for Horn Formulae and Computer Implementation. Inform. Process. Lett. 29, 1–12 (1988)
11. Monien, B., Speckenmeyer, E.: Solving satisfiability in less than 2^n steps. Discrete Appl. Math. 10, 287–295 (1985)
12. Porschen, S.: A CNF Formula Hierarchy over the Hypercube. In: Orgun, M.A., Thornton, J. (eds.) AI 2007. LNCS (LNAI), vol. 4830, pp. 234–243. Springer, Heidelberg (2007)
13. Porschen, S., Speckenmeyer, E.: Satisfiability of Mixed Horn Formulas. Discrete Appl. Math. 155, 1408–1419 (2007)
14. Porschen, S., Speckenmeyer, E., Randerath, B.: On linear CNF formulas. In: Biere, A., Gomes, C.P. (eds.) SAT 2006. LNCS, vol. 4121, pp. 221–225. Springer, Heidelberg (2006)

15. Schaefer, T.J.: The complexity of satisfiability problems. In: Proc. STOC 1978. ACM, pp. 216–226 (1978)
16. Schlipf, J., Annexstein, F.S., Franco, J., Swaminathan, R.P.: On finding solutions for extended Horn formulas. Inform. Process. Lett. 54, 133–137 (1995)
17. Tovey, C.A.: A Simplified NP-Complete Satisfiability Problem. Discrete Appl. Math. 8, 85–89 (1984)

How Many Conflicts Does It Need to Be Unsatisfiable?

Dominik Scheder and Philipp Zumstein*

Institute of Theoretical Computer Science, ETH Zürich
8092 Zürich, Switzerland
dscheder@inf.ethz.ch, zuphilip@inf.ethz.ch

Abstract. A pair of clauses in a CNF formula constitutes a conflict if there is a variable that occurs positively in one clause and negatively in the other. Clearly, a CNF formula has to have conflicts in order to be unsatisfiable—in fact, there have to be many conflicts, and it is the goal of this paper to quantify how many.

An unsatisfiable k-CNF has at least 2^k clauses; a lower bound of 2^k for the number of conflicts follows easily. We improve on this trivial bound by showing that an unsatisfiable k-CNF formula requires $\Omega(2.32^k)$ conflicts. On the other hand there exist unsatisfiable k-CNF formulas with $O(\frac{4^k \log^3 k}{k})$ conflicts. This improves the simple bound $O(4^k)$ arising from the unsatisfiable k-CNF formula with the minimum number of clauses.

Keywords: satisfiability, unsatisfiable formulas, conflict graph, Lovász Local Lemma.

1 Introduction

If you want to explain to your non-computer science friend what satisfiability of CNF formulas is all about, you will probably say it is about a list of constraints, all of which you want to satisfy simultaneously. Perhaps you will add that while each constraint is very easy to satisfy individually, the difficulty arises because many constraints conflict with each other. A natural guess is that if you cannot satisfy all your constraints, then there must be a lot of conflicts between them.

In our case, constraints are boolean clauses with no repetition of literals and no complementary literals, e.g., $x \vee \bar{y} \vee z$. A k-CNF formula is the conjunction of such clauses each containing *exactly* k literals. We use this notation for CNF formulas to be closer to the semantical viewpoint of formulas instead of using the notation of clause-sets which is closer to the syntactical viewpoint.

Two clauses have a *conflict* if one contains a positive literal, while the other contains its negation. Kullmann [1] introduced the notion of the symmetric conflict matrix, which has an entry for each pair of clauses counting the number of conflicts between them. We take the 0-1-version of this matrix interpreted as a graph. More formally, the *conflict graph* $CG(F)$ of a k-CNF formula F contains

---

* Research is supported by the SNF Grant 200021-118001/1.

H. Kleine Büning and X. Zhao (Eds.): SAT 2008, LNCS 4996, pp. 246–256, 2008.

the clauses as vertices, and two clauses are connected if there is a conflict between them. Any lower bound on the number of edges in the conflict graph, i.e., the number of conflicts, is also a lower bound on the number of multi-edges in the symmetric conflict matrix interpreted as a multi-graph.

We study the extremal values of several natural parameters of $CG(F)$ for unsatisfiable k-CNF formulas F, such as its minimum degree, maximum degree, and of course, above all, its number of edges. We will use the notations $d_{\min}, d_{\max}, e$, respectively, for these parameters and use concepts of graph theory also in the context of CNF formulas, e.g., the neighborhood of a clause C in a CNF formula F, denoted by $\Gamma_F(C)$, is the set of all clauses in F conflicting with C. To avoid notational confusion we use the e for the number of edges and $\mathbf{e}$ for the Eulerian constant.

Further, we introduce a notation that will come handy when defining CNF formulas: Let $F = C_1 \wedge \ldots \wedge C_m$ and $G = D_1 \wedge \ldots \wedge D_n$ be two CNF formulas over disjoint sets of variables. Define

$$F \veebar G := \bigwedge_{\substack{i=1,\ldots,m \\ j=1,\ldots,n}} C_i \vee D_j \, .$$

The formula $F \veebar G$ is a CNF formula with mn clauses. If F is a k-CNF formula and G an ℓ-CNF formula, $F \veebar G$ is a $(k + \ell)$-CNF formula. Moreover, by using distributivity it is easy to see that the two formulas $F \vee G$ and $F \veebar G$ are equivalent (describing the same boolean function) but $F \vee G$ is syntactically not a CNF formula. Note that a CNF formula is just *one* representation of a boolean function and there might exists other representations (logically equivalent formulas). The problem of deciding whether a given CNF formula is unsatisfiable is the same as to answer the question whether it is logically equivalent to the constant 0 function. Thus it is essential how a formula is represented.

Example 1. The *complete formula* K_k on the variables $x_1, \ldots, x_k$ is the k-CNF formula with all 2^k possible k-clauses on these variables. More formally

$$K_1 := x_1 \wedge \bar{x}_1, \quad K_{k+1} := (x_{k+1} \veebar K_k) \wedge (\bar{x}_{k+1} \veebar K_k) \, .$$

K_k is a k-CNF formula, and using induction and the fact that $\veebar$ is logically equivalent to $\vee$, one can easily see that it is unsatisfiable. Furthermore the conflict graph $CG(K_k)$ is a clique, has 2^k vertices and $\binom{2^k}{2}$ edges.

It is clear that an unsatisfiable 1-CNF formula contains two complimentary clauses and thereby at least one edge. After deleting all clauses of degree 0 in an unsatisfiable k-CNF formula, the so-obtained formula stays unsatisfiable. Therefore, we have at least 2^k clauses because this is the minimum number of clauses needed for a k-CNF formula to be unsatisfiable, and every clause has degree at least 1. Thus all unsatisfiable k-CNF formulas have $\Omega(2^k)$ many conflicts. Example 1 shows that there is an unsatisfiable k-CNF with $\Theta(4^k)$ many conflicts. What is the right order of magnitude for the number of conflicts needed for an unsatisfiable k-CNF formula?

There is a similar question in Ramsey theory: A graph G is H-Ramsey if every 2-coloring of its edges contains a monochromatic copy of H. The size Ramsey number $\hat{r}(H)$ asks for the minimum number of edges over all H-Ramsey graphs where the Ramsey number $r(H)$ is the minimum number of vertices over all H-Ramsey graphs. For which graphs does it hold that $\hat{r}(H) = \binom{r(H)}{2}$? It is known that equality holds for complete graphs [2] but clearly not for stars, cf. [3] for a survey about the size Ramsey number.

1.1   Results

The technical groundwork, namely the Lopsided Lovász Local Lemma is discussed in Section 2. We examine thereby also the maximum and minimum degree of an unsatisfiable k-CNF formula. The main theorem follows in Subsection 3.1 and states that every unsatisfiable k-CNF formula has at least $\Omega(2.32^k)$ conflicts. A construction by Hoory and Szeider [4] shows that there exists unsatisfiable k-CNF formulas with $O(\frac{4^k \log^3 k}{k})$ conflicts, discussed in Subsection 3.2. In the end we discuss the maximization versions of these parameters and formulate some open problems.

2   Maximum Degree and Minimum Degree

The Lovász Local Lemma can be used to show that k-CNF formulas where every clause depends only on a small subset of the other clauses are always satisfiable, compare for example [5] Section 2.2. As an implication k-CNF formulas where every variable occurs only a few times are always satisfiable, cf. Corrollary 5 and [6,7]. The Lovász Local Lemma can also be used to define a branching rule for a DPLL-algorithm on SAT as shown in [8]. Another applications is shown in [9] for the MAX-SAT problem.

Definition 2. *Let $A_1, \ldots, A_m$ be events in some probability space. A graph $G = (V, E)$ with $V = \{1, \ldots, m\}$ is called a* lopsided dependency graph *if for any A_i and any $U \subseteq V \setminus (\{A_i\} \cup \Gamma_G(A_i))$ with $\Pr[\bigcap_{j \in U} \bar{A}_j] > 0$, it holds that*

$$\Pr\left[A_i \,\middle|\, \bigcap_{j \in U} \bar{A}_j\right] \leq \Pr[A_i] \,.$$

Lemma 3 ([10],[11],[12]). *Let $A_1, \ldots, A_m$ be events in some probability space, and let G be a lopsided dependency graph for them. If there are numbers $0 \leq \gamma_i < 1$, $1 \leq i \leq m$, such that for any i,*

$$\Pr[A_i] \leq \gamma_i \prod_{j \in \Gamma_G(A_i)} (1 - \gamma_j) \,,$$

then

$$\Pr\left[\bar{A}_1 \cap \cdots \cap \bar{A}_m\right] > 0 \,.$$

Now think of a CNF formula $F = C_1 \wedge \cdots \wedge C_m$. Set each variable independently uniformly at random. Define A_i to be the event that C_i is *not* satisfied. It is not difficult to see that the conflict graph of F is a lopsided dependency graph for the events $A_1, \ldots, A_m$. Applying Lemma 3, we obtain the following result:

Theorem 4. *The maximum degree of any unsatisfiable k-CNF formula is at least $\frac{2^k}{e}$.*

Proof. The proof is basically given in [6] by Kratochvíl et al. Assume that the maximum degree of a k-CNF formula F is at most $d := \frac{2^k}{e} - 1$. Set each variable independently uniformly at random. Then for each clause C_i, the probability that it is not satisfied is 2^{-k}. Now apply Lemma 3 with $\gamma_i := \frac{1}{d+1}$ for all i, and use the fact that $(1 - \frac{1}{d+1})^d \geq e^{-1}$. □

The complete formula K_k (Example 1) has maximum degree $2^k - 1$ and shows that this bound is tight up to a constant factor.

Corollary 5 (Kratochvíl et al. [6]). *Suppose F is a k-CNF formula. If every variable occurs in at most $\frac{2^k}{ek}$ clauses, then F is satisfiable.*

Proof. Consider any clause C of F. Clearly, every literal in C causes at most $\frac{2^k}{ek} - 1$ conflicts and hence the maximum degree of F is at most $\frac{2^k}{e} - k$. Thus F is satisfiable by Theorem 4. □

Does this result implies anything about the number of conflicts in an unsatisfiable k-CNF formula? The number of conflicts is trivially at least the maximum degree, hence $\Omega\left(2^k\right)$. Further, some variable x occurs in many clauses of F. Assume that this variable is more or less *balanced*, i.e., it occurs equally often as a positive and negative literal. In this case this variable by itself induces $\Omega\left(\frac{4^k}{k^2}\right)$ conflicts. Does every unsatisfiable k-CNF formula have such a balanced high-frequency variable? The next example gives the most negative answer to this question: We will define an unsatisfiable formula in which every variable occurs *exactly once* negative.

Example 6. We set $F_1 := x_1 \wedge \bar{x}_1$, and for $k \geq 1$ define recursively

$$F_k := (\bar{x}_1 \vee \bar{x}_2 \vee \ldots \vee \bar{x}_k) \wedge \bigwedge_{i=1}^{k} \left(F_{k-1}^{(i)} \veebar x_i \right),$$

where $F_{k-1}^{(i)}$ are copies of F_{k-1} on different set of variables. By definition of the operator $\veebar$ one sees inductively that F_k is indeed a k-CNF formula. It is easy to see by using induction again that F_k is unsatisfiable for all $k \geq 1$ and that every negative literal occurs only once.

Let $a(k)$ be the number of clauses in F_k. We have $a(1) = 2$ and $a(k) = ka(k-1) + 1$. Solving this recurrence, we obtain $a(k) = \sum_{j=0}^{k} \frac{k!}{j!} = \lfloor ek! \rfloor$. Each "top level" variable x_i occurs once negatively and $\lfloor e(k-1)! \rfloor$ times positively.

Therefore, the number of conflicts in this formula is huge and does not give us any good upper bound. It seems that we have to pay for extreme non-balancedness by huge variable frequency. Still, the concept of balanced and non-balanced variables will be of great importance in the next section, when we prove a lower bound on the number of conflicts.

Let us make a second stab on proving a lower bound on $e(F)$ for F being an unsatisfiable k-CNF formula. Observe that $e(F) \geq \frac{d_{\min}(F)}{2} m(F)$, where $m(F)$ is the number of clauses. We know that $m(F) \geq 2^k$. The minimum degree $d_{\min}(F)$ gets 0 when we add clauses of degree 0. Thus to get any meaningful bound, we have to consider minimal unsatisfiable formulas. A CNF formula is called *minimal unsatisfiable* if it is unsatisfiable and deleting any clause makes it satisfiable. Minimal unsatisfiable formulas are interesting objects themselves, for they have many algorithmic aspects, which were studied for example in [13,14,15,16,17]. In this paper however, we use only some straightforward combinatorial properties. First, we can assume w.l.o.g. that F is a minimal unsatisfiable k-CNF formula, as every unsatisfiable CNF formula has a minimal unsatisfiable subformula (which does not have more conflicts). Second, we actually can state a lower bound on $d_{\min}(F)$ if F is minimal unsatisfiable:

Lemma 7. *If F is a minimal unsatisfiable k-CNF formula, then $d_{\min}(F) \geq k$.*

Proof. Let F be a minimal unsatisfiable k-CNF formula and assume for contradiction that there is a clause C such that its degree in the conflict graph is less than k. By minimality of F, $F - C$ is satisfiable. Take a satisfying assignment α of $F - C$. The neighbors $\Gamma_F(C)$ of C are satisfied by α, so we can assign to each $D \in \Gamma_F(C)$ a variable x_D such that the corresponding literal in D is satisfied. At least one of the k variables in C is not assigned, say ℓ is the corresponding literal in C. By changing α such that ℓ is true, we get a satisfying assignment for F, which is a contradiction. $\square$

This result is also tight: The k-CNF formula F_k defined above is minimal unsatisfiable, and has a minimum degree of k.

Corollary 8. *Every unsatisfiable k-CNF formula has at least $k \cdot 2^{k-1}$ conflicts.*

3 Number of Conflicts

3.1 A Lower Bound

In this section, we prove a lower bound on the number of conflicts in an unsatisfiable k-CNF formula. Our main technical tool will be a corollary of the Lopsided Lovász Local Lemma:

Corollary 9. *Let F be a CNF formula not containing the empty clause. If for all clauses C in F it holds that*

$$\sum_{D \in \Gamma_F(C)} \Pr\left[\, D \text{ not satisfied} \,\right] \leq \frac{1}{4} \,, \tag{1}$$

then F is satisfiable.

Proof. Write $F = C_1 \wedge \ldots \wedge C_m$. First, we can assume that there are no isolated clauses, i.e., every clause has some conflict. We apply Lemma 3 with $\gamma_i :=$ $2 \Pr[C_i \text{ not satisfied}]$. Note that (1) implies $\Pr[C_i \text{ not satisfied}] \leq \frac{1}{4}$ for each C_i, thus $\gamma_i \leq \frac{1}{2}$ for all i. A short calculation using the fact that $\prod(1-\gamma_j) \geq 1-\sum\gamma_j$ completes the proof. $\square$

Consider the function $f(p) = 1 - \log(1 - p) - \frac{\log(1-p)}{\log(p)}$ which has a unique $p^* \in (0, \frac{1}{2})$ maximizing $f(p)$. In fact, $p^* \approx 0.30$ and $f(p^*) > 1.218$.

Theorem 10. *Let F be an unsatisfiable k-CNF formula. For any $p < \frac{1}{2}$, F has at least*

$$\frac{2^{f(p)k}}{8 + 128k^2}$$

conflicts. Furthermore, plugging in the optimal value p^*, we obtain that F has $\Omega\left(2.32^k\right)$ conflicts.

Before we proceed to the proof, we explain the basic idea behind it. The Lovász Local Lemma implies that any unsatisfiable k-CNF formula contains a variable of high degree, namely around $\frac{2^k}{ek}$. Assume that this variable is more or less *balanced*, i.e., it occurs positively and negatively equally often. In this case there are already $\Omega\left(\frac{4^k}{k^2}\right)$ conflicts due to this variable, and we are done. Otherwise, we can assume that if a literal u occurs frequently, then $\bar{u}$ does not. A natural idea now comes to mind: In the probability space used in the Lovász Local Lemma, we set each variable independently, choosing 1 with probability $\frac{1}{2}$. Should we not bias u towards 1, if many clauses benefit from this, and only few suffer?

Proof (of Theorem 10). For a literal u, let $\mathrm{occ}_F(u)$ denote the number of clauses in F containing the literal u. Please note that $\mathrm{occ}_F(u)$ and $\mathrm{occ}_F(\bar{u})$ may differ. Choose parameters ℓ, θ as follows: $\ell := \frac{k\log(1-p)}{\log(p)}$ and $\theta = \frac{2^{k-\ell}}{8k}$.

We will color the literals of the formula and choose a probability with which they are set to 1. (i) If both $\mathrm{occ}_F(u) \geq \theta$ and $\mathrm{occ}_F(\bar{u}) \geq \theta$, color u and $\bar{u}$ red, and set each to 1 with probability $\frac{1}{2}$. (ii) If $\mathrm{occ}_F(u) \geq \theta$ and $\mathrm{occ}_F(\bar{u}) < \theta$, color u green and $\bar{u}$ red, and set u to 1 with probability $1 - p$ (and thus $\bar{u}$ to 1 with probability p). (iii) If both $\mathrm{occ}_F(u) < \theta$ and $\mathrm{occ}_F(\bar{u}) < \theta$, color both *black*,[1] and set each to 1 with probability $\frac{1}{2}$. What you should keep in mind is that every black literal is "not frequent", and the complement of each red literal is "frequent", and that green literals are likely to be satisfied, since $1 - p > \frac{1}{2}$.

Let $e(F)$ be number of conflicts/edges in F. As a first observation, note that any literal u causes $\mathrm{occ}_F(u)\mathrm{occ}_F(\bar{u})$ conflicts. Summing this up over all red literals u we obtain the number of conflicts caused by red literals. However, we might (i) count the pair $\{u, \bar{u}\}$ twice, if both are red, and (ii) count the same conflict k times, as two clauses can have up to k conflicting literals, e.g. $(u_1 \vee \cdots \vee u_k)$ and $(\bar{u}_1 \vee \cdots \vee \bar{u}_k)$. Still, we obtain

---

[1] Or white, if you are working on a blackboard...

$$e(F) \geq \frac{1}{2k} \sum_{u:\text{red}} \text{occ}_F(u)\text{occ}_F(\bar{u}) \geq \frac{\theta}{2k} \sum_{u:\text{red}} \text{occ}_F(u)$$

$$\geq \frac{\theta}{2k} |\{C \in F \mid C \text{ contains at least one red literal}\}| . \tag{2}$$

Our interpretation of this inequality is that if F has few conflicts, then only few clauses can contain red literals. Next, we define a new formula F' as follows: Start with F, and for each clause C in F that has fewer than ℓ green literals, and no red ones, remove all green literals. This formula F' is still unsatisfiable, and $e(F) \geq e(F')$. It is no k-CNF formula anymore, but each clause has at least $(k - \ell)$ literals. Further, $\text{occ}_{F'}(u) \leq \text{occ}_F(u)$ for any literal u.

Since F' is unsatisfiable, we can use the contrapositive of Corollary 9, i.e., there is a clause C^* in F' such that

$$\sum_{D \in \Gamma_{F'}(C^*)} \Pr[D \text{ not satisfied}] > \frac{1}{4} . \tag{3}$$

Let us partition $\Gamma_{F'}(C^*)$ into sets $\mathcal{B}$, $\mathcal{BG}$ and $\mathcal{BGR}$ as follows: Let $\mathcal{B}$ contain all clauses $D \in \Gamma_{F'}(C^*)$ containing only black literals, $\mathcal{BG}$ those containing at least one green literal, but no red ones, and $\mathcal{BGR}$ those containing at least one red literals. There are several useful observations: First, by construction of F', every clause in $\mathcal{BG}$ contains at least ℓ green literals. Second, clauses in $\mathcal{B}$ have at least $k - \ell$ literals, and $\mathcal{BGR}$ and $\mathcal{BG}$ contain only k-clauses. We can give certain bounds on $|\mathcal{BGR}|$ and $|\mathcal{B}|$. Every $D \in \mathcal{BGR}$ contains a red literal, hence by (2)

$$|\mathcal{BGR}| \leq \frac{2ke(F)}{\theta} . \tag{4}$$

To give a bound on $|\mathcal{B}|$, note that for every $D \in \mathcal{B}$ there is a black literal u in C^* such that $\bar{u}$ is in D. Since $\text{occ}_{F'}(\bar{u}) \leq \theta$ for each such literal u in C^*, and C^* contains at most k of them,

$$|\mathcal{B}| \leq k\theta . \tag{5}$$

To evaluate the sum in (3), let us estimate the probabilities with which the clauses in the three sets are unsatisfied. First, a clause $D \in \mathcal{BGR}$ has at least k literals, and each is satisfied with probability at least p (we pessimistically assume the worst case, namely that all literals in D are red). We obtain

$$\forall D \in \mathcal{BGR} : \Pr[D \text{ not satisfied}] \leq (1 - p)^k . \tag{6}$$

For $D \in \mathcal{B}$, observe that D has at least $k - \ell$ literals, which are all black, thus

$$\forall D \in \mathcal{B} : \Pr[D \text{ not satisfied})] \leq 2^{-k+\ell} . \tag{7}$$

For $D \in \mathcal{BG}$, we know that D contains at least ℓ green literals, each of which is satisfied with probability $1 - p$. All other literals in D are black. Therefore we obtain

$$\forall D \in \mathcal{BG} : \Pr[D \text{ not satisfied}] \leq 2^{-k+\ell}p^\ell . \tag{8}$$

We plug (4)–(8) into (3) and get

$$\frac{1}{4} < |\mathcal{BG}|2^{-k+\ell}p^\ell + 2ke(F)\theta^{-1}(1-p)^k + k\theta 2^{-k+\ell} .$$

This yields a lower bound on $|\mathcal{BG}|$. Since every clause in $\mathcal{BG}$ has a conflict with C, it yields a lower bound on $e(F')$, and thus on $e(F)$:

$$e(F) \geq |\mathcal{BG}| > p^{-\ell}\left(2^{k-\ell-2} - 2ke(F)\theta^{-1}(1-p)^k2^{k-\ell} - k\theta\right) .$$

Plugging in our values for θ and ℓ, a few calculations show that

$$e(F) > \frac{2^{f(p)k}}{8 + 128k^2} .$$

This completes the proof of the theorem.   □

We should make some comments on the proof above. It is a natural idea to choose a probability $\neq \frac{1}{2}$ for unbalanced variables. However, it is not clear why it makes sense to delete green (frequent) literals from clauses. Are these not exactly those literals making the clause more likely to be satisfied? What would happen if we re-did the proof without deleting green literals? We would have to assume the worst case that every $D \in \mathcal{BG}$ contains $k-1$ black literals and only one green, only marginally pushing up its probability of being satisfied. Overall, we would obtain a bound like $\Omega\left(\frac{k^2 2^k}{\ln k}\right)$, if we choose our parameters optimally. The benefit of crossing out some green literal becomes clear if one thinks the other way round: We are given a formula with fewer than $\frac{2^{f(p)k}}{8+128k^2}$ conflicts and want to prove that it is satisfiable. By deleting green literals from some clauses D, we sacrifice by decreasing the probability of D being satisfied. On the other hand, no clause C can have many of these shrunk, now completely black clauses D in its neighborhood, due to the bound $|\mathcal{B}|$. We are actually making the conflict graph much sparser, thus more amenable to the Local Lemma.

3.2   Upper Bound

For an upper bound construction we show that the construction given by Hoory and Szeider [4] gives an unsatisfiable k-CNF formula with $O(\frac{4^k \log^3 k}{k})$ conflicts. They used this construction to give an instance of a k-CNF formula with few occurrences of each variable.

For notational matter we write $K(v_1,\ldots,v_s)$ to denote the complete formula on the variables $v_1,\ldots,v_s$, and $K^-(v_1,\ldots,v_s)$ is used for the formula obtained by $K(v_1,\ldots,v_s)$ deleting the all-positive-clause $(v_1 \vee \ldots \vee v_s)$. The formula $K^-(v_1,\ldots,v_s)$ is equivalent to $(\bar{v}_1 \wedge \ldots \wedge \bar{v}_s)$. Let $k \geq 1$ and choose $\ell = \lfloor \log(k/\log^2 k) \rfloor$, $u = \lfloor k/l \rfloor$, and $v = k - lu$. Define the formula $F = \bigwedge_{i=0}^u F_i$ where

$$F_0 = K(z_1,\ldots,z_v) \veebar \bigvee_{i=1}^u K^-(x_1^{(i)},\ldots,x_\ell^{(i)}),$$
$$F_i = K(y_1^{(i)},\ldots y_{k-\ell}^{(i)}) \veebar (x_1^{(i)} \vee \ldots \vee x_\ell^{(i)}).$$

Replace the complete formulas by the constant 0 function and the "almost complete" formulas $K^-(x_1^{(i)}, \ldots, x_s^{(i)})$ by $(\bar{x}_1^{(i)} \wedge \ldots \wedge \bar{x}_s^{(i)})$ to see that this formula is unsatisfiable. It remains to calculate the number of conflicts in this formula:

$$e(F) = e(F_0) + ue(F_1) + ue(F_0, F_1) \leq |F_0|^2 + u|F_1|^2 + u|F_0||F_1|$$

$$\leq \frac{e^2 4^k}{k^2} + \frac{4^k \log^3 k}{k \log^2 e} + \frac{4^k e \log^3 k}{k \log^2 e} = O\left(\frac{4^k \log^3 k}{k}\right).$$

The calculations for $|F_0|, |F_1|$ are made in [4] and the better upper bound described there can also be used to get rid of one logarithm, i.e., this would yield an unsatisfiable k-CNF formula with $O(\frac{4^k \log^2 k}{k})$ conflicts. The details are omitted because the lower bound is far away from this upper bound and we cannot be sure what the right order of magnitude is.

4   Maximizing the Parameters

How big can the parameters $d_{\min}, d_{\max}, e$ of a minimal unsatisfiable k-CNF formula be? We will show that there exists minimal unsatisfiable k-CNF formulas with arbitrarily large minimum degree, for $k \geq 3$. From this it follows that the maximum degree, number of edges are arbitrarily big as well.

Example 11. The *cycle formula* C_ℓ on the variables $x_1, \ldots, x_\ell$ contains the clauses $(\bar{x}_i \vee x_{i+1}), i = 1, \ldots, \ell$ (the index is taken here and in the following modulo ℓ). We can interpret each clause as a logical implication $x_i \to x_{i+1}$ or as an inequality $x_i \leq x_{i+1}$ over the boolean values $0, 1$. It follows that C_ℓ is satisfied if and only if all the variables $x_1, \ldots, x_\ell$ are set to the same value. By adding the clauses $(x_1 \vee x_2), (\bar{x}_1 \vee \bar{x}_2)$ to C_ℓ we obtain a minimal unsatisfiable 2-CNF formula. Consider now

$$H_{k,\ell} := (C_\ell \wedge (x_1 \vee x_2) \wedge (\bar{x}_1 \vee \bar{x}_2)) \veebar K(y_1, \ldots, y_{k-2})$$

It is not difficult to see for two minimal unsatisfiable formulas F, G also $F \veebar G$ is minimal unsatisfiable. Thus $H_{k,\ell}$ is a minimal unsatisfiable k-CNF formula. It contains $2^{k-2}(\ell + 2)$ clauses, and each clause has at least $(2^{k-2} - 1)(\ell + 2) + 2$ conflicts. Therefore, the minimum degree of a minimal unsatisfiable k-CNF formula can be arbitrarily large, for each fixed $k \geq 3$.

5   Conclusion

Let e_k denote the minimum number of conflicts/edges over all unsatisfiable k-CNF formulas. Theorem 10 and the construction in Section 3.2 leads to the following asymptotic bounds

$$\Omega(2.32^k) \leq e_k \leq O\left(\frac{4^k \log^3 k}{k}\right).$$

This improves over the trivial bounds, but the gap between the lower and upper bound is still huge. We suspect that the true value of e_k is much closer to the upper bound, but it took us a considerable effort to get away from lower bounds like $\Omega(k^t 2^k)$ for some fixed t. The right magnitude of e_k is therefore still the main open question of this paper.

It is easy to change the definition of the complete formula in such a way that every pair of clauses has exactly one conflicting variable. By substituting this variant of the complete formula into the example given in Section 3.2 yields an unsatisfiable k-CNF formula with $\Theta\left(\frac{4^k \log^3 k}{k}\right)$ multi-edges.

As we pointed out in Example 6 there exists unsatisfiable k-CNF formulas where all variables of high degree are very unbalanced. But even there it holds that there is one variable u such that it induces almost all conflicts, i.e., $\mathrm{occ}(u) \cdot \mathrm{occ}(\bar{u})$ is already very large. Is it true in general that the maximum of $\mathrm{occ}(u) \cdot \mathrm{occ}(\bar{u})$ is very large?

Acknowledgment. We thank all the participants of Gremo Workshop on Open Problems, GWOP 2007, for the helpful discussions at the workshop.

References

1. Kullmann, O.: The combinatorics of conflicts between clauses. In: Giunchiglia, E., Tacchella, A. (eds.) SAT 2003. LNCS, vol. 2919, pp. 426–440. Springer, Heidelberg (2004)
2. Erdős, P., Faudree, R.J., Rousseau, C.C., Schelp, R.H.: The size ramsey number. Periodica Mathematica Hungarica 9(2-2), 145–161 (1978)
3. Faudree, R.J., Schelp, R.H.: A survey of results on the size Ramsey number. In: Paul Erdős and his mathematics, II (Budapest, 1999), János Bolyai Math. Soc. Budapest. Bolyai Soc. Math. Stud, vol. 11, pp. 291–309 (2002)
4. Hoory, S., Szeider, S.: A note on unsatisfiable k-CNF formulas with few occurrences per variable. SIAM Journal on Discrete Mathematics 20(2), 523–528 (2006)
5. Welzl, E.: Boolean satisfiability – combinatorics and algorithms (lecture notes) (2005), http://www.inf.ethz.ch/~emo/SmallPieces/SAT.ps
6. Kratochvíl, J., Savický, P., Tuza, Z.: One more occurrence of variables makes satisfiability jump from trivial to NP-complete. SIAM Journal of Computing 22(1), 203–210 (1993)
7. Berman, P., Karpinski, M., Scott, A.D.: Approximation hardness and satisfiability of bounded occurrence instances of SAT. Electronic Colloquium on Computational Complexity (ECCC) 10(022) (2003)
8. Hooker, J.N., Vinay, V.: Branching rules for satisfiability. Journal of Automated Reasoning 15, 359–383 (1995)
9. Czumaj, A., Scheideler, C.: A new algorithm approach to the general Lovász local lemma with applications to scheduling and satisfiability problems (extended abstract). In: Proceedings of the Thirty-Second Annual ACM Symposium on Theory of Computing, pp. 38–47. ACM, New York (2000); (electronic)
10. Erdős, P., Spencer, J.: Lopsided Lovász local lemma and Latin transversals. Discrete Appl. Math. 30, 151–154 (1991); ARIDAM III (New Brunswick, NJ, 1988)
11. Alon, N., Spencer, J.H.: The probabilistic method, 2nd edn. Wiley-Interscience Series in Discrete Mathematics and Optimization. Wiley-Interscience [John Wiley & Sons], New York (2000); With an appendix on the life and work of Paul Erdős

12. Lu, L., Székely, L.: Using Lovász local lemma in the space of random injections. Electron. J. Combin. 14(1), 13 (2007) Research Paper 63 (electronic)
13. Papadimitriou, C.H., Wolfe, D.: The complexity of facets resolved. J. Comput. Syst. Sci. 37(1), 2–13 (1988)
14. Szeider, S.: Minimal unsatisfiable formulas with bounded clause-variable difference are fixed-parameter tractable. J. Comput. Syst. Sci. 69(4), 656–674 (2004)
15. Kleine Büning, H.: On subclasses of minimal unsatisfiable formulas. Discrete Appl. Math. 107(1-3), 83–98 (2000); Boolean functions and related problems
16. Fleischner, H., Kullmann, O., Szeider, S.: Polynomial-time recognition of minimal unsatisfiable formulas with fixed clause-variable difference. Theoret. Comput. Sci. 289(1), 503–516 (2002)
17. Kleine Büning, H., Zhao, X.: The complexity of some subclasses of minimal unsatisfiable formulas. J. Satisf. Boolean Model. Comput. 3(1-2), 1–17 (2007)

Speeding-Up Non-clausal Local Search for Propositional Satisfiability with Clause Learning

Zbigniew Stachniak and Anton Belov[*]

Department of Computer Science and Engineering,
York University, Toronto, Canada
{zbigniew,antonb}@cse.yorku.ca

Abstract. In this paper we discuss search heuristics for non-clausal stochastic local search procedures for propositional satisfiability. These heuristics are based on a new method for variable selection as well as a novel clause learning technique for dynamic input formula simplification as well as for guiding the search for a model.

Introduction

The search heuristic of a typical stochastic local search method for propositional satisfiability is a two step selection procedure. First, some variables that occur in an input formula are chosen. In the second step, these *candidate variables* are subjected to an evaluation by a variable selection heuristic which picks one of them in order to modify the current truth-value assignment by changing the truth-value of the selected variable.

With a few notable exceptions, most clause-based stochastic local search SAT solvers select the candidate variables in the same way: they come from a random input clause false under the current assignment. The difference in performance of such solvers is therefore due mainly to differences in their respective variable selection heuristics.

In non-clausal case, the situation is different. The performance of non-clausal solvers depends critically not only on variable selection heuristics employed in them but also on the method for the selection of candidate variables. Clearly, the structure of an input formula has to play a role in the selection of the candidate variables and so should the current truth-value assignment. But the options are many.

In this paper we address the problem of the candidate variable selection for non-clausal stochastic local search solvers. This problem, that originated from our attempts at improving the performance of the polSAT procedure, first introduced in [14], led us to the definition of *wish lists* – special sets of variables that record the differences between models of an input formula. We have empirically confirmed that by modifying polSAT's candidate variable generation method to guarantee that the sets of candidate variables are wish lists, we obtain a significantly more efficient solver. Furthermore, by replacing the variable selection heuristic employed by the original polSAT procedure with a new one, modeled after *Adaptive Novelty+* (cf. [6]), we obtained an even more

---

[*] Research of both authors supported by grants from the Natural Sciences and Engineering Research Council of Canada.

H. Kleine Büning and X. Zhao (Eds.): SAT 2008, LNCS 4996, pp. 257–270, 2008.

dramatic improvement in performance (measured in terms of both the number of search steps as well CPU time). As shown in the Experimental Studies section of the paper, on some classes of problems, the new solver performs as well as, and in some cases better than, some clausal state of the art solvers.

Wish lists offer other opportunities for boosting the performance of stochastic local search solvers. In this paper we propose a novel technique for deriving information from wish lists recording failed attempts at finding a model for a Boolean formula. This technique, which we call *clause learning*, allows a SAT solver to derive and record such information from wish lists in the form of clauses. Learned clauses are logical consequences of input formulas and, hence, can be used in a number of ways to speed-up the performance of a solver – from input formula simplification and adjusting its search heuristic, to unsatisfiability detection and certification. In the present paper, we discuss the use of learned clauses for dynamic input formula simplification and for the refinement of search heuristics. The empirical results collected during our preliminary study on the incorporation of clause learning into non-clausal local search indicate that information provided by learned clauses can significantly improve the performance of a solver. However, our study also indicates that learned clauses have to be used in a rather subtle way to support rather than dominate a given search heuristic.

1   Logical Preliminaries

We assume that the reader is familiar with the syntax and semantics of classical propositional logic (CPL) and its clausal fragment. In this section we clarify some terminology and notation used in the paper that, otherwise, might result in unintended ambiguity.

Formulas of CPL (or simply *formulas*) are constructed in terms of countably infinite set of propositional variables, logical constants F (*false*) and T (*true*) and some selection of logical connectives (including $\neg$ (*negation*), $\vee$ (*disjunction*), and $\wedge$ (*conjunction*)). For reasons of simplicity of presentation, we restrict our discussion to formulas in negation normal form (NNF) only, i.e., to formulas in which the negation connective can only apply to propositional variables. However, all the definitions and results given in this paper extend to all the formulas of CPL without much difficulty.

Let α and β be formulas. We shall write $\alpha \vdash \beta$ when β is a logical consequence of α. $Var(\alpha)$ denotes the set of all the variables that occur in α. We shall write $\alpha(p)$ to explicitly indicate that a variable p occurs in α. If v is a logical constant, then $\alpha(p/v)$ denotes the result of a simultaneous replacement of every occurrence of the variable p in α by v.

We identify truth-value assignments for α with functions h mapping $Var(\alpha) \cup \{F, T\}$ into the set $\{0, 1\}$ of *truth-values* in such a way that $h(F) = 0$ and $h(T) = 1$. Since the CPL semantics allows to extend h, in the unique way, to the set of all sub-formulas of α, we shall be making no distinction between truth-value assignments and their extensions. Hence, α is *satisfiable* (or, $\alpha \in SAT$) if and only if there exists a truth-value assignment h for α such that $h(\alpha) = 1$. Every truth-value assignment h for α such that $h(\alpha) = 1$ is called a *model* for α. Finally, if h is a truth-value assignment, p is a variable, and if $v \in \{0, 1\}$, then $h[p/v]$ denotes the truth-value assignment which is defined in the same way as h with the exception that $h[p/v](p) = v$.

The polSAT procedure discussed in this paper is defined in terms of logical polarity which we review next.

Let $\alpha(p)$ be a NNF formula. An occurrence of p in $\alpha(p)$ is positive (resp. negative) if it is not in the scope (resp. it is in the scope) of the negation connective. This variable is *positive* (resp. *negative*) in $\alpha(p)$ if its every occurrence is positive (resp. negative) in $\alpha(p)$. Clearly, a formula may contain variables that are neither negative nor positive.

If one is not searching for all models for a formula α, then every positive and negative variable should be eliminated from α. This syntactic simplification of α is accomplished by, first, replacing every occurrence of every positive (resp., negative) variable with the constant T (resp. F) and, then by applying rewrite rules reflecting the semantics of the logical connectives (the rules: *rewrite a subformula $\beta \vee F$ as β and rewrite $F \wedge \beta$ as F*, can serve as examples).

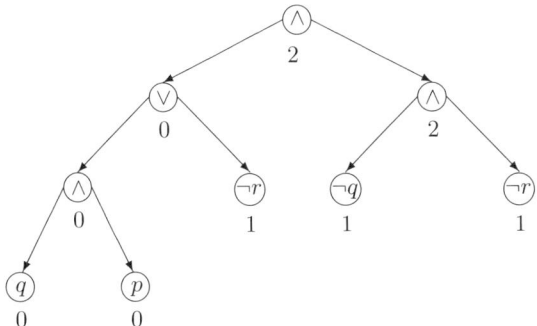

Fig. 1. Tree formula representation of $\alpha = ((q \wedge p) \vee \neg r) \wedge (\neg q \wedge \neg r)$

The following definition of the polarity clash of a formula comes from [14].

DEFINITION 1.1. Let α be a formula and let h be a truth-value assignment for α. The *polarity clash* of α with respect to h, denoted as $clash(\alpha, h)$, is defined as follows:

- if α is a literal, then $clash(\alpha, h) = 1 - h(\alpha)$;
- if $\alpha = \alpha_0 \vee \alpha_1$, then $clash(\alpha, h) = min\{clash(\alpha_0, h), clash(\alpha_1, h)\}$;
- if $\alpha = \alpha_0 \wedge \alpha_1$, then $clash(\alpha, h) = clash(\alpha_0, h) + clash(\alpha_1, h)$.

The $clash(\alpha, h)$ parameter was introduced in [14] as a measure of progress in the search for a satisfying truth-value assignment (the smaller the $clash(\alpha, h)$ value the better; it is easy to see that $h(\alpha) = 1$ if and only if $clash(\alpha, h) = 0$). If α is a conjunction of clauses, then $clash(\alpha, h)$ is simply the number of clauses in α false under h.

EXAMPLE 1.2. Figure 1 depicts a tree representation of the formula $\alpha = ((q \wedge p) \vee \neg r) \wedge (\neg q \wedge \neg r)$. The reader is asked to verify that p is positive, r is negative, and q is of no polarity in α. The label of every node N in the tree indicates the polarity clash of the subformula of α rooted in N with respect to the truth-value assignment h defined by $h(p) = h(q) = h(r) = 1$. ∎

2 polSAT Algorithm

polSAT is a non-clausal stochastic local search algorithm. In the preprocessing step, it converts an input formula into NNF in which $\vee$ and $\wedge$ are multi-argument rather than binary connectives. Similarly to clausal stochastic local search SAT methods, such as WalkSAT, polSAT performs a local search over the space of truth-value assignments (see Figure 2). Given an input formula α, polSAT generates a random truth-value assignment h and makes a fixed number (MaxFlip) of changes to h, each time selecting one of the variables $p \in Var(\alpha)$ and 'flipping' its truth-value, i.e., changing its truth-value from $h(p)$ to $1 - h(p)$. Such variable selections and flips are repeated until either h satisfies α or the allocated time to modify h into a satisfying assignment has elapsed ($MaxFlips$). The process is repeated (if needed) up to the specified $MaxTries$ times.

polSAT's variable selection is a two-step process. First, a subset $CandList(\alpha, h)$ of $Var(\alpha)$ is generated (line (vs1) in Figure 2). In the second step (line (vs2)), $CandList(\alpha, h)$ is searched for a variable v to flip using a variable selection heuristic.

---

procedure polSAT(α)
    for $i := 1$ to $MaxTries$ **do**
        $h :=$ *random truth-value assignment*
        for $j := 1$ to $MaxFlips$ **do**
            if $h(\alpha) = 1$ **then return** h **else**
(vs1)   compute $CandList(\alpha, h) \subseteq Var(\alpha)$;
(vs2)   pick $v \in CandList(\alpha, h)$ using a variable selection heuristic;
                $h(v) := 1 - h(v)$;
            end if
        end for
    end for
    return 'satisfying valuation for α not found'

Fig. 2. Generic polSAT algorithm

polSAT forms $CandList(\alpha, h)$ by collecting all the variables that can be reached from the root of α (when α is represented as a binary tree) branching into children β having the smallest but non-zero $clash(\beta, h)$ values. It's variable selection heuristic picks, with probability p, a variable from $CandList(\alpha, h)$ at random or, with probability $1 - p$, using the objective function to minimize the $clash(\alpha, h[v/1 - h(v)])$ value. The following example illustrates polSAT's computation of $CandList(\alpha, h)$.

EXAMPLE 2.1. Let α and h be as in Example 1.2. The search for the elements of $CandList(\alpha, h)$ begins with the root of α. The clash value of the root's leftmost child is 0. Therefore, the search continues by inspecting the rightmost subtree of the root (clash value 2). The root of this subtree, labeled with '$\wedge$', has two children with identical clash values equal to 1. Therefore, our search has to include both paths: one going through $\neg q$ and ending with q, and the second going through $\neg r$ and terminating at r. This gives us $CandList(\alpha, h) = \{q, r\}$.   ■

3   From Wish Lists to Learned Clauses

The original definition of $CandList(\alpha, h)$ discussed in the previous section has a draw-back. It may happen that every variable from $CandList(\alpha, h)$ could be a wrong choice for a flip. This is the case, for instance, when every model of α agrees with h on $CandList(\alpha, h)$. Indeed, the flip of any of the candidate variables would direct the search away from a model. For instance, if $\alpha = p \wedge (\neg p \vee (s \wedge s))$ and the truth-value assignment h is defined by $h(p) = 1$ and $h(s) = 0$, then $CandList(\alpha, h) = \{p\}$. Flipping p is certainly wrong since the only model for α assigns 1 to p as does h.

To remedy this problem, we may impose certain constraints on the $CandList$ sets such as the one proposed in the following definition.

DEFINITION 3.1. Let α be a formula and let h a truth-value assignment such that $h(\alpha) = 0$. A set $X \subseteq Var(\alpha)$ is said to be a *wish list* for α and h if for every model h' for α there exists a $p \in X$ such that $h(p) \neq h'(p)$.

Paraphrasing Definition 3.1, if X is a wish list for α and h, then no model for α as-signs truth-values to all the variables of X in the way h does. To guarantee that every candidate list of variables generated by polSAT is a wish list, we can adopt the follow-ing new definition of $CandList(\alpha, h)$:

DEFINITION 3.2 (**all–$\vee$–random–$\wedge$**). Let α be a formula in NNF and let h be a truth-value assignment. The set $CandList(\alpha, h)$ is defined as follows. If $h(\alpha) = 1$, then $CandList(\alpha, h) = \emptyset$. Else:

- If α is a literal containing a variable p, then $CandList(\alpha, h) = \{p\}$.
- If $\alpha = \beta_0 \vee \ldots \vee \beta_n$, then $CandList(\alpha, h) = \bigcup\{CandList(\beta_i, h) : i \leq n\}$.
- If $\alpha = \beta_0 \wedge \ldots \wedge \beta_n$, then $CandList(\alpha, h) = CandList(\beta_i, h)$, where β_i is a randomly selected conjunct of α false under h.

PROPOSITION 3.3. If α is a formula and h is a truth-value assignment h such that $h(\alpha) = 0$, then $CandList(\alpha, h)$, defined as in Definition 3.2, is a wish list for α and h.

PROOF (by induction on the syntactic complexity of α)
Let α (in NNF) and h be as stated. If $\alpha(p)$ is a literal, then $CandList(\alpha, h) = \{p\}$. Clearly, for every model h' for α, $h'(p) \neq h(p)$.

Next, let us assume that $\alpha = \beta_0 \vee \ldots \vee \beta_n$ and that h' is a model for α. So, there is $i \leq n$, such that $h'(\beta_i) = 1$. Since $h(\beta_i) = 0$, by inductive hypothesis, there exists $p \in CandList(\beta_i, h) \subseteq CandList(\alpha, h)$ such that $h'(p) \neq h(p)$.

Finally, suppose that $\alpha = \beta_0 \wedge \ldots \wedge \beta_n$. Let β_i be the randomly selected conjunct used to compute $CandList(\alpha, h)$. Since h' is also a model for β_i, by inductive hypoth-esis, there exists $p \in CandList(\beta_i, h) = CandList(\alpha, h)$ such that $h(p) \neq h'(p)$. ■

EXAMPLE 3.4. Let us use the **all–$\vee$–random–$\wedge$** method to compute $CandList(\alpha, h)$, where α and h are as in Example 1.2. Since α is a conjunction, by Definition 3.2, to compute $CandList(\alpha, h)$ we should randomly select one of the conjuncts of α which has non-zero clash value. Since there is only one such conjunct, $\neg q \wedge \neg r$,

(a) $CandList(\alpha, h) = CandList(\neg q \wedge \neg r, h)$.

Now, to compute $CandList(\neg q \wedge \neg r, h)$, we note that both $\neg q$ and $\neg r$ have non-zero clash values. If $\neg q$ is selected, then $CandList(\neg q \wedge \neg r, h) = CandList(\neg q, h) = \{q\}$ and, by (a), $CandList(\alpha, h) = \{q\}$. On the other hand, if $\neg r$ is selected, then $CandList(\alpha, h) = \{r\}$. To conclude, when computed using the **all–∨–random–∧** method, $CandList(\alpha, h)$ is a singleton rather a two-element set as computed in Example 1.2.   ∎

In Sections 4 and 5 we shall discuss two techniques for boosting the performance of non-clausal stochastic local search using information derived from failed attempts at local transformation of the current truth-value assignment into a model for an input formula. The notion of the learned clause derived from a wish list, given in the following definition, is at the core of these methods.

DEFINITION 3.5 (**learned clause**). Let α be a formula, h be a truth-value assignment such that $h(\alpha) = 0$, and let $CandList$ be a wish list for α and h. For every variable $p \in CandList$, let p^* be $\neg p$, if $h(p) = 1$ and $p^* = p$, otherwise. The learned clause associated with $CandList$ is the disjunction

$$\bigvee \{p^* : p \in CandList\}.$$

THEOREM 3.6. Let α, h, and $CandList$ be as in Definition 3.5, and let c_α be the learned clause associated with $CandList$. Then $\alpha \vdash c_\alpha$.

PROOF. Follows directly from Definitions 3.1 and 3.5.   ∎

THEOREM 3.7. Let α be a formula and let $\mathcal{C}_\alpha$ be the set of learned clauses obtained from all wish lists for α. Then $\alpha \in SAT$ iff $\mathcal{C}_\alpha$ is consistent.

PROOF. Let α and $\mathcal{C}_\alpha$ be as stated. The *only if* part of the theorem follows from Theorem 3.6 and the soundness of $\vdash$. The *if* part can be justified as follows. Suppose that h is a model for every clause in $\mathcal{C}_\alpha$ but $h(\alpha) = 0$. Let c be the learned clause obtained from the wish list for α and h. By the definition of a learned clause, we must have $h(c) = 0$. However, since $c \in \mathcal{C}, h(c) = 1$. Hence, $h(\alpha) = 1$, as required.   ∎

The derivation of learned clauses from wish lists defined using the **all–∨–random–∧** method is native to the non-clausal approach and is of no benefit to clausal SAT solvers. Indeed, when a formula α is in conjunctive normal form, then every $CandList(\alpha, h)$ coincides with the set of variables that occur in some randomly selected clause C in α that is false under h. Furthermore, the learned clause associated with such a clause C is C itself. On the other hand, in the non-clausal case, learned clauses are not necessarily subformulas of α. Furthermore, by inspecting the proof of Theorem 3.7 we can conclude that the set $\mathcal{C}_\alpha$ (defined as in Theorem 3.7) is a CNF representation of α, i.e., α is equivalent to $\bigwedge \mathcal{C}_\alpha$. This opens up a possibility of developing hybrid non-clausal solvers that utilize clausal methods for SAT.

4   Learned Clauses and Dynamic Formula Simplification

An input formula α can be periodically subjected to the syntactic simplification process during the search for its model. Learned clauses can be used for this purpose by, first, subjecting them to automated reasoning methods (such as binary resolution or implication graph reasoning) in an attempt to derive literals $l(p)$ such that

(a) $\alpha \vdash l(p)$,

or pairs l_0, l_1 of literals such that

(b) $\alpha \vdash l_0 \equiv l_1$.

Every literal $l(p)$ satisfying (a) (frequently referred to as a *backbone literal* in SAT literature) uniquely determines the truth-value of p in every model of α. We can use this knowledge to simplify α by reducing either $\alpha(p/T)$, if $l(p) = p$, or $\alpha(p/F)$, if $l(p) = \neg p$ using equivalence preserving rewrite rules discussed in Section 1.1. The resulting simplified formula may now have positive or negative variables that could be used to further simplify the formula.

Pairs l_0, l_1 of literals satisfying (b) should be used to reduce the variable count in α through literal renaming, that is by replacing every occurrence of l_0 in α by l_1.

For reasons of efficiency, the dynamic formula simplification may employ reasoning methods of limited inference power and be restricted to specific classes of clauses only (e.g., unary, binary, or Horn). Table 4 shows the results of detecting literals satisfying (a) and (b) on sample SAT instances. The results are obtained using the **all–∨–random–∧** method for wish list generation and using Boolean Constraint Propagation style reasoning on binary learned clauses.

5   Guiding Local Search Using Learned Clauses

Conflict clause learning technique for clausal backtracking SAT solvers is a method of derivation and retention of clauses which record reasons for inconsistencies as represented by failed search paths (cf. [10]). Learned clauses are conjoined with an input formula preventing the exploration of other search paths whose reasons for failure have already been encoded by some of these clauses.

The clause learning method presented in this paper derives additional clauses from wish lists. In polSAT, a search failure is indicated by $CandList(\alpha, h)$ being non-empty and that can be recorded as a learned clause (assuming, of course, that $CandList(\alpha, h)$ is a wish list). These clauses, in turn, can be used in a number of ways to constrain the search for the best greedy move.

One way to incorporate learned clauses into polSAT's search heuristic is to modify the way the cumulative clash $clash(\alpha, h)$ is computed. Since every model of an input formula has to be a model of every learned clause, we may take the clauses false under the current assignment into consideration. This, too, can be done in a number of ways. First, one can simply conjoin α with all the learned clauses generated and retained by a solver, i.e., the new clash value for α and h, $clash^*(\alpha, h)$, is calculated as follows

$$clash^*(\alpha, h) = clash(\alpha \wedge C, h),$$

where C is the set of learned clauses generated and retained by a solver. This *eager approach* to clause learning can be contrasted with a more *cautious approach* which we briefly discuss next.

As it was reviewed in Section 2, a variable selection heuristic of the original pol-SAT algorithm selects a variable from $CandList(\alpha, h)$ either at random or by making a greedy move which selects a variable p that minimizes the cumulative clash value $clash(\alpha, h[p/1 - h(p)])$. In the greedy move, the heuristic resolves ties by selecting a best variable at random. Learned clauses can provide another way to brake ties by looking at the number of learned clauses becoming false after the flip of a variable's truth-value. Clearly the eager and cautious approaches could be combined, refined, or incorporated into other variable selection heuristics.

As our preliminary study shows, making clause learning work well in a non-clausal stochastic local search solver requires a carefully crafted learning strategy to support the main search heuristic of the solver. Adding any of the clause learning methods defined above to polSAT may significantly improve the performance of the solver on some SAT instances. However, in other cases, these clause learning methods seem to 'hijack' the search process causing the solver to perform more search steps to find models.

6   Refining polSAT

We can now use the theoretical concepts and results discussed in the previous sections to evolve polSAT into a more efficient solver.

6.1   From polSAT-G to polSAT-N

First, we can replace the candidate variable generation method of the original polSAT procedure with the **all–∨–random–∧** method. This replacement alone improves pol-SAT's performance considerably (see Table 1). Let us refer to this version of polSAT as polSAT-G.

(vs2)	with probability wp pick v from $CandList(\alpha, h)$ at random;
(vs3*)	with probability $1 - wp$ do the following:

- sort the variables r in $CandList(\alpha, h)$, in ascending order, by the $clash(\alpha, h[r/1 - h(r)])$ value; ties are broken according to variables' age (oldest first); if there are several variables with the same clash value and age, the one that has been added to $CandList(\alpha, h)$ first is chosen;
- let v_1 be the first and v_2 the second variable in the ordering;
- if v_1 does not have minimal age among the variables in $CandList(\alpha, h)$), then return v_1;
- else, with probability p return v_1 and with probability $1 - p$ return v_2.

Fig. 3. Variable selection heuristic of polSAT-N

Having improved the selection of candidate variables (cf. line (vs1) in Figure 1), one may now experiment with new variable selection heuristics (line (vs2)). Our next solver–polSAT-N–is obtained from polSAT-G by modeling the variable selection heuristic after *Adaptive Novelty+* heuristic ([11,4,5,6]). The new heuristic's behavior is controlled by two parameters: dynamically adjusted noise setting p, and the pre-defined walk probability, wp (see Figure 3). At each local-search step, the variable to flip is selected from the candidate list based on its clash value and on its age, that is on the number of local search steps since this variable was last flipped.

The noise setting p is adjusted dynamically, using essentially the same mechanism as in [6], except that the step threshold is calculated using the initial clash value of the input formula, instead of the number of clauses. Furthermore, this new heuristic assumes that $CandList(\alpha, h)$ is not a singleton. When such a set has only one variable, say v, then v is a backbone variable which can be used to simplify α.

As we shall see shortly, there is an empirical evidence that polSAT-N performs significantly better than both polSAT and polSAT-G.

6.2   Clause Learning and Formula Simplification in polSAT-G and -N

Since every set of candidate variables generated by new versions of polSAT is a wish list, we can augment the solver with clause learning and dynamic formula simplification.

The dynamic formula simplification is implemented by periodically inspecting all the binary learned clauses generated by the solver and, then, attempting to deduce backbone variables and equivalent literals from them. To this end, the solvers maintain an implication graph to store the binary clauses and use this structure to perform a Boolean Constraint Propagation style reasoning. When either backbone variables or equivalent literals are deduced, the input formula is simplified in the way discussed in Section 4 and the search resumes.

Our experiments with clause learning concentrated on *eager* approach as applied to polSAT-N. To this end, in the variable selection heuristic described in Figure 3, the $CandList(\alpha, h)$ was sorted using the function $clash^*$ defined in Section 5 rather than $clash$. During its execution, polSAT-N collected all the learned clauses and performed only rudimentary operations to manage them such as the removal of subsumed clauses. In view of the experimental results obtained during our studies, this implementation of the learning method should be viewed as a proof of concept.

7   Experimental Results

We performed series of experiments with the following objectives: to confirm the effectiveness of the new candidate list generation algorithm; to evaluate the effect of replacement of the original polSAT's search strategy with Adaptive Novelty+, which is known to perform well in the clausal setting; to compare polSAT's performance with a state-of-the-art incomplete clausal solver; and, to evaluate the effects of the addition of learning and reasoning mechanisms described in Sections 4 and 5. The evaluation of the algorithms is based on the analysis of run-length and run-time distributions (RLDs and RTDs) on variety of instances from different benchmark classes. The methodology for RLD/RTD based analysis of stochastic local search algorithms is described in detail in [7]. Due to the limited resources, some of the experiments could not be performed with a high enough cutoff value to guarantee close to 100% success rate - in such cases, the reader should account for the fact that the descriptive statistics are taken over all runs, including the unsuccessful ones. Finally, all the experiments were performed on Intel Xeon X5355, 2.66GHz, 4MB cache, 4GB RAM.

7.1 Benchmarks

To facilitate our testing objectives, we implemented a formula generator capable of producing random and structured formulas from various classes [1]. One of the features of our generator is that when the generated formula is outputted in non-clausal format (DIMACS SAT and ISCAS formats), *the same* formula is converted to CNF and outputted in DIMACS CNF. The conversion to clauses is performed in two ways: using the equivalence preserving transformation, and using the well-known Plaisted-Greenbaum structure preserving CNF translation, optimized to avoid introduction of unnecessary variables (see [12]).

The benchmark classes used in our experimental studies are listed below:

fs-200-560-3-3-2. This class consists of the fixed-shape random formulas introduced in [12]. Each formula in this class consists of 560 200-variable *hyper-clauses* of shape $\langle 3, 3, 2 \rangle$. Each such hyper-clause can be viewed as a 3 level tree with the 3-child disjunction at the root, followed by 3-child conjunction at the second level, followed by a 2-child disjunction of literals at the third level. In [12], authors performed a detailed study of such formulas, and conjectured a phase transition at the ratio of hyper-clauses to variables at about 3.02. We took the formulas slightly below the phase-transition to ensure the satisfiability (to our knowledge, no complete solver, clausal or not, is capable of handling formulas from this class).

fsf-300-354-2-2-3-2. Similar to the above, the formulas in this class consist of 354 300-variable hyper-clauses of shape $\langle 2, 2, 3, 2 \rangle$. The main difference this time is that we forced satisfiability of the generated formulas by seeding them with two complimentary satisfying assignments – this technique was suggested in [1], and was shown to produce formulas that are very difficult for local-search based methods.

parity8. The instances in this class represent the standard encoding of 8-bit Minimal Disagreement Parity (MDP) problem, as described in [2].

2dlx-cc-mc-ex-bp-f2-bugXXX. These instances are taken from the sss-sat-1.0 processor verification benchmark available from [16]. The benchmarks contain instances in both non-clausal (ISCAS) and clausal formats.

For each of the benchmark classes, we generated 100 instances, and performed a set of preliminary experiments to determine a noise value for each of the solvers involved in our experimental study. Additionally, for CNF versions of benchmarks, we selected the CNF translation that produces instances that are as easy as possible for the clausal

---

[1] The generator and the benchmarks used in this paper are available at
http://www.cse.yorku.ca/~antonb

solver involved in our study. To obtain the benchmark hardness distribution, we measured the median number of search steps over 100 runs for each instance and for each of the solvers. To select representative instances for the experiments, we were careful to choose instances that were equally hard for each of the solvers involved in the study. Thus, for each class the easy representative falls within 5% of the easiest instances for *all* solvers, medium representative falls within 3% of the median, and the hard representative falls within the 5% of the hardest instances.

7.2   The Effect of all–∨–random–∧ Candidate List Generation Strategy

Our first test was set up to determine the effect of the new candidate list generation strategy. To this end, we compared the performance of the original version of polSAT with the performance of polSAT-G. The results presented in Table 1 clearly indicate the superiority of **all–∨–random–∧** strategy both in terms of the number of search steps and in terms of CPU time. The disproportionally large improvement in CPU time is due to the fact that the new strategy is susceptible to a more efficient implementation.

Table 1. Basic descriptive statistics of RLD and RTD for polSAT and polSAT-G; all results are based on 1000 tries; cutoff is set at $5 \cdot 10^5$ for fsf-300-354-2-2-3-2.easy and 10^6 for the rest

instance	algorithm	success rate	mean flips	stddev/ mean flips	median flips	mean CPU (ms)
fs-200-560-3-3-2.easy	polSAT	97.3	180128	1.23	97179	5416
	polSAT-G	99.9	28485	1.45	15275	366
fsf-300-354-2-2-3-2.easy	polSAT	80.7	218970	0.82	163957	10462
	polSAT-G	100	36988	0.76	28243	303
parity8.medium	polSAT	42.6	n/a	n/a	>5000000	n/a
	polSAT-G	77.9	2536741	0.70	2191527	7059

7.3   polSAT-G vs. polSAT-N

The goal of the second set of experiments was to test our hypothesis that due to the independence of polSAT from a particular search strategy, the modifications to search strategy that are known to work well in the clausal setting might work well in the context of non-clausal framework described in this paper. To that end we compared the performance of polSAT-G and polSAT-N. The results of this set of experiments, some of which are presented in Table 2 (next page), clearly indicate that polSAT-N significantly outperforms polSAT-G both in terms of the number of search steps, and in terms of the CPU time.

7.4   Comparison with Clausal Solver

Since the variable selection in polSAT-N is performed using Adaptive Novelty+ algorithm, we compared the performance of our solver with the performance of a clausal solver that employs the same algorithm. We chose ubcsat [15] for comparative study due to the fact

Table 2. Basic descriptive statistics of RLD and RTD for polSAT-G and polSAT-N; all results are based on 1000 tries; cutoff is set at 10^5 for 2dlx-cc-mc-ex-bp-f2-bug099, $5 \cdot 10^6$ for parity8 instances, 10^6 for the rest

instance	algorithm	success rate	mean flips	stddev/ mean flips	median flips	mean CPU (ms)
fs-200-560-3-3-2.medium	polSAT-G	99.7	123218	1.21	69350	1753
	polSAT-N	100	13351	0.91	9619	236
fsf-300-354-2-2-3-2.medium	polSAT-G	99.9	759746	1.00	513189	6930
	polSAT-N	100	51749	0.94	35865	698
parity8.medium	polSAT-G	77.9	2536741	0.70	2191527	7059
	polSAT-N	99.7	816693	0.98	582834	1667
2dlx-cc-mc-ex-bp-f2-bug099	polSAT-G	99.8	7977	1.76	1782	55
	polSAT-N	100	1317	0.80	1034	7

Table 3. Basic descriptive statistics of RLD and RTD for polSAT-N and ubcsat with Adaptive Novelty+ heuristic; most of the results, with the exception of the following, are for 1000 tries with 10^6 cutoff for polSAT-N and 10^8 cutoff for ubcsat; for ubcsat on fsf-300-354-2-2-3-2.medium the results are for 100 tries with 10^9 cutoff; for 2dlx-cc-mc-ex-bp-f2-bug049 the results are for 100 tries with 10^7 cutoff for polSAT-N and 10^9 cutoff for ubcsat.

instance	algorithm	success rate	mean flips	stddev/ mean flips	median flips	mean CPU (ms)
fs-200-560-3-3-2.hard	polSAT-N	99.6	156911	1.00	107737	2799
	ubcsat/anovelty+	100	170770	0.93	117011	1043
fsf-300-354-2-2-3-2.easy	polSAT-N	100	10540	0.75	8384	143
	ubcsat/anovelty+	76.2	37885144	1.06	17055099	14344
fsf-300-354-2-2-3-2.medium	polSAT-N	100	51749	0.94	35865	698
	ubcsat/anovelty+	44	n/a	n/a	$> 10^9$	n/a
parity8.hard	polSAT-N	97.7	2583752	0.93	1804016	5275
	ubcsat/anovelty+	99.9	1484619	0.96	1029140	374
2dlx-cc-mc-ex-bp-f2-bug049	polSAT-N	100	2231360	0.87	1873929	21195
	ubcsat/anovelty+	50	n/a	n/a	987001033	n/a

that it is both a high-performing solver (in terms of raw flip speed), and a very convenient tool for experimentation. The representative results of our study are presented in Table 3.

Our results clearly demonstrate that there are classes of problems that can be solved very efficiently in non-clausal representation. Even though some classes, like parity-8, can be handled more efficiently in CNF, for others translation to CNF is extremely harmful: equivalence preserving transformation of fsf-300-354-2-2-3-2 formulas produces 5-6-CNF instances with approximately 88000 clauses which can not be solved by ubcsat. On the other hand, Plaisted Greenbaum structure preserving CNF translation of these formula introduces over 3000 extra variables, many of which are interdependent, which makes these formulas very difficult for local search (see [13]). Similarly, introduction of extra variables during translation of processor verification instances to

CNF (over 1600 extra variables for `2dlx-cc-mc-ex-bp-f2-bug049`) makes them very hard for clausal local search.

7.5   Learning and Formula Simplification

In this section we discuss the results of evaluation of the effects of the addition of clause learning and formula simplification techniques to polSAT-N, as described in Sections 4 and 5. As it has been already mentioned, making clause learning work well in a non-clausal stochastic local search solver requires additional study of the interaction between the main heuristic of the solver and the clause learning engine.

Table 4 shows that in some cases even a very rudimentary form of clause learning, such as the eager approach discussed in Section 5, may result in a significant improvement in the performance of a solver. For example, the results for `parity8` instances demonstrate that the savings in the number of search steps could be significant enough to offset the additional computational effort required to perform learning, and could result in the improvement in the CPU time. However, in other cases, we observed that the clause learning interfered negatively with the polSAT-N's search heuristic and that resulted in a diminished performance of polSAT-N with learning.

Our results also indicate that even restricting reasoning to binary learned clauses polSAT-N was still able to simplify input formulas (see Table 4)

Table 4. Basic descriptive statistics of RLD and RTD for polSAT-N with and without the addition of learning and formula simplification techniques

instance	reasoning mechanism	success rate	mean flips	stddev/ mean flips	median flips	mean CPU (ms)
`fs-200-560-3-3-2.hard`	none	99.6	156911	1.00	107737	2799
	learning	100	121883	0.82	94283	4228
`fsf-300-354-2-2-3-2.medium`	none	100.0	51749	0.94	35865	698
	learning	99.6	67826	1.64	36456	5526
`parity8.hard`	none	99.7	2583752	0.93	1804016	5275
	learning	100	1354522	0.98	944137	3612
	simplification	100	1740627	0.92	1309759	4491
`2dlx-cc-mc-ex-bp-f2-bug049`	none	100	2231360	0.87	1873929	21195
	simplification	99.7	2152581	1.00	1451799	28152

8   Concluding Remarks

The notions of a wish list and a learned clause introduced and studied in this paper offer other promising directions for future studies. One of them is the possibility of building complete hybrid clausal/non-clausal SAT solvers. Learned clauses deduced during the execution of a non-clausal local search solver can be subjected to automated reasoning using a variety of clause-based tools from SAT solvers to refutational proof systems. In Section 4 we discussed the use of clause-based automated deduction for the purpose of input formula simplification. Clausal reasoning can also be used to detect and certify the unsatisfiability of input formulas. Indeed, if an input formula α is unsatisfiable then, by Theorem 3.7, the set of all learned clauses that can be derived from α is inconsistent. Achieving completeness of a stochastic local search SAT procedure in such a way has been suggested by other authors. A framework for complete clause-based local search proposed in [3] uses a clause generator to derive implied clauses of an input formula α and to conjoin them with α. The generation of the empty clause indicates that $\alpha \notin SAT$.

Another research direction, that has already been initiated by the authors, is the extension of the polSAT framework to directly handle other non-clausal representations of input formulas such as DAGs. There have been some previous work on DAG-based stochastic local search, most notably that by Kautz, McAllester, and Selman [9], which identifies classes of Boolean satisfiability problems that could benefit from such an extended framework.

Acknowledgments. We thank the anonymous referees for helpful comments.

References

1. Achlioptas, D., Jia, H., Moore, C.: Hiding Satisfying Assignments: Two are Better than One. J. of Artificial Intelligence Research 24, 623–639 (2005)
2. Crawford, J.M., Kearns, M.J., Shapire, R.E.: The Minimal Disagreement Parity Problem as Hard Satisfiability Problem. Computational Intell. Research Lab and AT&T Bell Labs TR (1994)
3. Fang, H., Ruml, W.: Complete Local Search for Propositional Satisfiability. In: AAAI, pp. 161–166 (2004)
4. Hoos, H.H.: Local Search – Methods, Models, Applications. TU Dermstadt, FB Informatik, Darmstadt, Germany (1998)
5. Hoos, H.H.: On the Run-Time Behavior of Stochastic Local Search Algorithms for SAT. In: AAAI/IAAI, pp. 661–666 (1999)
6. Hoos, H.H.: An Adaptive Noise Mechanism for WalkSAT. In: AAAI, pp. 655–660 (2002)
7. Hoos, H.H., Stutzle, T.: Local Search Algorithms for SAT: An Empirical Evaluation. Journal of Automated Reasoning 24, 421–481 (2000)
8. Hoos, H.H., Stutzle, T.: Stochastic Local Search: Foundations and Applications. Elsevier, Amsterdam (2005)
9. Kautz, H., Selman, B., McAllester, D.: Exploiting Variable Dependency in Local Search. In: IJCAI (1997)
10. Lynce, I., Marques-Silva, J.P.: An Overview of Backtrack Search Satisfiability Algorithms. Annals of Mathematics and Artificial Intelligence, 307–326 (2003)
11. McAllester, D., Selman, B., Kautz, H.: Evidence for Invariants in Local Search. In: AAAI, pp. 321–326 (1997)
12. Navarro, J.A., Voronkov, A.: Generation of Hard Non-Clausal Random Satisfiability Problems. In: AAAI, pp. 436–442 (2005)
13. Prestwich, S.D.: Variable Dependency in Local Search: Prevention Is Better Than Cure. In: Marques-Silva, J., Sakallah, K.A. (eds.) SAT 2007. LNCS, vol. 4501, pp. 107–120. Springer, Heidelberg (2007)
14. Stachniak, Z.: Going Non-clausal. In: SAT, pp. 316–322 (2002)
15. Tompkins, D.A., Hoos, H.: UBCSAT: An Implementation and Experimentation Environment for SLS Algorithms for SAT and MAX-SAT. In: H. Hoos, H., Mitchell, D.G. (eds.) SAT 2004. LNCS, vol. 3542, pp. 306–320. Springer, Heidelberg (2005)
16. Velev, M.: Miroslav Velev's SAT Benchmarks,
   http://www.miroslav-velev.com/sat_benchmarks.html

Local Restarts

Vadim Ryvchin[1,2] and Ofer Strichman[1]

[1] Information Systems Engineering, IE, Technion, Haifa, Israel
[2] Design Technology Solutions Group, Intel Corporation, Haifa, Israel
rvadim@tx.technion.ac.il, ofers@ie.technion.ac.il

Abstract. Most or even all competitive DPLL-based SAT solvers have a "restart" policy, by which the solver is forced to backtrack to decision level 0 according to some criterion. Although not a sophisticated technique, there is mounting evidence that this technique has crucial impact on performance. The common explanation is that restarts help the solver avoid spending too much time in branches in which there is neither an easy-to-find satisfying assignment nor opportunities for fast learning of strong clauses. All existing techniques rely on a global criterion such as the number of conflicts learned as of the previous restart, and differ in the method of calculating the threshold after which the solver is forced to restart. This approach disregards, in some sense, the original motivation of focusing on 'bad' branches. It is possible that a restart is activated right after going into a good branch, or that it spends all of its time in a single bad branch. We suggest instead to localize restarts, i.e., apply restarts according to measures local to each branch. This adds a dimension to the restart policy, namely the decision level in which the solver is currently in. Our experiments with both Minisat and Eureka show that with certain parameters this improves the run time by 15% - 30% on average (when applied to the 100 test benchmarks of SAT-race'06), and reduces the number of time-outs.

1 Global vs. Local Restarts

Most or even all competitive DPLL SAT solvers have a "restart" policy, a strategy initially proposed by Gomes et. al [3]. The solver is restarted after a certain number of conflict clauses have been learned. The fact that new clauses have been added to the clause database deviates the search from one restart to the next. In those solvers that is relevant, the search is changed also owing to randomness.

Different restart policies are used by different solvers. A recent survey by Huang [4] includes several types of restart policies. We briefly describe various types of popular restart techniques based on that survey and on some new developments.

1. *Arithmetic (or fixed) series.* Parameters: x, y. A policy in which there is a restart every x conflicts, which is increased by y every restart. Some sample values are: in zchaff 2004 $x = 700$, in Berkmin $x = 550$, in Siege $x = 16000$ and in Eureka $x = 2000$. In all of these solvers the series is in fact fixed (i.e.,

H. Kleine Büning and X. Zhao (Eds.): SAT 2008, LNCS 4996, pp. 271–276, 2008.

$y = 0$), owing to the observation that completeness is meaningless in the realm of timeouts.

2. *Geometric series.* Parameters: x, y. A policy in which the initial interval is x, which is then multiplied by a factor of y in each restart, for some $y > 1$. This policy is used in Minisat-2 with $x = 100$ conflicts and $y = 1.5$.

3. *Inner-Outer Geometric series.* Parameters: x, y, z. An idea suggested by Biere and implemented in PicoSAT [1], by which restarts follow what can be seen as a two dimensional pattern that increases geometrically in both dimensions. The inner loop multiplies a number initialized to x, by z, at each restart. When this number is larger than a threshold y, it is reset back to x and the threshold y is also multiplied by z (this is the outer loop). Hence, both the inner and outer loops follow a geometric series, and the whole series creates an oscillating pattern.

4. *Luby et al. series* [5]. Parameter: x. A policy in which restarts are performed according to the following series of numbers: 1,1,2,1,1,2,3,1,1,2,1,1,2,3,4,... multiplied by the constant x (called the *unit-run*). Formally, let t_i denote the i-th number in this series. Then t_i is defined recursively:

$$t_i = \begin{cases} 2^{k-1} & \text{if } \exists k \in \mathbb{N}.\ i = 2^k - 1 \\ t_{i-2^{k-1}+1} & \text{if } \exists k \in \mathbb{N}.\ 2^{k-1} \le i < 2^k - 1 \end{cases}$$

This is a well-defined series, as the two conditions are mutually-exclusive. This policy has some nice theoretical characteristics in a class of randomized algorithms called Las Vegas algorithms[1], but the relevance of these results to DPLL has only been empirical so far – it is not clear what is the reason that it works well in practice. The experiments reported in [4] show that it outperforms the other restart strategies, and indeed this is now the restart method of choice of several state-of-the-art solvers, such as TinySAT [4] and RSAT [8].

For completeness of this list, we should also mention that there is a family of techniques in which 'restart' does not entail backtracking to level 0, but rather to some decision level which is lower than what is computed as the backtracking level by a conflict analysis procedure. Such a procedure was proposed, for example, by Lynch [6]. We did not experiment with these techniques, however.

All of the strategies listed above are based on a global counter of conflict clauses, and therefore they measure progress over many branches together. Assuming that the motivation for restarts is to prevent the solver from getting stuck in a bad branch (which can, informally, be defined as a branch which neither contains an easy-to-find satisfying assignment nor leads to efficient learning that directs the solver to a different search-space or to a proof of unsatisfiability), such a global policy may miss the point.

For example, it is possible that the solver spent a significant amount of time searching in a branch, eventually left it, and very soon after that it restarts (since

---

[1] Algorithms that use randomness, but the quality of the result is not affected by it. Typically randomness in such algorithms only affects run-times.

the global threshold was reached), although there is no knowledge yet about the potential of the current branch. It is also possible that the restart is too late, for example if it spends all its time between restarts in a single bad branch.

A possibly better strategy is to localize the measure of difficulty of branches, and restart when the branch is more difficult than some threshold. Each of the global strategies mentioned above can be applied locally, because we can count the number of conflicts under each branch easily, as follows. For each decision level d we maintain a counter $c(d)$, which is initially (when a decision is made at that level) set to the global number of conflicts. When backtracking back to that level, we examine the difference between the current global number of conflicts, and $c(d)$. This difference reflects the number of conflicts that were encountered above level d, since the last time a decision was made at this level. If this difference is larger than some strategy-dependent threshold, we restart.

Locality opens a new dimension, namely that of the decision level. In other words, the threshold can be a function of the level in which the solver is currently in. We call such strategies *dynamic*. It can be expected that the work done between two visits to a decision level (from decision to backtracking back to that level) will be smaller as the level increases. Also, we collected statistics regarding the size of learned clauses at each level, and it shows that conflict clauses at low decision levels are smaller on average. Hence giving less chance to deeper levels forces the solver to learn stronger facts first. Each of the strategies above can be made dynamic, although in strategies in which the series oscillates as in Luby et al. and the Inner-Outer strategy, it is not clear how to add this new dimension. We focused, then, on the following strategy:

5. Dynamic-fix. Parameters: x, y, d, min. A policy in which at decision level i there is a restart every $\max(x - i \cdot d, min)$ conflicts, which is increased by y every restart.

Making the strategy local instead of global requires re-tuning of the parameters – there is no reason to believe that parameters that optimize a global restart policy also optimize a local one. Hence a major empirical evaluation is needed in order to check the effect of locality on each of these strategies.

2   Experimental Results and Conclusions

The table in Fig. 1 shows results with 40 different restart configurations, when implemented on top of MINISAT 2007 [2], and ran on the 100 industrial benchmarks that were used as preparation for SAT-race'06 (divided evenly to the two test-sets TS1 and TS2). A similar table for the latest version of Eureka [7], with 41 configurations, appears in Fig. 2. The set of configurations is not identical, but close, because we chose them dynamically: when a good strategy was found, we tried to change it incrementally. The tables are sorted according to the type of strategy, local/global, and parameters. The third column indicates whether this strategy is implemented globally or locally. Timeout was set to 30 minutes. Instances that timed-out are included and contribute 30 minutes (we added them

Place	Strategy	G/L	Parameters	TS1				TS2				Overall	
				SAT	UNSAT	TO	Total	SAT	UNSAT	TO	Total	TO	Time
✓3	Arith	L	100,10	1.12	2.06	4	3.18	2.17	2.59	6	4.75	10	7.93
26	Arith	L	10,1	2.12	2.62	6	4.74	2.42	2.99	6	5.41	12	10.15
8	Arith	L	100,1	1.89	1.96	4	3.85	2.37	2.84	6	5.21	10	9.05
6	Arith	L	100,20	2.49	1.99	6	4.48	2.32	2.21	5	4.53	11	9.02
12	Arith	L	100,40	2.51	1.95	6	4.47	2.11	2.74	6	4.86	12	9.33
10	Arith	L	1000,0.1	2.3	2.05	4	4.35	1.89	2.85	6	4.74	10	9.09
9	Arith	L	1000,1	2.15	1.93	5	4.08	2.07	2.9	6	4.97	11	9.05
32	Arith	L	1000,10	2.76	2.13	7	4.89	2.72	2.99	8	5.71	15	10.6
34	Arith	L	1000,20	3.13	2.07	8	5.2	2.61	2.93	5	5.54	13	10.74
21	Arith	L	2500,1	2.11	2.38	6	4.49	2.37	3.03	7	5.39	13	9.89
24	Arith	L	3,1	2.47	1.87	3	4.34	2.88	2.81	9	5.69	12	10.03
29	Arith	L	3,10	2.69	1.92	6	4.61	2.95	2.92	9	5.87	15	10.48
14	Arith	L	5,0.2	2.41	1.62	6	4.04	2.59	2.85	8	5.43	14	9.47
15	Arith	L	5000,1	2.33	2.48	7	4.81	2.13	2.56	4	4.69	11	9.5
18	Arith	L	6,1	2.02	2.23	5	4.25	2.61	2.86	8	5.46	13	9.71
27	Geom.	L	10,1.1	2.53	2.03	6	4.56	2.5	3.18	8	5.68	14	10.24
37	Geom.	L	10,1.5	2.46	2.63	7	5.08	2.62	3.29	6	5.91	13	10.99
40	Geom.	L	10,2	2.89	2.77	9	5.65	3.03	3.39	9	6.42	18	12.07
16	Geom.	L	100,1.1	1.71	2.16	3	3.86	2.55	3.14	8	5.69	11	9.56
38	Geom.	L	100,1.5	3.33	2.71	9	6.03	2.94	2.77	6	5.71	15	11.75
36	Geom.	L	100,2	2.33	2.86	7	5.19	2.42	3.35	7	5.76	14	10.95
33	Geom. *	G	100,1.5	1.6	2.76	6	4.36	3.06	3.22	8	6.28	14	10.64
11	IO	G	100,1000,1.1	2.68	2.07	6	4.75	1.72	2.86	7	4.57	13	9.32
4	IO	G	100,1000,1.5	1.81	2.04	4	3.86	2.04	2.97	6	5	10	8.86
39	IO	G	100,1000,2	2.81	2.16	8	4.97	3.33	3.48	10	6.81	18	11.78
✓1	IO	L	100,1000,1.1	1.59	2	4	3.59	1.27	2.51	4	3.78	8	7.38
7	IO	L	100,1000,1.5	2.22	2.02	5	4.24	1.92	2.88	6	4.8	11	9.04
30	IO	L	100,1000,2	2.89	2.22	8	5.11	2.6	2.79	7	5.39	15	10.5
22	Luby	G	32	2.22	1.49	3	3.71	3.06	3.15	10	6.21	13	9.91
23	Luby	G	128	3.08	1.76	6	4.84	2.21	2.89	7	5.1	13	9.94
13	Luby	G	512	2.84	1.93	7	4.77	1.92	2.64	5	4.56	12	9.33
5	Luby	G	1024	2.26	1.97	5	4.22	2.02	2.74	6	4.76	11	8.98
✓2	Luby	L	32	1.6	1.15	3	2.75	2.22	2.92	6	5.14	9	7.89
25	Luby	L	128	2.75	2.01	7	4.76	2.29	3.02	7	5.32	14	10.08
17	Luby	L	512	2.18	2.08	5	4.26	2.33	3.1	6	5.43	11	9.69
19	Luby	L	1024	2.71	2.02	4	4.73	1.94	3.05	7	5	11	9.73
28	D-arith	L	1000,0.1,10,10	3.45	1.02	6	4.47	2.7	3.13	8	5.84	14	10.31
20	D-arith	L	1000,0.1,20,10	2.92	0.99	4	3.91	2.77	3.1	8	5.87	12	9.78
31	D-arith	L	1000,10,10,10	3.5	2	8	5.51	1.64	3.41	7	5.05	15	10.56
35	D-arith	L	1000,10,20,10	3.22	2.02	8	5.24	2.25	3.4	8	5.65	16	10.89

Fig. 1. Results, in hours, based on MINISAT 2007. The original configuration of MIN-ISAT 2007 is marked with *.

to the SAT or UNSAT column according to our prior knowledge of the expected result). Instances that none of our configurations nor any SAT'06-race competitor can solve are not included. The overall number of timeouts and total run time are given in the last two columns, where time is measured in hours. All together the two tables represent over 40 days of CPU time.

The first column indicates the position of each solver when measured by the total run time, and the best three configurations according to this measure are preceded by '✓'. With both solvers, the best three configurations that we tried are local (also when measured by time-outs).

To the extent that the benchmark set is representative of industrial problems, and that MiniSat 2007 and Eureka represent state-of-the-art solvers, it

Place	Strategy	G/L	Parameters	TS1				TS2				Overall	
				SAT	UNSAT	TO	Total	SAT	UNSAT	TO	Total	TO	Time
39	Arith	L	10,0.1	2.34	1.26	4	3.6	2.78	4.22	11	7	15	10.59
38	Arith	L	10,1	1.92	1.67	4	3.59	2.93	4.06	10	6.98	14	10.58
41	Arith	L	100,1	2.19	1.63	3	3.81	3.24	4.04	10	7.28	13	11.09
17	Arith	L	100,10	1.78	1.11	2	2.89	2.8	3.44	7	6.24	9	9.13
√2	Arith	L	1000,1	1.6	1.04	2	2.64	2.74	2.72	6	5.46	8	8.09
5	Arith	L	1000,10	1.63	0.96	2	2.59	3.05	2.68	5	5.72	7	8.31
√1	Arith	L	1000,20	1.83	0.92	2	2.75	2.57	2.67	5	5.24	7	7.98
40	Arith	L	20,0.1	2.47	1.35	4	3.82	2.65	4.23	11	6.87	15	10.69
31	Arith	L	20,1	2.4	1.32	3	3.72	2.63	3.69	9	6.32	12	10.04
14	Arith	L	2000,1	1.76	1.1	2	2.86	3.4	2.81	6	6.21	8	9.08
32	Arith	L	3,1	2.04	1.19	3	3.23	3.4	3.43	9	6.83	12	10.06
8	Arith	L	3,10	1.63	1	2	2.63	2.66	3.24	6	5.89	8	8.52
4	Arith	L	3,20	1.7	0.9	2	2.6	2.47	3.21	7	5.68	9	8.28
21	Arith	L	3,40	1.79	0.92	2	2.71	3.54	3.39	8	6.93	10	9.64
37	Arith	L	5,0.2	2.29	1.23	3	3.53	3.17	3.85	10	7.02	13	10.55
18	Arith	L	5000,1	1.71	1.08	2	2.79	3.01	3.44	7	6.45	9	9.24
19	Arith*	G	2000,0	2.15	1.07	3	3.22	3.17	3	6	6.17	9	9.39
29	Geom.	L	10,1.1	2.2	1.07	3	3.26	3.27	3.49	9	6.76	12	10.03
36	Geom.	L	10,1.5	1.89	1.1	2	2.99	3.17	4.23	10	7.4	12	10.39
25	Geom.	L	10,2	1.96	1.32	2	3.28	3.14	3.38	9	6.52	11	9.80
11	Geom.	L	100,1.1	1.98	0.9	2	2.88	2.8	3.1	7	5.9	9	8.78
28	Geom.	L	100,1.5	1.73	0.95	2	2.68	3.46	3.78	9	7.24	11	9.93
30	Geom.	L	100,2	2.11	1.01	2	3.12	3.16	3.75	7	6.91	9	10.04
10	IO	G	100,1000,1.1	1.54	0.93	2	2.47	3.05	3.12	7	6.17	9	8.64
15	IO	G	100,1000,1.5	1.59	0.9	1	2.49	3.01	3.57	8	6.58	9	9.08
26	IO	G	100,1000,2	2.12	0.87	3	2.99	3.34	3.48	8	6.83	11	9.82
√3	IO	L	100,1000,1.1	1.72	0.88	2	2.6	2.82	2.7	6	5.52	8	8.12
22	IO	L	100,1000,1.5	2.19	0.86	3	3.05	3.14	3.55	8	6.68	11	9.73
34	IO	L	100,1000,2	2.34	1.1	3	3.44	3.13	3.76	8	6.88	11	10.32
16	Luby	G	32	1.83	1.03	3	2.86	2.97	3.29	7	6.26	10	9.12
12	Luby	G	128	2.17	0.87	2	3.05	2.92	2.94	7	5.86	9	8.90
13	Luby	G	512	1.59	1	2	2.59	3.18	3.27	7	6.46	9	9.05
23	Luby	G	1024	2.22	1.09	3	3.31	3.58	2.88	6	6.46	9	9.76
9	Luby	L	32	1.67	0.94	1	2.61	2.75	3.17	7	5.92	8	8.53
7	Luby	L	128	1.71	0.91	1	2.62	2.84	2.96	6	5.79	7	8.41
6	Luby	L	512	1.6	0.94	2	2.54	3.14	2.72	6	5.86	8	8.40
27	Luby	L	1024	2.33	1.1	3	3.43	3.6	2.87	7	6.47	10	9.90
24	D-arith	L	1000,0.1,10,10	1.91	1.34	3	3.25	3.26	3.27	8	6.53	11	9.77
35	D-arith	L	1000,0.1,20,10	1.86	1.71	4	3.57	3.15	3.66	9	6.81	13	10.38
20	D-arith	L	1000,10,10,10	1.88	1.2	2	3.08	3.25	3.28	5	6.53	7	9.61
33	D-arith	L	1000,10,20,10	1.82	1.31	2	3.13	3.25	3.74	8	6.98	10	10.11

Fig. 2. Results, in hours, based on EUREKA. The original configuration of EUREKA is marked with *.

seems that locality can help with the four types of strategies that we tried. The following table shows, for the Luby and Inner-Outer strategies, the figures corresponding to the best local and best global configurations that we could find.

Strategy	Minisat				Eureka			
	Global		Local		Global		Local	
	TO	Time	TO	Time	TO	Time	TO	Time
Luby	11	8.98	9	7.89	9	8.90	8	8.40
IO	10	8.86	8	7.38	9	8.64	8	8.12

There seems to be such an advantage for the local geometric and local arithmetic strategies as well, but more global configurations of these strategies need to be tested in order to draw concrete conclusions. If we take the default parameters of Minisat and Eureka as best of their respective global strategies, then this can be said with some confidence.

What about the dynamic strategy? it does not seem to score well in general, at least not with the 4 parameters set that we tried, but it performs well with unsatisfiable instances. In the case of the first table (Minisat), the dynamic strategies with parameters 1000,0.1,20,10 and 1000,0.1,10,10 arrive at the second and third places, respectively, if we measure only unsatisfiable instances. More parameters and variations of this strategy are necessary in order to see if it can become competitive in the general case.

We are currently trying more configurations and looking for other measures for the quality of the branch that can be checked with a marginal cost in run-time. It is possible that measures such as the size of the backtrack can be factored in the restart policy.

References

1. A. Biere. PicoSAT essentials. JSAT (2008) (to be published)
2. Een, N., Sorensson, N.: Minisat v2.0 (beta). In: Solvers description, SAT-race (2006), http://fmv.jku.at/sat-race-2006/descriptions/27-minisat2.pdf
3. Gomes, C.P., Selman, B., Kautz, H.A.: Boosting combinatorial search through randomization. In: AAAI/IAAI, pp. 431–437 (1998)
4. Huang, J.: The effect of restarts on the efficiency of clause learning. In: IJCAI, pp. 2318–2323 (2007)
5. Luby, M., Sinclair, A., Zuckerman, D.: Optimal speedup of Las Vegas algorithms. In: ISTCS, pp. 128–133 (1993)
6. Baptista, L.L., Marques Silva, J.P.: Stochastic systematic search algorithms for satisfiability. In: LICS Workshop on Theory and Applications of Satisfiability Testing, pp. 190–204 (2001)
7. Nadel, A., Gordon, M., Palti, A., Hana, Z.: Eureka-2006 SAT solver. In: Solvers description, SAT-race (2006)
8. Pipatsrisawat, K., Darwiche, A.: Rsat 2.0: SAT solver description. In: SAT competition 2007 (2007)

Regular and General Resolution: An Improved Separation

Alasdair Urquhart[*]

Department of Computer Science
University of Toronto
Toronto, Ontario M5S 1A4,
Canada
urquhart@cs.toronto.edu

Abstract. This paper gives an improved separation between regular and unrestricted resolution. The main result is that there is a sequence $\Pi_1, \Pi_2, \ldots, \Pi_i, \ldots$ of sets of clauses for which the minimum regular resolution refutation of Π_i has size $2^{\Omega(R_i/(\log R_i)^7)}$, where R_i is the minimum size of an unrestricted resolution refutation of Π_i. This improves earlier lower bounds for which the separations proved were of the form $2^{\Omega(\sqrt[3]{R})}$ and $2^{\Omega(\sqrt[4]{R}/(\log R)^3)}$.

1 Introduction

1.1 The Regularity Restriction

This paper proves an improved separation between the size of regular and unrestricted resolution refutations of sets of clauses. This provides a nearer approach to an optimal separation between these two propositional proof systems than earlier results.

The regularity restriction was first introduced by Grigory Tseitin in a groundbreaking article [1], the published version of a talk given in 1966 at a Leningrad seminar. This restriction is very natural, in the sense that algorithms such as that of Davis, Logemann and Loveland [2] (the prototype of almost all satisfiability algorithms used in practice today) can be understood as a search for a regular refutation of a set of clauses. If refutations are represented as trees, rather than directed acyclic graphs, then minimal-size refutations are regular, as can be proved by a simple pruning argument [3, p. 436].

The main result of Tseitin's paper [1] is an exponential lower bound for regular resolution refutations of contradictory CNF formulas based on graphs. Tseitin makes the following remarks about the heuristic interpretation of the regularity restriction:

> The regularity condition can be interpreted as a requirement for not proving intermediate results in a form stronger than that in which they

---

[*] The author gratefully acknowledges the support of the Natural Sciences and Engineering Research Council of Canada.

H. Kleine Büning and X. Zhao (Eds.): SAT 2008, LNCS 4996, pp. 277–290, 2008.
© Springer-Verlag Berlin Heidelberg 2008

are later used (if A and B are disjunctions such that $A \subseteq B$, then A may be considered to be the stronger assertion of the two); if the derivation of a disjunction containing a variable ξ involves the annihilation of the latter, then we can avoid this annihilation, some of the disjunctions in the derivation being replaced by "weaker" disjunctions containing ξ.

These remarks of Tseitin suggest that there is always a regular resolution refutation of minimal size, as in the case of tree resolution. Consequently, some researchers tried to extend Tseitin's results to general resolution by showing that regular resolution can simulate general resolution efficiently. However, these attempts were doomed to failure.

The first example of a contradictory CNF formula whose shortest resolution refutation is irregular was given by Wenqi Huang and Xiangdong Yu [4]. Subsequently, Andreas Goerdt [5] gave the first super-polynomial separation between regular resolution and unrestricted resolution by constructing a family of formulas that have polynomial-size resolution refutations, but require super-polynomial size regular resolution refutations.

Goerdt's results were improved to an exponential separation in a paper by Alekhnovich, Johannsen, Pitassi and Urquhart [6]. The paper in fact contains two separate proofs of an exponential separation. The first presents a sequence $GT'_{n,\rho}$ of sets of clauses that have general resolution refutations with size $O(n^3)$, but require regular resolution refutations of size $2^{\Omega(n)}$. The second gives an infinite sequence of sets of clauses $Stone(G, S)$ based on a pebbling problem that have general resolution refutations with size $O(n^4)$, but require regular resolution refutations of size $2^{\Omega(n/(\log n)^3)}$.

Hence, the best separations so far between regular and general resolution are of the form $2^{\Omega(\sqrt[3]{R})}$, and $2^{\Omega(\sqrt[4]{R}/(\log R)^3)}$, where R is the size of the smallest general resolution refutation of the set of clauses in question. It is natural to ask whether we can improve these separations. In fact, we know that we cannot do better than a $2^{\Omega(R \log \log R/ \log R)}$ separation. This is because Ben-Sasson, Impagliazzo and Wigderson [7] showed that if R is the size of a general resolution refutation of a set of clauses, then there is a tree resolution refutation with size $2^{O(R \log \log R/ \log R)}$. Since a tree resolution refutation of minimal size is regular, it follows that the same upper bound holds for regular resolution.

The present paper makes a closer approach to a matching lower bound. The main result is that there is a sequence of contradictory sets of clause Π_i, with an associated unbounded parameter $n = n(i)$, so that Π_i has a general resolution refutation with size $O(n(\log n)^7)$, but any regular resolution refutation has size $2^{\Omega(n/[(\log n)^2 \log \log n])}$ The proof of this result is an amalgamation and extension of ideas underlying the two previous separation results.

1.2   Preliminaries

A *literal* is a propositional variable x or its negation $\neg x$. A *clause* is a set of literals, interpreted as the disjunction of the set. For clauses containing exactly one positive literal, we use the implication $p_1, \ldots, p_k \to q$ as alternative notation for the clause $\neg p_1 \lor \cdots \lor \neg p_k \lor q$. The *resolution rule* allows us to derive the *resolvent* $C \lor D$ from the clauses $C \lor x$ and $D \lor \neg x$ by *resolving* on the variable x; a clause $C \lor D$ can also be derived from a clause C by *weakening*. A *resolution derivation* of a clause C from a set of clauses Σ consists of a sequence of clauses in which each clause is either a clause of Σ, or derived from earlier clauses by resolution or weakening, and C is the last clause in the sequence; it is a *refutation* of Σ if C is the empty clause Λ. The *size* $|\mathcal{R}|$ of a refutation $\mathcal{R}$ is the number of resolvents in it. We can represent it as a directed acyclic graph (dag) where the nodes are the clauses in the refutation, each clause of F has out-degree 0, and any other clause has one or two arcs pointing to the clause or clauses from which it is derived. Resolution is a *sound* and *complete* propositional proof system, that is to say, a set of clauses Σ is unsatisfiable if and only if there is a resolution refutation for Σ.

A resolution refutation is *regular* if on any path from Λ to a clause in F (in the directed acyclic graph associated with the refutation), each variable is resolved on at most once along the path.

It is sometimes helpful to view a regular resolution refutation as a branching program. Representing the refutation as a dag, let us say that a variable x is *queried at a node* q in the dag if q is labelled with a clause $C \lor D$, derived from parent clauses $C \lor x$ and $D \lor \neg x$ by resolving on x. Starting from the empty clause Λ at the root of the dag, we can construct a path in the refutation by answering the queries occurring in the path; the answers determine an assignment to the variables queried along the path. The path is chosen so that the assignment falsifies all the clauses in the path. Thus, if the variable x is queried at a node, and the answer is "false," then the next node in the path is labelled with the parent clause containing the literal x; similarly for the answer "true." If $C \lor D$ is derived by weakening from C, then the path continues to C. The path constructed in this way must end with an initial clause falsified by this assignment.

An *assignment* (*restriction*) for a set of clauses is a Boolean assignment to some of the variables in the set; the assignment is *total* if all the variables in the set are assigned values. If C is a clause, and σ an assignment, then we write $C \restriction \sigma$ for the result of applying the assignment to C, that is, $C \restriction \sigma = 1$ if $\sigma(l) = 1$ for some literal l in C, otherwise, $C \restriction \sigma$ is the result of removing all literals set to 0 by σ from C. If Σ is a set of clauses, then $\Sigma \restriction \sigma$ is the set of clauses $C \restriction \sigma$, C a clause in Σ.

If $\mathcal{R}$ is a resolution refutation of Σ, and σ a restriction for Σ, then we define the *restriction* $\mathcal{R} \restriction \sigma$ of $\mathcal{R}$ to be the sequence of clauses resulting from $\mathcal{R}$ by replacing all of the clauses C in $\mathcal{R}$ by $C \restriction \sigma$, and then removing all of the clauses set to 1. It is easy to verify that $\mathcal{R} \restriction \sigma$ is a resolution refutation of $\Sigma \restriction \sigma$, and that $\mathcal{R}$ is regular, if $\mathcal{R}$ is regular.

If Σ is a set of clauses, and x, y are variables in Σ, or the propositional constant $\perp$, then we say that there is an *implicational chain from x to y in Σ* if there is a sequence $x = x_0, \ldots, x_k = y$ of variables (or constants) and a sequence $C_1, \ldots, C_k$ of clauses so that for all i, $0 < i \leq k$, x_{i-1} occurs negatively and x_i positively in C_i.

The notation $\log x$ stands for the base two logarithm of x, and $\ln x$ the natural logarithm of x.

2   Pebbling Games and Pebbling Formulas

2.1   The Pebbling Game

A *pointed graph* G is a directed acyclic graph where all nodes have indegree at most two, having a unique sink, or target node, to which there is a directed path from all the nodes in G. It is *binary* if all nodes except for the source nodes have indegree two. If v is a node in a pointed graph G, then $G \restriction v$ is the subgraph of G restricted to the nodes from which there is a directed path to v.

The *pebbling game* played on a pointed graph G is a one-player game in the course of which pebbles are placed on or removed from nodes in G. The rules of the game are as follows;

1. A pebble may be placed on a source node at any time.
2. If all predecessors of a node u are marked with pebbles, then a pebble may be placed on node u.
3. A pebble may be removed from a node at any time.

A *move* in the game consists of the placing or removing one of the pebbles in accordance with one of the three rules. The *configuration* at a given stage in the game is the set of nodes in G that are marked with a pebble. The goal of the game is to place a pebble on the sink node t, while minimizing the number of pebbles used (that is, minimizing the number of pebbles on the graph at any stage of the game). Thus a successful play of the game can be presented as a sequence of configurations $C_0, \ldots, C_k$, where $C_0 = \emptyset$ and $t \in C_k$.

A *strategy* for the game is a sequence of moves following the rules of the game that ends in pebbling the target node. The *cost* of such a strategy is the minimum number of pebbles required in order to execute it, that is to say, the size of the largest configuration in the sequence of configurations produced by following the strategy. The *pebbling number* of G, written as $\sharp G$, is the minimum cost of a strategy for the pebbling game played on G.

2.2   Pebbling Formulas

We associate a contradictory set of clauses with every pointed graph G. Each node in G except the target t is assigned a distinct variable; to simplify notation, we identify a node with the variable associated with it, and use the notation $\mathrm{Var}(G)$ for the set of these variables. We associate the constant $\perp$ (falsum) with the target node t, and make the identification $t = \perp$.

Definition 1. *If G is a pointed graph, $Peb(G)$ is a set of clauses expressed in terms of the variables $Var(G)$, so that $Peb(G) = \{ Clause(v) : v \in G \setminus \{t\} \}$.*

1. If v is a source node of G, then $Clause(v) = v$.

2. If v is a node in G, with predecessor u, then Clause(v) = u → v.
3. If v is a node in G, with predecessors u, w, then Clause(v) = u, w → v.

If we set some variables in $\mathrm{Peb}(G)$, then the resulting set of clauses is not necessarily of the form $\mathrm{Peb}(G')$, where G' is a subgraph of G. We shall focus on a family of special assignments, called *pebbling assignments*, that preserve this property. If $v \in G$, $v \neq t$, then we define the assignment $[\![v := 1]\!]$ to be the assignment defined by first setting the variable v to 1, and then setting to 1 any variable u for which there is no implicational chain from u to $\bot$ in the resulting clause set. The assignment $[\![v := 0]\!]$ is defined as follows: first, choose a directed path $\pi = (v, \ldots, t)$ from v to the target t, set all the nodes in the path to 0, and in addition set any node from which v is not reachable, but not in the path π, to 1. The assignment $[\![v := 0]\!]$ is not uniquely determined by this construction, since it depends on the path chosen – however, this is not important, since the set of clauses $\mathrm{Peb}(G) \restriction [\![v := 0]\!]$ resulting from the restriction is independent of the path. A *pebbling assignment* results from a sequence of restrictions of the form $[\![v := 0]\!]$ and $[\![w := 1]\!]$.

The effect of the restrictions just defined can be described directly as an operation on the underlying graph. If G is a pointed graph, and $v \in G$, $v \neq t$, $G[v := 1]$ is the graph resulting from G by first removing v, together with all edges entering or leaving v, and then restricting the resulting graph to the nodes from which the target node t is accessible. $G[v := 0]$ is the pointed graph $G \restriction v$.

Lemma 1.   *1. For $b = 0, 1$, $\mathrm{Peb}(G) \restriction [\![v := b]\!] = \mathrm{Peb}(G[v := b])$.*
2. If G is a pointed graph, and $v \in G$, then

$$\sharp G \leq \max\{\sharp G[v := 0], \sharp G[v := 1] + 1\}.$$

Proof. The first part of the lemma follows straightforwardly from the definitions. For the second part, we employ the following strategy in the pebble game on G. First, follow a minimum cost strategy to pebble v in $G[v := 0]$. Second, leaving a pebble on v, but removing all other pebbles, follow a minimum cost strategy in the pebbling game on $G[v := 1]$ to pebble the target node in G, using the extra pebble for any moves where a pebble is needed on v to justify a placement. The cost is at most $\max\{\sharp G[v := 0], \sharp G[v := 1] + 1\}$.   □

If Σ is a set of clauses, then a *C-critical assignment* is an assignment to the variables in Σ that makes all the clauses true, except C. In the case of $\mathrm{Peb}(G)$, we are interested in a particular family of critical assignments. Let v be a vertex in G, and $\pi = (v, \ldots, t)$ a directed path in G from v to the target node t. Set all the nodes in the path π to 0, and all other nodes in G to 1. This assignment makes all of the clauses in $\mathrm{Peb}(G)$ true, except for $\mathrm{Clause}(v)$. An assignment determined by the path π we shall call a *v-critical assignment*, since the clause that it falsifies is associated with the node v. Since we have assumed that G is a pointed graph, such v-critical assignments exist for all the nodes v in G, so that $\mathrm{Peb}(G)$ is minimally inconsistent.

Lemma 2. *If G is a pointed graph with $\sharp G = p$, then there are at least p vertices v in G for which there is a v-critical assignment for $\mathrm{Peb}(G)$.*

Proof. Every pebbling strategy for G must contain a configuration with p pebbles, so there must be at least p vertices in G. For every vertex v in G, we can construct a v-critical assignment for Peb(G) by choosing a path from v to the target node.   □

If G is a binary pointed graph, then the clause set Peb(G) contains both 3-literal clauses and unit clauses. It is convenient, in view of a later construction, to convert it into a set of 3-literal clauses.

Definition 2. *Let Σ be a set containing both 3-literal clauses and unit clauses. Then Σ^* is the set of clauses obtained from Σ by the following construction. First, introduce for each unit clause l in Σ, a pair of new auxiliary variables x_l and y_l. Second, replace the unit clause l by the set of four 3-literal clauses $\{l \vee x_l \vee y_l, l \vee \overline{x_l} \vee y_l, l \vee x_l \vee \overline{y_l}, l \vee \overline{x_l} \vee \overline{y_l}\}$.*

We write Peb$^*(G)$ for Peb(G)*. Let G be a binary pointed graph. If v is a node in G that is not a source node, then we write Clauses(v) for $\{u, w \rightarrow v\}$, where u, w are the predecessors of v, and if v is a source node, then Clauses(v) is defined to be the set of four clauses $\{v \vee x_v \vee y_v, v \vee \overline{x_v} \vee y_v, v \vee x_v \vee \overline{y_v}, v \vee \overline{x_v} \vee \overline{y_v}\}$. If X is a subset of the nodes in G, then Clauses(X) is defined to be $\bigcup\{\text{Clauses}(v) : v \in X\}$.

3   Constructing Hard Problems

3.1   Earlier Constructions

In this section, we construct the problems that produce our improved separation between general and regular resolution. The overall approach is derived from the first separation result described above in §1; the proof of this result is based on the following idea. The construction begins with a sequence of problems GT_n that are hard for tree resolution but not for regular resolution. The set of clauses GT_n asserts that there is a directed acyclic ordering on n nodes that has no sink; these problems were introduced in the proof complexity literature by Krishnamurthy [8], who conjectured that they require superpolynomial-size resolution refutations. That conjecture was refuted by Stålmarck, who showed that they in fact have linear size resolution refutations [9]. However, Bonet and Galesi [10] showed that they require exponentially large tree resolution refutations, thus showing an exponential separation between general and tree resolution.

The exponential lower bound for tree resolution shows that any tree refutation for GT_n must contain exponentially many paths starting from the root of the tree. Although this fact does not force regular resolution refutations to be large, as Stålmarck showed, nevertheless we can convert the GT_n examples into hard problems for regular resolution by making a small modification. The idea is to add new literals to certain clauses in such a way as to force the exponentially many paths in a tree refutation not to overlap, at least in their initial segments. The new sets of clauses $GT'_{n,\rho}$ require exponentially large regular resolution refutations, though the general resolution size remains linear, as in the case of the original GT_n problems.

The construction in the present paper follows the outline above, but this time starting from the pebbling formulas. The second lower bound proof in [6] also began from the pebbling formulas, but used a somewhat different construction to convert them into hard examples for regular resolution. The present result combines features of both proofs; the construction proceeds in two stages.

3.2 Xorification of Clause Sets

The construction starts from $\mathrm{Peb}(G)$, for G a pointed graph. The first stage applies to $\mathrm{Peb}(G)$ a construction of Alekhnovich and Razborov.

Definition 3. *Let Σ be a set of clauses, and $k > 1$ a positive integer. For each variable x in Σ, introduce a set of k distinct variables $\{x_1, \cdots, x_k\}$. Then the set of clauses $\Sigma^{k\oplus}$, the k-xorification of Σ, is defined as follows: first, substitute the formula $x_1 \oplus \cdots \oplus x_k$ for all of the variables x occurring in Σ, second, convert the resulting formula into conjunctive normal form.*

If C is a clause containing m literals, then $\{C\}^{k\oplus}$ contains $2^{m(k-1)}$ clauses, each of length mk. Hence, when G is a binary pointed graph with n nodes, $\mathrm{Peb}^*(G)^{k\oplus}$ contains nk variables, and $n2^{3(k-1)}$ clauses, each of length $3k$.

The special case of Definition 3 where $k = 2$ is the original construction of Alekhnovich and Razborov [11]. They observed that it could be used to produce hard problems for resolution from clause sets requiring refutations of large width. Let $\mathrm{Width}(\Sigma)$ be the size of the largest clause in Σ, and $\mathrm{Width}(\Sigma \vdash 0)$ the minimum width of a resolution refutation of Σ.

Theorem 1. (Alekhnovich and Razborov) *If Σ is contradictory, then any resolution refutation of $\Sigma^{2\oplus}$ has size $\exp[\Omega(\mathrm{Width}(\Sigma \vdash 0) - \mathrm{Width}(\Sigma))]$.*

More important in the present context is the fact that the construction can be used to produce examples that separate tree resolution from general resolution.

Theorem 2. *If G is a pointed graph with n nodes and pebbling number p, then the set of clauses $\mathrm{Peb}(G)^{2\oplus}$ has general resolution refutations of size $O(n)$, but every tree resolution refutation of $\mathrm{Peb}(G)^{2\oplus}$ has size $2^{\Omega(p)}$.*

Proof. The theorem can be proved by imitating the proof of Ben-Sasson, Impagliazzo and Wigderson [7]. Their result involves clause sets that are the "orification" $\mathrm{Peb}(G)^\vee$ of $\mathrm{Peb}(G)$ rather than the xorification $\mathrm{Peb}(G)^{2\oplus}$; however, the steps in their proof can be imitated almost word for word in the case of $\mathrm{Peb}(G)^{2\oplus}$ to produce essentially the same result as their main theorem. □

An assignment μ for $\mathrm{Peb}^*(G)^{k\oplus}$ is defined to be *full* if whenever v is a vertex in G, and μ assigns a value to some variable v_j associated with v, then all the variables attached to v are assigned values by μ. If μ is such a full assignment, then we can construct an assignment for $\mathrm{Peb}(G)$ from μ by setting $\sigma(v) = \mu(v_1 \oplus \cdots \oplus v_k)$. In this case, we say that the constructed assignment is the *projection* of μ, written $\pi(\mu)$. We shall say that an assignment μ for $\mathrm{Peb}^*(G)^{k\oplus}$ is a *pebbling assignment* if its projection $\pi(\mu)$ is a pebbling assignment for $\mathrm{Peb}(G)$.

Lemma 3. *Let G be a binary pointed graph. If σ is a v-critical assignment for $Peb(G)$, $v \in G$, and C is in $Clauses(v)^{k\oplus}$, then there is a C-critical assignment μ for $Peb^*(G)^{k\oplus}$ so that $\pi(\mu) = \sigma$.*

Proof. If v is a source node in G, and D is a clause in $Clauses(v)$, then we can construct a D-critical assignment for $Peb^*(G)$ by giving the appropriate values to the auxiliary variables x_l and y_l. If v is not a source node, then σ is already a v-critical assignment for $Peb^*(G)$.

Starting from a v-critical assignment for $Peb^*(G)$, we can construct an assignment μ that assigns values to $u_1, \ldots, u_k$, for all nodes $u \in G$, so as to make C false, but all other clauses in $Peb^*(G)^{k\oplus}$ true, and in addition, this assignment μ satisfies $\pi(\mu) = \sigma$. $\qquad\square$

3.3   Adding Random Literals

The second stage of the construction starts from $Peb^*(G)^{k\oplus}$, for a binary pointed graph G and suitable k, and replaces each clause C in the set with a pair of clauses, $C \vee \rho(C)$ and $C \vee \neg\rho(C)$, where $\rho(C)$ is a variable associated with C by the function ρ. For the second stage to work (that is to say, for the resulting sets of clauses to require exponentially large regular resolution refutations), it is essential that ρ have a special property, namely that the image of a large set of clauses has a large intersection with a large set of variables. The easiest way to construct such a function is by a probabilistic argument, given in the following lemma.

Lemma 4. *If G is a binary pointed graph with n nodes, $\delta = 5/3$ and $k = \lceil \delta \log \log n \rceil + 1$, define $\Sigma = Peb^*(G)^{k\oplus}$, and $V = Var(\Sigma)$. Then for sufficiently large n, there exists a map ρ from Σ to V satisfying the condition: For all $A \subseteq G$ with $|A| = \lfloor n/4 \log n \rfloor$, and $B \subseteq V$, with $|B| = \lfloor n/4 \log n \rfloor$, $|\rho(Clauses(A)) \cap B| \geq n/8 \log n$.*

Proof. If $A \subseteq G$ with $|A| = \lfloor n/4 \log n \rfloor$, and $x \in A$, then $Clauses(x)$ contains $2^{3(k-1)} = 2^{5\lceil \log \log n \rceil} \geq (\log n)^5$ clauses, so that $|Clauses(A)|$ contains $\Theta(n(\log n)^4)$ clauses, $|\Sigma| = \Theta(n(\log n)^5)$, and $|V| = nk = n(\lceil \delta \log \log n \rceil + 1) \leq 2n \log \log n$, for sufficiently large n.

Consider the space $\mathcal{R}$ of all random maps from Σ to V; that is to say, for each $C \in \Sigma$, a variable $\rho(C) \in V$ is chosen uniformly at random. For $A \subseteq G$ with $|A| = \lfloor n/4 \log n \rfloor$, and $B \subseteq V$, with $|B| = \lfloor n/4 \log n \rfloor$, we say that ρ is *bad* for A and B if $|\rho(Clauses(A)) \cap B| < n/8 \log n$.

We establish the existence of the map ρ by a probabilistic argument; to accomplish this, we need to prove exponentially small upper bounds on the probability that a random map is bad for some sets A and B. In proving this, it helps to view the construction of a random map as resulting from a series of independent experiments, each of them consisting in the construction of a random map from a subset of Σ.

We partition $Clauses(x)$ as $\Xi_1(x), \ldots, \Xi_q(x)$, where $q = \lfloor (\log n)^3 \rfloor$, so that each set $\Xi_j(x)$ in the sequence contains at least $(\log n)^2$ clauses. For fixed j,

$1 \leq j \leq q$, let Ξ_j be the union of all the $\Xi_j(x)$, for $x \in G$, and for $A \subseteq G$, $|A| = \lfloor n/4 \log n \rfloor$, let Clauses$_j(A)$ be the union of all the $\Xi_j(x)$, for $x \in A$. Then Clauses$_j(A)$ contains $\Theta(n \log n)$ clauses. Let ρ_j be a random map from Ξ_j to V; we take ρ to be the union of the sequence $\rho_1, \ldots, \rho_q$ of independently constructed random maps.

For a given j, where $1 \leq j \leq q$, let Z be the random variable representing the number of variables in B not in the image of Clauses$_j(A)$ under ρ_j:

$$Z(\rho_j) = |\{x \in B | \ x \notin \rho_j(\text{Clauses}_j(A)) \}|.$$

For $B = \{b_1, b_2, \ldots, b_i, \ldots, b_m\}$, where $m = \lfloor n/4 \log n \rfloor$, define an indicator random variable Θ_i by:

$$\Theta_i(\rho_j) = \begin{cases} 1, & \text{if } b_i \notin \rho_j(\text{Clauses}_j(A)) \\ 0, & \text{if } b_i \in \rho_j(\text{Clauses}_j(A)), \end{cases}$$

so that $Z = \Theta_1 + \cdots + \Theta_m$. We estimate the expected value of Θ_i by

$$E(\Theta_i) = \left(1 - \frac{1}{|V|}\right)^{|\text{Clauses}_j(A)|}$$

$$\leq \left(1 - \frac{1}{2n \log \log n}\right)^{\Theta(n \log n)}$$

$$\leq \exp\left(-\Omega\left(\frac{\log n}{\log \log n}\right)\right),$$

showing that

$$E(Z) \leq m \cdot \exp\left(-\Omega\left(\frac{\log n}{\log \log n}\right)\right) = m \cdot o(1).$$

It follows that for any given positive γ, $E(Z) < \gamma m$, for sufficiently large n. For the remainder of the proof, we assume that n is chosen sufficiently large so that $E(Z) < m/8$.

We need to show that the random variable Z is tightly concentrated around its mean. To do this, we employ a large deviation bound for martingales, following [12].

Order Clauses$_j(A)$ as $\{C_1, \ldots, C_p\}$. For $\rho \in \mathcal{R}$, and $1 \leq j \leq p$, define $\rho \restriction j$ to be the restriction of ρ to the set $\{C_1, \ldots, C_j\}$. Define an equivalence relation $\equiv_j$ on $\mathcal{R}$ by setting

$$\rho \equiv_j \sigma \iff \rho \restriction j = \sigma \restriction j,$$

for $1 \leq j \leq p$, and let $\equiv_0$ be the universal relation on $\mathcal{R}$. Let $\mathcal{F}_j$ be the finite Boolean algebra whose atoms are the blocks of the partition of $\mathcal{R}$ induced by $\equiv_j$, for $0 \leq j \leq p$. Now define a sequence of random variables $Z_0, \ldots, Z_p$ by setting $Z_j = E(Z|\mathcal{F}_j)$. Then $Z_0 = E(Z)$, $Z_p = Z$, and the sequence $Z_0, \ldots, Z_p$ forms a martingale, with $|Z_{j+1} - Z_j| \leq 1$. Consequently, by the martingale tail inequality of Hoeffding and Azuma [13, p. 221],

$$P(Z \geq m/2) \leq P(Z - E(Z) > 3m/8)$$
$$< \exp(-(3m/8)^2/2p)$$
$$\leq \exp(-\Omega(n/(\log n)^3)).$$

Let W be the random variable representing the number of variables in B not in the image of Clauses(A) under ρ:

$$W(\rho) = |\{x \in B| \, x \notin \rho(\text{Clauses}(A)) \,\}|.$$

Since the maps $\rho_1, \ldots, \rho_q$ are constructed independently, it follows that

$$P(W \geq m/2) \leq [\exp(-\Omega(n/(\log n)^3))]^q = \exp(-\Omega(n)).$$

We can now complete the proof of the existence of a map ρ satisfying the condition of the lemma. The probability that a random map $\rho \in \mathcal{R}$ is bad for some A and B is bounded by

$$\binom{n}{\lfloor n/4 \log n \rfloor} \binom{n \log \log n}{\lfloor n/4 \log n \rfloor} \exp(-\Omega(n)).$$

Let $H(x) = x \log(1/x) + (1 - x) \log(1/(1 - x))$ be the binary entropy function. Then the first binomial coefficient above can be bounded by

$$\binom{n}{\lfloor n/4 \log n \rfloor} \leq \exp(O(nH(n/\lfloor n/4 \log n \rfloor)))$$
$$= \exp(O(nH(1/\log n)))$$
$$= \exp(O(n \log \log n/\log n)).$$

A similar computation shows that the second binomial coefficient has the same upper bound. Hence, the probability can be bounded above by

$$\exp(O(n \log \log n/\log n)) \exp(-\Omega(n)) = \exp(-\Omega(n)).$$

Consequently, the probability that a random map ρ is bad for some A and B is exponentially small for sufficiently large n, showing that a map satisfying the condition of the lemma must exist. $\square$

3.4   Construction of the Hard Problems

Let's say that for $\Sigma = \text{Peb}^*(G)^{k\oplus}$, a map ρ is *good for* Σ if it satisfies the condition of Lemma 4. This lemma states that for $\delta = 5/3$, $k = \lceil \delta \log \log n \rceil + 1$, and sufficiently large n, there is a map that is good for $\Sigma = \text{Peb}^*(G)^{k\oplus}$. This enables us to construct our set of hard problems for regular resolution. The construction is based on the following result of Paul, Celoni and Tarjan.

Theorem 3. *[14] There is a sequence of binary pointed graphs $G_1, G_2, \ldots, G_i, \ldots$ with pebbling number $\Omega(n(i)/\log n(i))$, where $n(i) = |G_i| = O(i2^i)$.*

We construct our sequence $\Pi_1, \Pi_2, \ldots, \Pi_i, \ldots$ by applying the earlier constructions to $G_1, G_2, \ldots, G_i, \ldots$.

Definition 4. *Let $G_1, G_2, \ldots, G_i, \ldots$ be the sequence of graphs of Theorem 3, $k(i) = \lceil \delta \log \log n(i) \rceil + 1$, and ρ_i a map that is good for $\Sigma_i = \mathrm{Peb}^*(G_i)^{k(i)\oplus}$. Then Π_i is defined to be the set of clauses*

$$\{C \vee \rho_i(C) : C \in \Sigma_i\} \cup \{C \vee \neg\rho_i(C) : C \in \Sigma_i\}.$$

The set of clauses Π_i contains $\Theta(n(\log n)^5)$ clauses, and $\Theta(n \log \log n)$ variables, where $n = n(i)$ is the size of the pointed graph G_i. By a "pebbling assignment for Π_i" we mean a pebbling assignment for $\Sigma_i = \mathrm{Peb}^*(G_i)^{k(i)\oplus}$.

4   Lower Bound for Regular Resolution

4.1   Destroying Large Clauses by Restrictions

In this section, to avoid notational clutter, we adopt the following conventions. We assume that we are dealing with the set of clauses $\Pi = \Pi_i$, for sufficiently large i, write G for G_i, and n for $n(i) = |G_i|$. Define a clause to be *large* if it contains at least $n/8 \log n$ literals.

Lemma 5. *If Σ is a set of clauses in the language of Π, containing fewer than $2^{n/\lceil 64(\log n)^2 \log \log n \rceil}$ clauses, then there is a pebbling assignment μ so that:*

1. *$\Sigma\!\restriction\!\mu$ contains no large clauses.*
2. *$G\!\restriction\!\pi(\mu)$ has pebbling number at least $n/2 \log n$.*

Proof. There are at most $2n \log \log n$ variables in Π, and so at most $4n \log \log n$ literals involving those variables. If we choose a literal at random and set it to 1, then the probability this assignment sets a large clause C to 1 is at least $1/r$, where $r = 32 \log n \log \log n$. Hence, the average number of large clauses in Σ set to 1 is at least $|\Sigma|/r$.

Choose a literal l achieving at least this average, and set it to 1. Suppose that l contains a variable v_j, where $v \in G$. Set the remaining variables in the set of variables $\{v_1, \ldots, v_k\}$ so as to maximize $\sharp G[v := b]$, where $b = v_1 \oplus \cdots \oplus v_k$. Now extend this assignment to produce a pebbling assignment for Π whose projection to $\mathrm{Peb}(G)$ is $[\![v := b]\!]$. Then the set Σ' resulting from this restriction contains at most $(1 - 1/r)|\Sigma|$ large clauses, and by Lemma 1, $\sharp G[v := b]$ is at most one less than $\sharp G$.

If we repeat this procedure $\lfloor n/2 \log n \rfloor$ times, resulting in a restriction μ, then the set contains at most

$$(1 - 1/r)^{\lfloor n/2 \log n \rfloor} 2^{n/\lceil 64(\log n)^2 \log \log n \rceil}$$

large clauses. However, this last expression is bounded above by

$$\exp\left[-\frac{n}{(\log n)^2} \left(\frac{0.9 - \ln 2}{64 \log \log n} \right) \right] < 1,$$

showing that $\Sigma\!\restriction\!\mu$ contains no large clauses. By construction, the pebbling number of $G\!\restriction\!\pi(\mu)$ is at least $n/2 \log n$. $\qquad\square$

4.2   Large Clauses

Lemma 6. *Let $\mathcal{R}$ be a regular resolution refutation of $\Pi \restriction \mu$, where μ is a pebbling assignment, and $G \restriction \pi(\mu)$ has pebbling number $\geq n/2 \log n$. Then $\mathcal{R}$ contains a clause with at least $n/8 \log n$ literals.*

Proof. Viewing $\mathcal{R}$ as a branching program, we describe a strategy for constructing a path in $\mathcal{R}$, starting with the root, and concurrently constructing a full assignment to certain variables in π. The strategy is as follows. We suppose that the path has been constructed as far as a node p, and that σ is the current full assignment. We extend the path and the assignment according to these rules:

1. If the clause labelling p is derived by weakening, then continue the path to the unique parent node; the assignment remains unchanged.
2. If the variable queried at p is already assigned a value by σ, answer the query according to σ, and continue the path according to this answer.
3. If the variable queried at p is not assigned a value by σ, then it must be associated with a node $v \in G \restriction \pi(\sigma)$. Extend σ to a pebbling assignment σ' so that $\pi(\sigma') = [\![v := b]\!]$, choosing b so as to maximize the pebbling number of $(G \restriction \pi(\mu \cup \sigma)) \restriction [\![v := b]\!])$. Then extend the path in accordance with σ'.

Continue according to these rules until $\lfloor n/4 \log n \rfloor$ nodes in G have been queried (that is to say, variables attached to the nodes have been queried), let C be the clause at the end of the resulting path, and τ the resulting assignment.

By Lemma 2, there are at least $n/4 \log n$ vertices $v \in G \restriction \pi(\tau)$ for which there is a v-critical assignment for $\mathrm{Peb}(G \restriction \pi(\tau))$. If ϕ is such a critical assignment, then $\pi(\tau) \cup \phi$ is a v-critical assignment for $\mathrm{Peb}(G)$. Let A be the set of all nodes in G satisfying this condition, and B the set of variables assigned values by τ. Since $|A|, |B| \geq \lfloor n/4 \log n \rfloor$, by Lemma 4, $|\rho(\mathrm{Clauses}(A)) \cap B| \geq n/8 \log n$.

Let x be a variable in $\rho(\mathrm{Clauses}(A)) \cap B$. We claim that x must occur in C. Suppose not. By assumption, there is a $D \in \mathrm{Clauses}(v)$, for some $v \in A$, so that $\rho(D) = x$, and $D \vee x, D \vee \overline{x} \in \Pi$. Let's assume that $\tau(x) = 0$ (the case $\tau(x) = 1$ is symmetrical). By Lemma 3, there is a D-critical assignment ϕ for $\mathrm{Peb}^*(G)^{k\oplus}$ that extends τ, and so is a $D \vee x$-critical assignment for Π. Extend the path in $\mathcal{R}$ from C by answering queries in accordance with ϕ. This path must terminate in a node labelled with $D \vee x$. But since x does not occur in C, it follows that it must have been resolved on twice along the path, violating regularity. This contradiction proves that x must occur in C, showing that C contains at least $n/8 \log n$ literals.   □

4.3   Lower Bound

Theorem 4. *Let $\Pi_1, \Pi_2, \ldots, \Pi_i, \ldots$ be the sequence of contradictory sets of clauses based on the pointed graphs $G_1, G_2, \ldots, G_i, \ldots$ of Paul, Celoni and Tarjan, where $n = n(i)$ is the size of the graph G_i. Then:*

1. *There are resolution refutations of Π_i with size $O(n(\log n)^7)$.*
2. *Every regular resolution refutation of Π_i has size $2^{\Omega(n/[(\log n)^2 \log \log n])}$*

Proof. The set of clauses $\text{Peb}(G)$ has a refutation using unit resolution (where at least one of the premisses in every resolution step is a unit clause), with $O(n)$ steps and in which every clause contains at most three literals. We can imitate this refutation to produce a refutation of $\text{Peb}(G)^{k\oplus}$; in this refutation, a single resolution step in the original refutation corresponds to multiple resolution steps in the new refutation. Let us suppose that in the original refutation of $\text{Peb}(G)$, the clause $b \vee c$ was inferred from a and $\bar{a} \vee b \vee c$, where a, b, c are literals. Then in the new refutation, we infer $\{b \vee c\}^{k\oplus}$ from $\{\bar{a} \vee b \vee c\}^{k\oplus}$ and $\{a\}^{k\oplus}$. The set of clauses $\{a\}^{k\oplus} \cup \{\bar{a}\}^{k\oplus}$ consists of all the clauses in a fixed set of k variables, so it takes $O(2^k) = O((\log n)^{5/3})$ steps to deduce the empty clause from this set. Hence, a single clause in $\{b \vee c\}^{k\oplus}$ can be derived in $O((\log n)^{5/3})$ steps. It follows that the derivation of $\{b \vee c\}^{k\oplus}$ takes $O((\log n)^5 (\log n)^{5/3})$ resolution steps, showing that the entire refutation has size $O(n(\log n)^{20/3})$. By adding some extra resolution inferences, we can produce a resolution refutation of Π_i with the same size bound.

For the second part of the theorem, let us assume that $\mathcal{R}$ is a regular resolution refutation of Π_i, with size less than $2^{n/[64(\log n)^2 \log\log n]}$. By Lemma 5 there is a pebbling assignment μ so that $\mathcal{R} \upharpoonright \mu$ contains no large clauses, but $G_i \upharpoonright \pi(\mu)$ has pebbling number at least $n/2 \log n$. However, Lemma 6 shows that $\mathcal{R} \upharpoonright \mu$ must contain a large clause, showing that a regular refutation of this size cannot exist. $\square$

It is interesting to ask how close Theorem 4 comes to the optimum. We already observed in §1 that if R is the minimum size of a resolution refutation of a set of clauses, then the size of a regular refutation is bounded above by $2^{O(R \log\log R/\log R)}$. If we express the lower bound in these terms, then we find that the lower bound on regular refutations has the form $2^{\Omega(R/(\log R)^7)}$. So, the separation we have proved is certainly much closer to the optimum than previous bounds, but there is definitely room for improvement.

References

1. Tseitin, G.: On the complexity of derivation in propositional calculus. In: Slisenko, A.O. (ed.) Studies in Constructive Mathematics and Mathematical Logic, Part 2, pp. 115–125. Consultants Bureau, New York (1970); Reprinted in[15], vol. 2, pp. 466-483
2. Davis, M., Logemann, G., Loveland, D.: A machine program for theorem proving. Communications of the Association for Computing Machinery 5, 394–397 (1962); Reprinted in [15], vol. 1, pp. 267-270
3. Urquhart, A.: The complexity of propositional proofs. The Bulletin of Symbolic Logic 1, 425–467 (1995)
4. Huang, W., Yu, X.: A DNF without regular shortest consensus path. SIAM Journal on Computing 16, 836–840 (1987)
5. Goerdt, A.: Regular resolution versus unrestricted resolution. SIAM Journal on Computing 22, 661–683 (1993)
6. Alekhnovich, M., Johannsen, J., Pitassi, T., Urquhart, A.: An exponential separation between regular and general resolution. Theory of Computing 3, 81–102 (2007); Preliminary version. In: Proceedings of the 34th Annual ACM Symposium on Theory of Computing, May 19-21, 2002, Montréal, Québec, Canada (2002)

7. Ben-Sasson, E., Impagliazzo, R., Wigderson, A.: Near optimal separation of tree-like and general resolution. Combinatorica, 585–603 (2004); Preliminary version, ECCC TR00-005 (2000)
8. Krishnamurthy, B.: Short proofs for tricky formulas. Acta Informatica 22, 253–275 (1985)
9. Stålmarck, G.: Short resolution proofs for a sequence of tricky formulas. Acta Informatica 33, 277–280 (1996)
10. Bonet, M.L., Galesi, N.: Optimality of size-width tradeoffs for resolution. Computational Complexity 10, 261–276 (2001); Preliminary version: Proceedings 40th FOCS (1999)
11. Ben-Sasson, E.: Size Space Tradeoffs For Resolution. In: Proceedings of the 34th ACM Symposium on the Theory of Computing, pp. 457–464 (2002)
12. Kamath, A., Motwani, R., Palem, K., Spirakis, P.: Tail bounds for occupancy and the satisfiability threshold conjecture. Random Structures and Algorithms 7, 59–80 (1995)
13. McDiarmid, C.: Concentration. In: Habib, M., McDiarmid, C., Ramirez-Alfonsin, J., Reed, B. (eds.) Probabilistic Methods for Algorithmic Discrete Mathematics, vol. 16, pp. 195–248. Springer, Heidelberg (1998); Algorithms and Combinatorics 16
14. Paul, W., Tarjan, R., Celoni, J.: Space bounds for a game on graphs. Mathematical Systems Theory 10, 239–251 (1977)
15. Siekmann, J., Wrightson, G. (eds.): Automation of Reasoning. Springer, New York (1983)

Finding Guaranteed MUSes Fast

Hans van Maaren[1] and Siert Wieringa[2,*]

[1] Delft University of Technology
Faculty of EWI
Mekelweg 4, 2628 CD, Delft, The Netherlands
h.vanmaaren@tudelft.nl
[2] Helsinki University of Technology (TKK)
Department of Information and Computer Science
P.O. Box 5400, FI-02015 TKK, Finland
Siert.Wieringa@tkk.fi

Abstract. We introduce an algorithm for finding a *minimal unsatisfiable subset* (MUS) of a CNF formula. We have implemented and evaluated the algorithm and found that its performance is very competitive on a wide range of benchmarks, including both formulas that are close to minimal unsatisfiable and formulas containing MUSes that are only a small fraction of the formula size.

In our simple but effective algorithm we associate assignments with clauses. The notion of *associated assignment* has emerged from our work on a Brouwer's fixed point approximation algorithm applied to satisfiability. There, clauses are regarded to be entities that order the set of assignments and that can select an assignment to be associated with them, resulting in a *Pareto optimal agreement*.

In this presentation we abandon all terminology from this theory which is superfluous with respect to the recent objective and make the paper self contained.

1 Introduction

Solvers for instances of the Boolean satisfiability problem, so called SAT solvers, have found their way into numerous applications including *electronic design automation* (EDA) [1], formal verification [2,3] and artificial intelligence [4]. In many of those applications we would like to have an explanation of the *cause* of unsatisfiability in case a formula is unsatisfiable. For example if an FPGA routing problem is translated to a Boolean formula a satisfying assignment corresponds to a valid routing, and unsatisfiability means no such routing exists [1]. In the latter case the user might want to know which part of the design caused the unroutability.

An *unsatisfiable subset* or *core* of an unsatisfiable formula is a subset of clauses from that formula the conjunction of which is unsatisfiable. A *minimal unsatisfiable subset* (MUS) is an unsatisfiable subset that becomes satisfiable if any of its

---

* Supported by the Academy of Finland project #112016.

H. Kleine Büning and X. Zhao (Eds.): SAT 2008, LNCS 4996, pp. 291–304, 2008.

clauses is removed. There can be multiple MUSes in one formula. An unsatisfiable subset might help to understand at least one cause of a formula's unsatisfiability as the clauses that have been left out were not necessary to maintain unsatisfiability. Multiple algorithms have been presented for finding unsatisfiable cores that are not guaranteed to be MUSes (i.e. [5,6,7]). In this paper we present an algorithm for finding a MUS in an unsatisfiable formula. Our algorithm does not guarantee finding the minimum unsatisfiable core, which is the MUS with the least number of clauses [8].

In [6] a nice example of a small unsatisfiable FPGA routing problem can be found. It shows the use of finding multiple MUSes and is used to argue that the minimum unsatisfiable core is not necessarily the most useful core for diagnostic purposes. Other work has focused on algorithms for finding all MUSes [9,10] or the minimum unsatisfiable core [8]. Even if finding all MUSes is not feasible those algorithms might be very useful to find one or multiple MUSes under some time constraint.

Important notions for our algorithm emerged from our work [11] on applying a Brouwer's fixed point approximation algorithm [12] to satisfiability. In that work a clause is regarded to be an entity that can select an assignment from the set of all 2^n possible assignments on which it imposes a complete ordering in which all satisfying assignments are preferred over all unsatisfying assignments. Unsatisfiability is represented by the possibility to find a subset, or *coalition*, of clauses that form a *Pareto optimal agreement*. In such an agreement all clauses have chosen a unique assignment that does not satisfy themselves, while they would all prefer the chosen assignment of all other clauses in the coalition and there is no single assignment that all those clauses prefer over their own choice. The existence of this agreement proves that the preferences of the clauses are contradictionary from which the inconsistency of the formula follows. As the clauses in the coalition prove unsatisfiability of the formula they form an unsatisfiable subset. We implemented an algorithm for finding unsatisfiable subsets using this theory [11], but as it remains far from competitive we let go of most of this background and present a simple and efficient algorithm.

2   Extracting a MUS

We will represent a CNF formula $\mathcal{F}$ by a *sequence* of clauses. Furthermore a subscript will denote an element index in a sequence, so clause C_i from $\mathcal{F} = \langle C_1, C_2, ..., C_m \rangle$ is the ith clause in the sequence.

The most straightforward approach to reducing a CNF formula of m clauses to a MUS is given in Algorithm 1. This algorithm requires solving m SAT problem instances to find a MUS in the unsatisfiable set of clauses $\mathcal{F}$. In case the input set $\mathcal{F}$ is a MUS itself all the SAT problem instances have $m - 1$ clauses.

Given a CNF formula $\mathcal{F}$ Algorithm 2 finds a formula $\mathcal{F}' \subseteq \mathcal{F}$ such that $\mathcal{F}'$ and $\mathcal{F}$ have the same set of satisfying assignments. The conditional addition of the clauses from $\mathcal{F}$ to $\mathcal{F}'$ is performed one by one in the order of occurrence in the sequence $\mathcal{F} = \langle C_1, C_2, ..., C_m \rangle$. A clause C_i is added to $\mathcal{F}'$ iff there exists a

---

Algorithm 1. NAIVEFINDMUS($\mathcal{F}$)

1: $\mathcal{F}' := \mathcal{F}$
2: **for** $i = 1$ *to* $|\mathcal{F}|$ **do**
3:      **if** $\mathcal{F}' \setminus \{C_i\}$ is UNSATISFIABLE **then**
4:          $\mathcal{F}' := \mathcal{F}' \setminus \{C_i\}$
5:      **end if**
6: **end for**
7: **return** $\mathcal{F}'$

Algorithm 2. FINDEQUSUBSET($\mathcal{F}$)

1: $\mathcal{F}' := \langle\rangle$
2: **for** $i = 1$ *to* $|\mathcal{F}|$ **do**
3:      **if** $\neg C_i \wedge \mathcal{F}'$ is SATISFIABLE **then**
4:          **append** C_i to $\mathcal{F}'$
5:      **end if**
6: **end for**
7: **return** $\mathcal{F}'$

---

truth assignment that is satisfying all the clauses that have already been added to $\mathcal{F}'$ while not satisfying C_i.

Lemma 1. *The formula $\mathcal{F}'$ returned by Algorithm 2 has the same set of satisfying truth assignments as the input formula $\mathcal{F}$.*

Proof. Each clause $C_i \in \mathcal{F}$ is added to the initially empty sequence $\mathcal{F}'$ if there is an assignment satisfying the clauses already added to $\mathcal{F}'$ while not satisfying the clause C_i. If no such assignment exists then either $\mathcal{F}'$ is unsatisfiable or $\mathcal{F}' \models C_i$. If $\mathcal{F}'$ is unsatisfiable $\mathcal{F}$ must be unsatisfiable and thus both formulas have zero satisfying truth assignments. If $\mathcal{F}' \models C_i$ then C_i does not restrict the number of satisfying truth assignments in $\mathcal{F}$ as it implied by, and therefore can be derived from, the subset of $\mathcal{F}$ that was already added to $\mathcal{F}'$.   □

Corollary 1. *If the input $\mathcal{F}$ of Algorithm 2 is unsatisfiable so is the formula $\mathcal{F}'$ it returns*

Although it is possible to use Algorithm 2 to reduce the number of clauses in satisfiable formulas in this paper we limit ourselves to its applications for unsatisfiable formulas.

Definition 1. *A critical clause of an unsatisfiable formula $\mathcal{F}$ is a clause that belongs to every unsatisfiable subset of the formula $\mathcal{F}$.*

Proposition 1. *In a MUS every clause is a critical clause.*

Note that an unsatisfiable formula does not have to contain any critical clauses. An example of a formula without critical clauses would be a formula with two completely disjoint unsatisfiable subsets. A critical clause is called a *necessary clause* in [13].

Lemma 2. *The last clause appended to the sequence forming the unsatisfiable formula $\mathcal{F}' = \langle C_1', C_2', ..., C_m' \rangle$ which is a subset of the unsatisfiable formula $\mathcal{F}$ returned by Algorithm 2 is critical for $\mathcal{F}'$.*

Proof. By construction it holds for each clause C_i' in $\mathcal{F}'$ that there exists an assignment that satisfies all the clauses C_j' in $\mathcal{F}'$ with $j < i$. So without the clause $C_{|\mathcal{F}'|}'$ there is a satisfying assignment for the $|\mathcal{F}'| - 1$ other clauses in $\mathcal{F}'$. Consequently, every subset of $\mathcal{F}'$ that does not contain clause $C_{|\mathcal{F}'|}'$ is satisfiable.   □

Algorithm 3 reduces an unsatisfiable CNF formula to a MUS. It proceeds in multiple *rounds*, proving one clause critical in every round. The lines 3 to 8 in this algorithm are similar to the pseudo code of Algorithm 2 except for the addition of a sequence M to which $\mathcal{F}'$ is initialised. This sequence M consists of all clauses that have already been proven to be critical for the unsatisfiability of the MUS we are constructing.

Algorithm 3. REDUCETOMUS($\mathcal{F}$)

1: $M := \langle \rangle$
2: **while** $|M| < |\mathcal{F}|$ **do**
3:      $\mathcal{F}' := M$
4:      **for** $i = 1$ to $|\mathcal{F}|$ **do**
5:          **if** $(C_i$ does not appear in $M)$ **and** $(\neg C_i \wedge \mathcal{F}'$ is SATISFIABLE) **then**
6:              append C_i to $\mathcal{F}'$
7:          **end if**
8:      **end for**
9:      append LAST($\mathcal{F}'$) to M
10:     $\mathcal{F} := \mathcal{F}'$
11: **end while**

Lemma 3. *In each round of Algorithm 3 it finds a new critical clause unless the set of critical clauses is unsatisfiable.*

Proof. The first round proceeds just like Algorithm 2 would. At the end of the round the last clause is added to M as according to Lemma 2 it is critical. In every following round the sequence $\mathcal{F}'$ is initialised to contain the clauses in M, which are all clauses that have already been proven critical. If M is satisfiable at the start of a round more clauses will be added to $\mathcal{F}'$ which will cause the last clause added in that round to be a clause that was not yet in M. According to Lemma 2 it is critical and as it was not proven critical before we have found a new critical clause. If M is unsatisfiable at the start of a round then at the end of the round $\mathcal{F}'$ will still be equal to M and the algorithm will end.   □

Instead of using a sequence M to keep track of the critical clauses one can also reshuffle the sequence of clauses before each round in such a way that all the clauses already proven critical precede the other clauses in the sequence. One might also want to add a preliminary exit to the for loop as soon as $\mathcal{F}'$ becomes unsatisfiable but from experimental results this did not seem to result in an overall performance gain.

Example 1. Consider the following unsatisfiable CNF formula.

$$(x_1 \vee x_2) \wedge (x_3 \vee x_4) \wedge (x_1 \vee \neg x_2) \wedge (x_1 \vee x_2 \vee x_3) \wedge (\neg x_1) \ . \tag{1}$$

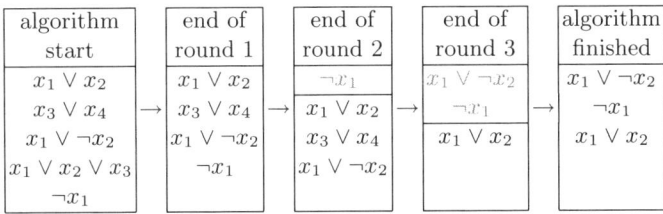

Fig. 1. Finding a MUS in Formula (1)

Each of the rectangles in Fig. 1 show the contents of the sequence $\mathcal{F}$ at some point in the execution of Algorithm 3 with the example formula as input.

- After the first round the clause $x_1 \vee x_2 \vee x_3$ is removed as there is no satisfying assignment to $\neg(x_1 \vee x_2 \vee x_3) \wedge (x_1 \vee x_2) \wedge (x_3 \vee x_4) \wedge (x_1 \vee \neg x_2)$. This is because clause $x_1 \vee x_2 \vee x_3$ subsumes (is a superset of) clause $x_1 \vee x_2$ which has already been added to the sequence. After this round the clause $\neg x_1$ is proven to be a critical clause.
- The first clause in the rectangle in Fig. 1 holding the contents of the sequence $\mathcal{F}$ after the second round is separated from the other clauses by a horizontal line because it is a critical clause and $\mathcal{F}$ is therefore initialised to hold that clause at the start of this round. In this round no clauses are removed from the sequence. After this round clause $x_1 \vee \neg x_2$ is proven to be a critical clause.
- At the start of the third round the sequence $\mathcal{F}$ is initialised to hold the two clauses that were proven to be critical so far. Clause $x_1 \vee x_2$ can be added to the sequence as $\neg(x_1 \vee x_2) \wedge (x_1 \vee \neg x_2) \wedge \neg x_1$ is satisfiable. After adding that clause the sequence $\mathcal{F}$ becomes unsatisfiable and therefore clause $x_3 \vee x_4$ will not be added. After this round the clause $x_1 \vee x_2$ is proven to be a critical clause. As all clauses in $\mathcal{F}$ are now proven to be critical we have found a MUS. ■

We will now describe a way to find more than one critical clause per round of the algorithm. This will reduce the number of rounds the algorithm needs to find a MUS and therefore the number of SAT problem instances that need to be solved. Recall that in each round a clause C_i from $\mathcal{F}$ is only added to the sequence $\mathcal{F}'$ if there is a satisfying assignment for what is already in $\mathcal{F}'$ that is not satisfying C_i. To find multiple critical clauses in one round we store that satisfying assignment with each clause $C_i' \in \mathcal{F}'$ with $\mathcal{F}' = \langle C_1', C_2', ..., C_m' \rangle$ and name it the *associated assignment* of the clause C_i'. This associated assignment does not have to define a truth assignment for all variables in $\mathcal{F}$.

Definition 2. *The associated assignment of a clause $C_i' \in \mathcal{F}'$ is an assignment that satisfies all $C_j' \in \mathcal{F}'$ with $j < i$ and does not satisfy C_i'.*

Lemma 4. *If the associated assignment of a clause $C_i' \in \mathcal{F}'$ satisfies all clauses $C_j' \in \mathcal{F}'$ with $j > i$ then clause C_i' is critical for $\mathcal{F}'$.*

Proof. By definition the associated assignment of a clause $C_i' \in \mathcal{F}'$ satisfies all clauses $C_j' \in \mathcal{F}'$ with $j < i$. If it also satisfies all clauses $C_j' \in \mathcal{F}'$ with $j > i$ then it is a satisfying assignment for all C_j' in $\mathcal{F}'$ with $j \neq i$ and therefore every subset of $\mathcal{F}'$ not containing C_i' is satisfiable. ☐

Example 2. Let us reconsider the unsatisfiable Formula (1) and feed it to the improved version of the algorithm. Figure 2 is constructed similarly to Fig. 1 with the exception that it shows the associated assignments for the clauses in $\mathcal{F}$ after round one. Let us assume that the associated assignments define a truth assignment for all variables in Formula (1). In this example the first, third and fourth clause are proven to be critical after the first round as their associated assignments satisfy all clauses succeeding them in the sequence. At the start of the second round $\mathcal{F}'$ is initialised to contain those three clauses and therefore it becomes unsatisfiable right away. This means that clause $x_3 \vee x_4$ will not be added and we have found a MUS as all clauses in $\mathcal{F}'$ are critical.

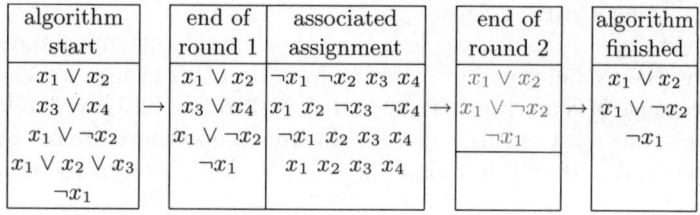

algorithm start	end of round 1	associated assignment	end of round 2	algorithm finished
$x_1 \vee x_2$	$x_1 \vee x_2$	$\neg x_1 \ \neg x_2 \ x_3 \ x_4$	$x_1 \vee x_2$	$x_1 \vee x_2$
$x_3 \vee x_4$	$x_3 \vee x_4$	$x_1 \ x_2 \ \neg x_3 \ \neg x_4$	$x_1 \vee \neg x_2$	$x_1 \vee \neg x_2$
$x_1 \vee \neg x_2$	$x_1 \vee \neg x_2$	$\neg x_1 \ x_2 \ x_3 \ x_4$	$\neg x_1$	$\neg x_1$
$x_1 \vee x_2 \vee x_3$	$\neg x_1$	$x_1 \ x_2 \ x_3 \ x_4$		
$\neg x_1$				

Fig. 2. Finding a MUS in Formula (1), using improved algorithm ∎

The associated assignment a of a clause $C_i' \in \mathcal{F}'$ is the result of solving a SAT problem instance that only contained those literals that occurred in the clauses $C_j' \in \mathcal{F}'$ with $j \leq i$. This means that the associated assignment a of C_i' might not include a truth assignment for a literal in a clause $C_k' \in \mathcal{F}'$ with $k > i$. It is possible that none of the literals of C_k' are satisfied by the truth assignments defined in a but there is a literal in C_k' for which no truth assignment is defined in a. In an implementation of the algorithm one must either always regard such a as an unsatisfying assignment or extend a to hold a satisfying truth assignment for at least one of the literals of C_k' for which no truth assignment was previously defined in a.

Consider the case where this algorithm is given a MUS of m clauses as input for which the algorithm manages to find only one critical clause per round. In this worst case $m\frac{m+1}{2}$ SAT problem instances must be solved to prove all clauses critical, whereas the naive strategy of Algorithm 1 only requires solving m problem instances. However, first of all these SAT problem instances are varying in size from 1 to m clauses so a large number of them are easy to solve, or even trivial, while all the instances that need to be solved using the naive approach have m clauses and thus will often all be hard. Secondly, it is possible to implement all of the SAT solver calls performed in one round in one *incremental* problem, thus requiring only m incremental problems. This can be implemented easily. Adding a clause to $\mathcal{F}'$ in the pseudo code is adding a clause to the problem instance in the solver. Satisfying $\neg C_i$ without having to add it to the solvers problem instance can be guaranteed by forcing the solver to assign all literals from C_i the value false. Using the SAT solver MiniSat [14] such assignments can be passed as so called *assumptions*. Thirdly, our technique for finding multiple critical clauses per round will usually reduce the number of required rounds to significantly less than m as will become clear from the experimental results further on in this document.

3   Implementation

In order to obtain an efficient implementation of our algorithm we have merged it with the code of the state of the art SAT solver MiniSat 2.0 [14] without the optional simplifier [15]. The simplifier is not used as we are interested in finding a MUS consisting of the original input clauses. It might be possible to modify the simplifier for our application but we will not discuss that here. Our MUS finder is called MiniUnsat. Each round is handled as one incremental SAT problem.

The problem mentioned earlier of an assignment that might not be complete for a succeeding clause is handled by extending the assignment with a satisfying truth assignment for a literal of the clause. In case there are multiple literals in the clause for which no truth assignment is defined by the assignment the literal that occurs last in the clause is chosen.

As the algorithm is greedy in the sense that it adds a clause to the subset unless it is proven to be unnecessary clause sorting has a great effect on performance in terms of speed, but also on which MUS the algorithm finds. By manually sorting the clauses before executing the program the user might give preference to one clause over the other and can thereby influence which MUS is found.

Two simple optional automatic clause sorting methods are also implemented. The first automatic sorting method, *sort by weight*, sorts clauses by the sum of the number of occurrences in the formula of all of the literals of the clause. This sorting method is by default enabled as it has positive influence on the speed with which MUSes are found.

The second automatic sorting method, *sort by length*, sorts the clauses with the shortest ones first and thereby focuses on removing subsumed clauses. A *subsumed clause* is a clause that is a superset of another clause in the formula. If the clauses are sorted using this method then the conditional addition of the shorter subsumed clauses will always precede the conditional addition of the subsuming clauses. The latter will never be added as the shorter clause implies the longer clause. Note that it is possible that a formula contains a MUS in which a clause C_i subsuming a clause C_j occurs as long as C_j does not, and that MUSes that have this property will not be found if clauses are sorted in this way. Sorting by length is also by default enabled and if both described clause sorting methods are used together then sorting by length has priority over sorting by weight.

To improve the speed with which the SAT problems are solved a heuristic was added which influences MiniSat's branch direction heuristic. At each variable decision MiniSat branches to the negative side by default. In the first round of the algorithm this is left untouched. After the first round every clause that has not been removed has had an associated assignment in the previous round. As the clauses will remain in the same order in each round apart from the new critical clauses moving to the front this assignment can be seen as an estimate to the assignment we are looking for. Our program therefore sets the variable branch directions to those that would lead the solver in the direction of the associated assignment found in the previous round when it is looking for a new assignment.

4   Related Work

The argument of Lemma 2 is also used in work on finding subsets of infeasible linear programmes [16,17] and in recent work on the *Constraint Satisfaction Problem* (CSP), which is more generalised than SAT. In the latter work it is credited to [18]. The constraint that is critical due to Lemma 2 is called the *transition constraint* in the work on CSP, as it is on the transition from satisfiability to unsatisfiability.

Where we have focused on proving more clauses critical than only this transition constraint, or transition clause in our case, they focused on more efficient approaches to finding the transition constraint. Instead of adding a constraint to a sequence until it becomes unsatisfiable they do a binary search for the transition constraint.

The authors call their approach the *dichotomic approach*. It has the advantage of a logarithmic, rather than linear, worst case number of required SAT problems to solve per round. The authors applied this algorithm successfully to instances of CSP. Although its worst case required number of SAT problems to be solved is lower we reckon it has some disadvantages as an incremental SAT implementation is not immediate, at least not without adding clause selector variables that will make the problems considerably harder. Besides that only one critical clause will be found per round. Still, their approach is interesting and might perform well when implemented efficiently for SAT problems or even combined with some of the ideas presented here.

Another approach to finding an unsatisfiable core is applied by zcore [5]. It records which clauses are necessary to derive the empty clause in a resolution proof of unsatisfiability generated using a SAT solver (zchaff in this case). Those clauses form an unsatisfiable subset, but it might be far from minimal. To approximate a MUS closer the authors suggest to iterate zcore until the size of the unsatisfiable subset no longer reduces. We will refer to this approach by the name of the script supplied with zchaff for this purpose, which is zruntillfix. For finding a guaranteed MUS the authors of zcore supply zminimal which is an implementation of the naive MUS finding algorithm (1).

In a recent publication [7] an unsatisfiable core extractor called *Approximate One MUS* (AOMUS) was introduced. Our experimental results support the claim of the authors that it has good performance at the useful task of reducing benchmarks that describe FPGA routing problems. As the name indicates the "approximated MUSes" found by AOMUS are not guaranteed to be MUSes. The algorithm OMUS, which does guarantee the output of exactly one MUS, is the AOMUS algorithm with the addition of a post processor with the slightly understated name "fine tune". This procedure is in fact an implementation of the naive MUS finding algorithm (1).

The core extractor AMUSE [6] is best suited for use with formulas that contain unsatisfiable cores that are small compared to the formula size. It does not guarantee finding a MUS. AMUSE's preference for small cores is supported by our experimental results and caused by the fact that it builds up the unsatisfiable subset by adding clauses to a satisfiable subset until it becomes unsatisfiable.

In [19] an approach for finding a MUS by using the resolution proof of the unsatisfiability of the input formula is presented. The algorithm presented there tests for every clause in the input formula whether the empty clause can still be derived from the original resolution proof after removing that clause and all clauses derived from it.

The *minimal unsatisfiability prover* (MUP) [20] is targeted at proving the input formula to be a MUS rather than at extracting MUSes from the input formula. It is meant as a post processor to core extractors that are able to give close approximation of MUSes, like AMUSE, AOMUS or zruntillfix.

5   Results

We have tested our implementation using a computing cluster with 20 nodes that each have 2 Intel Xeon 5130 (2Ghz) Dual Core processors, making up for a total of 80 processor cores. None of the tested MUS finders is based on a parallelised algorithm so each run was executed on a single processor core. All tests were run with an 1800 seconds time limit and a 2GB memory limit.

We used benchmarks describing various sorts of problems. We generated random 3-SAT formulas with 50 variables and 215 clauses, 100 variables and 430 clauses and 200 variables and 860 clauses. From each of the three sets we took 50 unsatisfiable formulas. We also added all unsatisfiable instances found in the DaimlerChrysler benchmark set which describes problems from automotive product configuration[1]. In the DaimlerChrysler benchmarks the unsatisfiable cores

---

[1] http://www-sr.informatik.uni-tuebingen.de/~sinz/DC

are only a small fraction of the formula size. Next we added the Bevan family from the handmade category of the SAT Competition[2] held in 2003. All benchmarks in the Bevan family are already MUSes. Finally, we put the benchmark set used in the paper presenting AOMUS together[3] we will refer to that set as *FPGA + Various*. Note that some FPGA routing problems occur both in original form and shuffled in that set [21].

Besides testing the MUS extraction qualities of MiniUnsat we have also tested it as a post processor to the output of AMUSE, AOMUS and zruntillfix. Those three programs were also tested using the naive MUS proving approach as a post processor. For AMUSE and zruntillfix a naive MUS prover implementation was found in zminimal. OMUS is the implementation of AOMUS followed by an internal naive MUS prover.

Although AOMUS outputs a core that is only an approximation to a MUS it may prove some clauses of that core critical. Those clauses are not tested again by the post processor implemented in OMUS. With permission of the author of AOMUS we have modified it to pass the information about those clauses to MiniUnsat when we used it with AOMUS.

We also tested MUP as a post processor to the three core extractors. We do not present the results here as MUP does not seem to be robust enough. The dtree generator supplied with MUP for generating the required *binary decision diagram* (BDD) often crashes. Fortunately, the c2d generator [22], which was suggested as an alternative by the author of MUP in a personal communication, works better. However, the number of successful runs using MUP with either of the two generators is much smaller than that of the other tested approaches, and where it is succesful it is not significantly faster either. The author of MUP uses a BDD variable reordering tool called MINCE [23] as a preprocessor for some of his benchmarks results, which might explain the difference between his and our results.

The results presented in Table 1 are meant to give a general impression of the performance of different approaches. Please note that only those benchmarks in which a MUS was found using all seven presented approaches were included in the calculations of the average MUS sizes. The run times and MUS sizes for each tested benchmark on all tested approaches, including MUP, are available on the internet[4]. Table 2 shows the average number of clauses that are proven critical in one round by making use of the associated assignment technique we described. From the results in this table one can easily see that checking if a clause is critical by testing if its successors are satisfied by its associated assignment leads to a significant reduction in the number of required rounds.

The six scatter plots that together form Fig 3 give an impression of the run times in seconds of the various approaches to extracting a MUS. Each scatter plot compares two program versions. In each scatter plot there is a data point for every benchmark, with the position along the horizontal axes indicating the run time of the approach labelled on the horizontal axis, and the vertical position

---

[2] http://www.satcompetition.org

[3] SAT Competitions and http://www.aloul.net/benchmarks.html

[4] http://www.tcs.hut.fi/~swiering/musfinding

Table 1. Results summary

A+zmin	AMUSE + zminimal
A+M	AMUSE + MiniUnsat
z+zmin	zruntillfix + zminimal
z+M	zruntillfix + MiniUnsat
O	OMUS
AO+M	AOMUS + MiniUnsat
M	MiniUnsat

Number of formulas a MUS was extracted from within 1800 seconds								
Set	#	A+zmin	A+M	z+zmin	z+M	O	AO+M	M
3-SAT 50 vars	50	50	50	50	50	50	50	50
3-SAT 100 vars	50	48	50	50	50	50	50	50
3-SAT 200 vars	50	6	50	8	50	48	50	50
DaimlerChrysler	84	84	84	84	84	84	84	84
Bevan	56	29	29	31	31	43	41	56
FPGA+Various	36	21	24	20	25	28	29	23
Sum	326	238	287	243	290	303	304	313

Average MUS size for formulas a MUS was extracted from by all approaches								
Set	#	A+zmin	A+M	z+zmin	z+M	O	AO+M	M
3-SAT 50 vars	50	101.4	96.2	95.2	94.8	92.4	92.3	101.3
3-SAT 100 v.	48	268	243.1	247.6	233.7	234	232.2	252.9
3-SAT 200 v.	3	624	563.3	639.3	576.3	546.3	546.7	578.3
DaimlerChr.	84	78.4	78.4	76.8	76.8	77.8	76	76.4
Bevan	28	186.4	186.4	186.4	186.4	186.4	186.4	186.4
FPGA+Various	15	225.5	220.4	231.9	217.3	226.4	220.6	221.5

Table 2. Average number of clauses proven critical per round

3-SAT 50 vars	2.6
3-SAT 100 vars	2.7
3-SAT 200 vars	2.8
DaimlerChrysler	10.2
Bevan	45.1
FPGA+Various	4.6

indicating the run time of the approach labelled on the vertical axis. Note that in all six plots both axes have a logarithmic scale. A value of 1800 seconds corresponds to a timeout.

The first three scatter plots show the gains the core extractors AMUSE (a), zruntillfix (b) and AOMUS (c) have from using MiniUnsat rather than a naive MUS proving approach as a post processor. The improvement caused by using MiniUnsat when regarding the combination of core extractor and prover as one program is quite remarkable, especially for the random 3-SAT formulas with 200

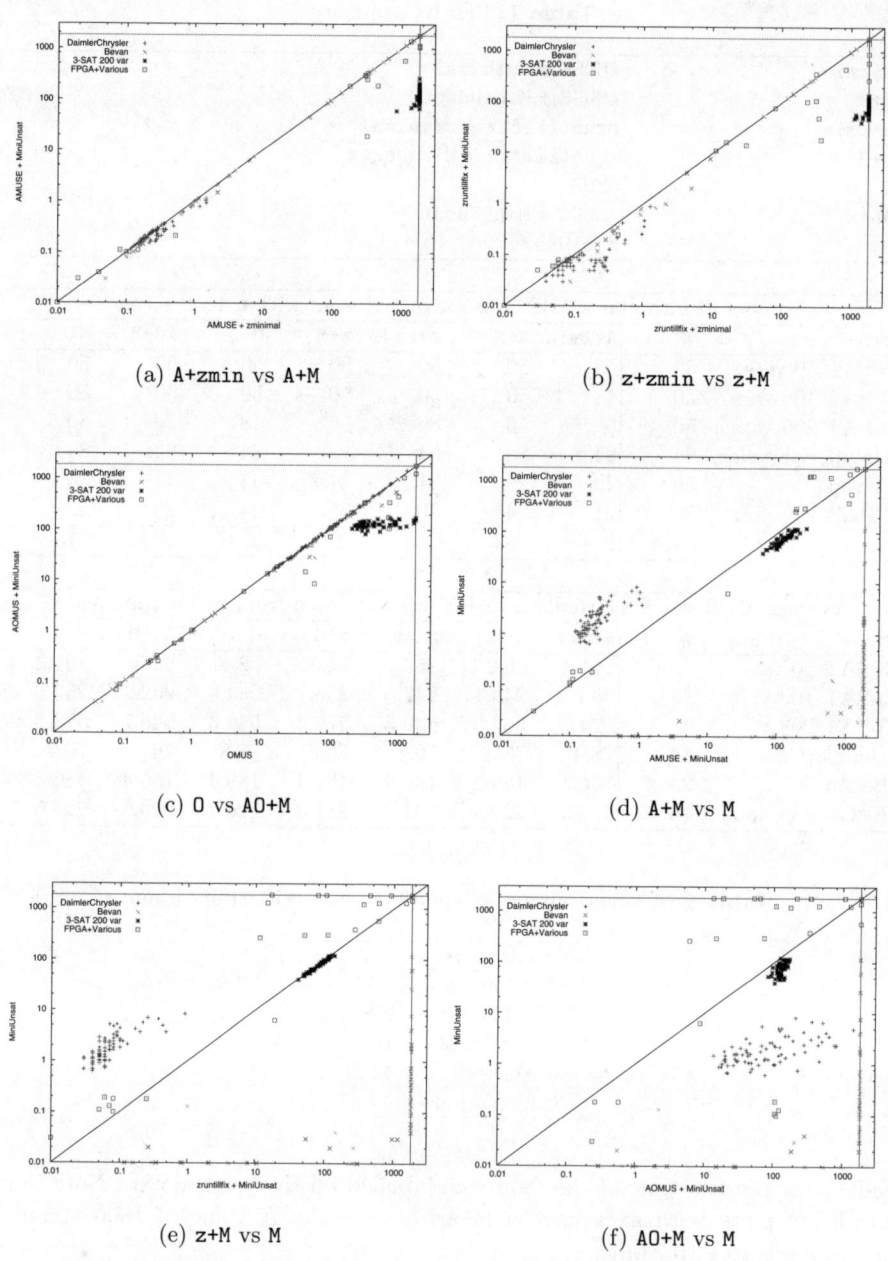

(a) A+zmin vs A+M

(b) z+zmin vs z+M

(c) O vs AO+M

(d) A+M vs M

(e) z+M vs M

(f) AO+M vs M

Fig. 3. Results

variables. While AMUSE+zminimal and zruntillfix+zminimal are not capable
of extracting a MUS in half an hour from the majority of those formulas that
goal can always reached when using MiniUnsat instead of zminimal.

The other scatter plots show how the three core extractors combined with `MiniUnsat` and regarded as one program perform against `MiniUnsat` without a preprocessor. From those scatter plots it can be seen that for the set of Daimler Chrysler benchmarks it pays off to use `AMUSE` (d) or `zruntillfix` (e) as a preprocessor. However, for the Bevan benchmarks using those preprocessors will mean most benchmarks will not be solved because the preprocessor times out. The performance of the `AOMUS` core extractor (f) as a preprocessor for the Bevan benchmarks is, unsurprisingly, similar to the other two tested preprocessors. However, in `AOMUS`'s scatter plot, on the horizontal line indicating a timeout for `MiniUnsat` we see a number of benchmarks from the set *FPGA+Various*. The benchmarks from that set that `MiniUnsat` fails to extract a MUS from without the help of `AOMUS` are FPGA routing problems, a domain in which `AOMUS` excels.

6   Conclusion

Although over the last years attention has been paid to the development of tools for extracting unsatisfiable subsets from unsatisfiable Boolean formulas most existing tools do not guarantee that the extracted unsatisfiable subsets are minimal. The `MiniUnsat` MUS finder we presented is capable of extracting a MUS from a wide range of unsatisfiable formulas at very competitive speeds.

In case the user wants to tune performance for a specific set of benchmarks several interesting combinations with existing software are recommendable. For example, if the program is applied in a setting were a MUS is often only a small fraction of the size of the formula it is extracted from, it is wise to use `AMUSE` as a preprocessor to `MiniUnsat`. For finding minimal unsatisfiable subsets in FPGA routing problems the use of `AOMUS` as a preprocessor is recommended.

Acknowledgements. Thanks to all developers of the tested tools for making their software available, either on the web or at request. Thanks to all colleagues that supported this work by showing their interest or commenting. Special thanks to Keijo Heljanko for all his help throughout this project.

References

1. Nam, G.J., Sakallah, K.A., Rutenbar, R.A.: Satisfiability-based layout revisited: Detailed routing of complex FPGAs vis search-based Boolean SAT. In: FPGA, pp. 167–175 (1999)
2. Biere, A., Cimatti, A., Clarke, E.M., Zhu, Y.: Symbolic model checking without BDDs. In: Cleaveland, W.R. (ed.) ETAPS 1999 and TACAS 1999. LNCS, vol. 1579, pp. 193–207. Springer, Heidelberg (1999)
3. McMillan, K.L.: Interpolants and symbolic model checking. In: Cook, B., Podelski, A. (eds.) VMCAI 2007. LNCS, vol. 4349, pp. 89–90. Springer, Heidelberg (2007)
4. Rintanen, J., Heljanko, K., Niemelä, I.: Planning as satisfiability: parallel plans and algorithms for plan search. Artif. Intell. 170(12-13), 1031–1080 (2006)

5. Zhang, L., Malik, S.: Extracting Small Unsatisfiable Cores from Unsatisfiable Boolean Formula. In: Theory and Applications of Satisfiability Testing, 6th International Conference, SAT 2003. Santa Margherita Ligure, Italy, May 5-8 (2003)
6. Oh, Y., Mneimneh, M.N., Andraus, Z.S., Sakallah, K.A., Markov, I.L.: AMUSE: A minimally-unsatisfiable subformula extractor. In: Malik, S., Fix, L., Kahng, A.B. (eds.) DAC, pp. 518–523. ACM, New York (2004)
7. Grégoire, É., Mazure, B., Piette, C.: Local-search extraction of MUSes. Constraints 12(3), 325–344 (2007)
8. Lynce, I., Silva, J.P.M.: On computing minimum unsatisfiable cores. In: H. Hoos, H., Mitchell, D.G. (eds.) SAT 2004. LNCS, vol. 3542, Springer, Heidelberg (2005)
9. Liffiton, M.H., Sakallah, K.A.: On finding all minimally unsatisfiable subformulas [25], 173–186
10. Liffiton, M.H., Sakallah, K.A.: Algorithms for computing minimal unsatisfiable subsets of constraints. Journal of Automated Reasoning 40(1), 1–33 (2008)
11. Wieringa, S.: Finding cores using a Brouwer's fixed point approximation algorithm. Master's thesis, Delft University of Technology, Faculty of EWI (2007)
12. van Maaren, H.: Pivoting algorithms based on Boolean vector labeling. Acta Mathematica Vietnamica 22(1), 183–198 (1997)
13. Kullmann, O., Lynce, I., Marques-Silva, J.: Categorisation of clauses in conjunctive normal forms: Minimally unsatisfiable sub-clause-sets and the lean kernel. [24] 22–35
14. Eén, N., Sörensson, N.: An extensible SAT-solver. In: Giunchiglia, E., Tacchella, A. (eds.) SAT 2003. LNCS, vol. 2919, pp. 502–518. Springer, Heidelberg (2004)
15. Eén, N., Biere, A.: Effective preprocessing in SAT through variable and clause elimination. [15] 61–75
16. Tamiz, M., Mardle, S.J., Jones, D.F.: Detecting IIS in infeasible linear programmes using techniques from goal programming. Computers & OR 23(2), 113–119 (1996)
17. Galinier, P., Hertz, A.: Solution techniques for the large set covering problem. Discrete Applied Mathematics 155(3), 312–326 (2007)
18. de Siqueira, N.J.L., Puget, J.-F.: Explanation-based generalisation of failures. In: ECAI, pp. 339–344 (1988)
19. Dershowitz, N., Hanna, Z., Nadel, A.: A scalable algorithm for minimal unsatisfiable core extraction. [24] 36–41
20. Huang, J.: MUP: A minimal unsatisfiability prover. In: Tang, T.A. (ed.) ASP-DAC, pp. 432–437. ACM Press, New York (2005)
21. Aloul, F.A., Ramani, A., Markov, I.L., Sakallah, K.A.: Solving difficult instances of Boolean satisfiability in the presence of symmetry. IEEE Trans. on CAD of Integrated Circuits and Systems 22(9), 1117–1137 (2003)
22. Darwiche, A.: New advances in compiling CNF into decomposable negation normal form. In: de Mántaras, R.L., Saitta, L. (eds.) ECAI, pp. 328–332. IOS Press, Amsterdam (2004)
23. Aloul, F.A., Markov, I.L., Sakallah, K.A.: MINCE: A static global variable-ordering heuristic for SAT search and BDD manipulation. J. UCS 10(12), 1562–1596 (2004)
24. Biere, A., Gomes, C.P. (eds.): SAT 2006. LNCS, vol. 4121. Springer, Heidelberg (2006)
25. Bacchus, F., Walsh, T. (eds.): SAT 2005. LNCS, vol. 3569. Springer, Heidelberg (2005)

Author Index

Printing: Mercedes-Druck, Berlin
Binding: Stein + Lehmann, Berlin

Lecture Notes in Computer Science

Sublibrary 1: Theoretical Computer Science and General Issues

For information about Vols. 1– 4669
please contact your bookseller or Springer

Vol. 4854: L. Bougé, M. Forsell, J.L. Träff, A. Streit, W. Ziegler, M. Alexander, S. Childs (Eds.), Euro-Par 2007 Workshops: Parallel Processing. XVII, 236 pages. 2008.

Vol. 4851: S. Boztaş, H.-F.(F.) Lu (Eds.), Applied Algebra, Algebraic Algorithms and Error-Correcting Codes. XII, 368 pages. 2007.

Vol. 4848: M.H. Garzon, H. Yan (Eds.), DNA Computing. XI, 292 pages. 2008.

Vol. 4847: M. Xu, Y. Zhan, J. Cao, Y. Liu (Eds.), Advanced Parallel Processing Technologies. XIX, 767 pages. 2007.

Vol. 4846: I. Cervesato (Ed.), Advances in Computer Science – ASIAN 2007. XI, 313 pages. 2007.

Vol. 4838: T. Masuzawa, S. Tixeuil (Eds.), Stabilization, Safety, and Security of Distributed Systems. XIII, 409 pages. 2007.

Vol. 4835: T. Tokuyama (Ed.), Algorithms and Computation. XVII, 929 pages. 2007.

Vol. 4818: I. Lirkov, S. Margenov, J. Waśniewski (Eds.), Large-Scale Scientific Computing. XIV, 755 pages. 2008.

Vol. 4800: A. Avron, N. Dershowitz, A. Rabinovich (Eds.), Pillars of Computer Science. XXI, 683 pages. 2008.

Vol. 4783: J. Holub, J. Žďárek (Eds.), Implementation and Application of Automata. XIII, 324 pages. 2007.

Vol. 4782: R. Perrott, B.M. Chapman, J. Subhlok, R.F. de Mello, L.T. Yang (Eds.), High Performance Computing and Communications. XIX, 823 pages. 2007.

Vol. 4771: T. Bartz-Beielstein, M.J. Blesa Aguilera, C. Blum, B. Naujoks, A. Roli, G. Rudolph, M. Sampels (Eds.), Hybrid Metaheuristics. X, 202 pages. 2007.

Vol. 4770: V.G. Ganzha, E.W. Mayr, E.V. Vorozhtsov (Eds.), Computer Algebra in Scientific Computing. XIII, 460 pages. 2007.

Vol. 4769: A. Brandstädt, D. Kratsch, H. Müller (Eds.), Graph-Theoretic Concepts in Computer Science. XIII, 341 pages. 2007.

Vol. 4763: J.-F. Raskin, P.S. Thiagarajan (Eds.), Formal Modeling and Analysis of Timed Systems. X, 369 pages. 2007.

Vol. 4759: J. Labarta, K. Joe, T. Sato (Eds.), High-Performance Computing. XV, 524 pages. 2008.

Vol. 4746: A. Bondavalli, F. Brasileiro, S. Rajsbaum (Eds.), Dependable Computing. XV, 239 pages. 2007.

Vol. 4743: P. Thulasiraman, X. He, T.L. Xu, M.K. Denko, R.K. Thulasiram, L.T. Yang (Eds.), Frontiers of High Performance Computing and Networking ISPA 2007 Workshops. XXIX, 536 pages. 2007.

Vol. 4742: I. Stojmenovic, R.K. Thulasiram, L.T. Yang, W. Jia, M. Guo, R.F. de Mello (Eds.), Parallel and Distributed Processing and Applications. XX, 995 pages. 2007.

Vol. 4739: R. Moreno Díaz, F. Pichler, A. Quesada Arencibia (Eds.), Computer Aided Systems Theory – EUROCAST 2007. XIX, 1233 pages. 2007.

Vol. 4736: S. Winter, M. Duckham, L. Kulik, B. Kuipers (Eds.), Spatial Information Theory. XV, 455 pages. 2007.

Vol. 4732: K. Schneider, J. Brandt (Eds.), Theorem Proving in Higher Order Logics. IX, 401 pages. 2007.

Vol. 4731: A. Pelc (Ed.), Distributed Computing. XVI, 510 pages. 2007.

Vol. 4728: S. Bozapalidis, G. Rahonis (Eds.), Algebraic Informatics. VIII, 291 pages. 2007.

Vol. 4726: N. Ziviani, R. Baeza-Yates (Eds.), String Processing and Information Retrieval. XII, 311 pages. 2007.

Vol. 4719: R. Backhouse, J. Gibbons, R. Hinze, J. Jeuring (Eds.), Datatype-Generic Programming. XI, 369 pages. 2007.

Vol. 4711: C.B. Jones, Z. Liu, J. Woodcock (Eds.), Theoretical Aspects of Computing – ICTAC 2007. XI, 483 pages. 2007.

Vol. 4710: C.W. George, Z. Liu, J. Woodcock (Eds.), Domain Modeling and the Duration Calculus. XI, 237 pages. 2007.

Vol. 4708: L. Kučera, A. Kučera (Eds.), Mathematical Foundations of Computer Science 2007. XVIII, 764 pages. 2007.

Vol. 4707: O. Gervasi, M.L. Gavrilova (Eds.), Computational Science and Its Applications – ICCSA 2007, Part III. XXIV, 1205 pages. 2007.

Vol. 4706: O. Gervasi, M.L. Gavrilova (Eds.), Computational Science and Its Applications – ICCSA 2007, Part II. XXIII, 1129 pages. 2007.

Vol. 4705: O. Gervasi, M.L. Gavrilova (Eds.), Computational Science and Its Applications – ICCSA 2007, Part I. XLIV, 1169 pages. 2007.

Vol. 4703: L. Caires, V.T. Vasconcelos (Eds.), CONCUR 2007 – Concurrency Theory. XIII, 507 pages. 2007.

Vol. 4700: C.B. Jones, Z. Liu, J. Woodcock (Eds.), Formal Methods and Hybrid Real-Time Systems. XVI, 539 pages. 2007.

Vol. 4699: B. Kågström, E. Elmroth, J. Dongarra, J. Waśniewski (Eds.), Applied Parallel Computing. XXIX, 1192 pages. 2007.

Vol. 4698: L. Arge, M. Hoffmann, E. Welzl (Eds.), Algorithms – ESA 2007. XV, 769 pages. 2007.

Vol. 4697: L. Choi, Y. Paek, S. Cho (Eds.), Advances in Computer Systems Architecture. XIII, 400 pages. 2007.

Vol. 4688: K. Li, M. Fei, G.W. Irwin, S. Ma (Eds.), Bio-Inspired Computational Intelligence and Applications. XIX, 805 pages. 2007.

Vol. 4684: L. Kang, Y. Liu, S. Zeng (Eds.), Evolvable Systems: From Biology to Hardware. XIV, 446 pages. 2007.

Vol. 4683: L. Kang, Y. Liu, S. Zeng (Eds.), Advances in Computation and Intelligence. XVII, 663 pages. 2007.

Vol. 4681: D.-S. Huang, L. Heutte, M. Loog (Eds.), Advanced Intelligent Computing Theories and Applications. XXVI, 1379 pages. 2007.

Vol. 4672: K. Li, C. Jesshope, H. Jin, J.-L. Gaudiot (Eds.), Network and Parallel Computing. XVIII, 558 pages. 2007.

Vol. 4671: V.E. Malyshkin (Ed.), Parallel Computing Technologies. XIV, 635 pages. 2007.